Studies in Classification, Data Analysis, and Knowledge Organization

Studies in Classification, Data Analysis, and Knowledge Organization is a book series which offers constant and up-to-date information on the most recent developments and methods in the fields of statistical data analysis, exploratory statistics, classification and clustering, handling of information and ordering of knowledge. It covers a broad scope of theoretical, methodological as well as application-oriented articles, surveys and discussions from an international authorship and includes fields like computational statistics, pattern recognition, biological taxonomy, DNA and genome analysis, marketing, finance and other areas in economics, databases and the internet. A major purpose is to show the intimate interplay between various, seemingly unrelated domains and to foster the cooperation between mathematicians, statisticians, computer scientists and practitioners by offering well-based and innovative solutions to urgent problems of practice.

Giuseppe Giordano • Michele La Rocca •
Marcella Niglio • Marialuisa Restaino •
Maurizio Vichi
Editors

Statistical Models and Learning Methods for Complex Data

Springer

Editors

Giuseppe Giordano
Department of Political and Social Studies
University of Salerno
Fisciano, Salerno, Italy

Michele La Rocca
Department of Economics and Statistics
University of Salerno
Fisciano, Salerno, Italy

Marcella Niglio
Department of Economics and Statistics
University of Salerno
Fisciano, Salerno, Italy

Marialuisa Restaino
Department of Economics and Statistics
University of Salerno
Fisciano, Salerno, Italy

Maurizio Vichi
Department of Statistical Sciences
Sapienza University of Rome
Rome, Italy

14th Scientific Meeting of the Classification and Data Analysis Group of the Italian Statistical Society (CLADAG 2023) Salerno (Italy), September 11–13, 2023 https://www.statlab-unisa.it/cladag2023/

ISSN 1431-8814 ISSN 2198-3321 (electronic)
Studies in Classification, Data Analysis, and Knowledge Organization
ISBN 978-3-031-84701-1 ISBN 978-3-031-84702-8 (eBook)
https://doi.org/10.1007/978-3-031-84702-8

Mathematics Subject Classification: 62-06, 62Jxx, 62Mxx, 62R10, 62Hxx, 62R07, 62Pxx

This Springer imprint is published by the registered company Springer Nature Switzerland AG
The registered company address is: Gewerbestrasse 11, 6330 Cham, Switzerland

Preface

This book comprises a selection of post-conference papers presented at the 14th meeting of the Classification and Data Analysis Group (CLADAG) of the Italian Statistical Society (SIS), which is also a member of the International Federation of Classification Societies (IFCS). Organized biennially by international scientific committees, the CLADAG meetings advance methodological research in multivariate statistics, with a strong focus on Data Analysis and Classification. The CLADAG group facilitates the exchange of ideas in these fields, promoting the dissemination of concepts, numerical methods, algorithms, and computational and applied results.

In 2023, the CLADAG conference was organized by the Department of Economics and Statistics and the Department of Social and Political Studies at the University of Salerno in Italy. It took place at the University Campus in Fisciano, Salerno, from September 11 to 13. The conference welcomed 280 participants from various countries and featured 5 Keynote Talks, 32 Invited Sessions, and 27 Contributed Sessions. A Special Event involved university students who took part in the data analysis competition, "Data (in) Context: Thinking with Data."

The diversity and depth of the Conference program are attributed to the efforts of the Conference Scientific Committee, chaired by Carla Rampichini. We extend our gratitude to the anonymous referees for their invaluable contributions, which significantly enhanced the quality of this book.

The 22 contributions featured in this book span a variety of topics, including different approaches to clustering and classification, multidimensional data analysis, panel data, social networks, time series, statistical inference, and mixture models. These methodologies are applied to a range of empirical domains such as economics, finance, hydrology, the social sciences, education, and sports.

We would like to express our appreciation to the administrative staff of the Department of Economics and Statistics and the Department of Social and Political Studies at the University of Salerno, as well as the staff of the Italian Statistical Society, for their support to the Local Organizing Committee.

We sincerely thank all the authors and participants, without whom the Conference would not have been possible. Above all, we are grateful to all who chose this book to share their research findings. We hope this book will contribute to advancing knowledge in the field.

Fisciano, Salerno, Italy	Giuseppe Giordano
Fisciano, Salerno, Italy	Michele La Rocca
Fisciano, Salerno, Italy	Marcella Niglio
Fisciano, Salerno, Italy	Marialuisa Restaino
Rome, Italy	Maurizio Vichi

Contents

Exploring Latent Evolving Ability in Test Equating and Its Effects on Final Rankings

Silvia Bacci, Bruno Bertaccini, Carla Galluccio, Leonardo Grilli, and Carla Rampichini

Abstract This paper investigates to what extent test equalisation methods are affected by changes in the population distribution of the ability. This problem arises when the subjects can repeat the test. The scenario considered here entails two test administrations at different times and increasing ability of the subjects. Additionally, this research question becomes especially relevant when the final output of this process is a unique merit ranking. Indeed, building a merit ranking can be interpreted as a classification problem, in which the goal is to correctly classify a subject in the ranking based on their true ability. To answer these questions, we conduct a simulation study comparing concurrent calibration with the classical item response theory (IRT) linking parameter methods.

1 Introduction

In large-scale assessments, multiple test forms are usually constructed to increase test security and allow tests to be implemented on different exam dates and times [15]. This organisation requires using many items (to construct parallel versions of the same test for each group) and applying equalisation methods to ensure the comparability of scores obtained on different test forms [6], which is a relevant concern when the final output of the procedure is a unique merit ranking. Indeed, building a merit ranking can be interpreted as a classification problem, in which the goal is to correctly classify a subject in the ranking based on his/her true ability.

An example is given by the Italian admission test for the faculty of medicine. Over the past few years, Italian universities introduced admission tests to manage students' enrollment in different courses, including medicine and health sciences.

S. Bacci · B. Bertaccini · C. Galluccio · L. Grilli (✉) · C. Rampichini
University of Florence, Department of Statistics, Computer Science, Applications 'G. Parenti', Firenze, Italy
e-mail: silvia.bacci@unifi.it; bruno.bertaccini@unifi.it; carla.galluccio@unifi.it; leonardo.grilli@unifi.it; carla.rampichini@unifi.it

G. Giordano et al. (eds.), *Statistical Models and Learning Methods for Complex Data*, Studies in Classification, Data Analysis, and Knowledge Organization,
https://doi.org/10.1007/978-3-031-84702-8_1

In this vein, until 2022, a printed test was administered once a year to all students on a single occasion for each type of degree course. Starting from the test scores, a national merit ranking was built, based on which the top students were allowed to enrol on the courses.

However, in 2023, the government introduced an experimental assessment for the medicine admission test with computer-based tests administered two times a year, in April and July, on different dates and times. The test forms were different to each other within each session and date/time, built by experts in order to make them parallel. Students starting from the fourth year of secondary school could answer the test in each session. A unique final merit ranking was provided in September. If a student attended both test sessions of the year, the highest score was used to build the merit ranking.

It is worth understanding that this organisation requires equalisation methods to compare the scores obtained on the different test forms [6]. Moreover, the populations that answer the test at different times may differ in terms of average ability and variance. Indeed, in this case, it is reasonable to suppose that students who took the test at the baseline have lower abilities than those who took it later, at least because they had more time to study and became familiar with the type of test. Then, the populations between which the item parameters are linked through equalisation methods at the baseline may differ significantly in the average ability and variance from those at subsequent administrations [1].

Drawing inspiration from this example, we designed a simulation study for a simplified framework considering a test administered in two sessions at different times with the following features: (i) a single form is administered in each session; (ii) the test aims at assessing a unidimensional latent trait, i.e. there are no testlets; and (iii) the test items conform to the Rasch model [10]. As for the Italian medicine admission test, we do not assume a pre-calibration of the items.

In the literature, equalisation methods have been widely studied, highlighting their strengths and weaknesses. For example, [13] reviewed traditional and IRT-based equating methods, highlighting how commonly used methods like the mean-mean method have drawbacks as they fail to take into account all of the item parameter estimates simultaneously in the estimation process. However, little has been done about the consequences of using these methods in terms of building merit rankings. This paper aims to fill this gap, providing an analysis of the behaviour of some common equalisation methods in this assessment context through a simulation study. In particular, we will investigate how the merit rankings derived from the test equalisation methods are affected by the fraction of subjects repeating the test and to what extent the rankings are different for subjects repeating and not repeating the test.

The remaining part of the contribution is organized as follows. In Sect. 2, we present the statistical model. Section 3 illustrates the basics of test equating, focusing on the description of the equating design and the IRT linking parameter methods used in this work, while in Sect. 4, we describe the simulation study. Section 5 is devoted to the results. Finally, the conclusions in Sect. 6 end the paper.

2 Statistical Model

In the IRT framework, the two-parameter logistic (2PL) model [2] defines the probability that a subject i ($i = 1, \dots, N$) taking a test with J items answers the binary item Y_{ij} ($j = 1, \dots, J$) as a function of her ability θ_i and a set of item-specific parameters:

$$P(Y_{ij} = 1|\theta_i) = \frac{\exp[\alpha_j(\theta_i - \beta_j)]}{1 + \exp[\alpha_j(\theta_i - \beta_j)]}, \tag{1}$$

where α_j and β_j denote the discrimination and difficulty parameters of item j, respectively. Note that, when $\alpha_j = 1$ for all items, the 2PL model corresponds to the Rasch model [10].

3 Test Equating

Test equating is a statistical procedure used to make the scores obtained on different test forms comparable, i.e. having equivalent meaning. Equalising two or more forms of a test is a practical matter, requiring the use of subjects' scores as a starting point. The scores used to perform equalisation are collected according to established principles known as equating designs. In this work, we considered the case of the non-equivalent groups with anchor test design [6].

Regarding equalisation methods, in the literature, it is possible to find methods associated with classical test theory (CTT) and others related to IRT methods. Among the latter, herein we used the concurrent calibration [5] and the mean-mean [9], mean-sigma [7], Haebara [4], and the Stocking-Lord [14] IRT linking parameter methods.

3.1 *Equating Design*

In the Non-Equivalent Groups with Anchor Test (NEAT) design [6], it is assumed that the groups of subjects responding to different test forms come from different populations. In this case, to equalise the scores provided by different test forms, it is necessary to define a subset of items, named *anchor items* [6], that are in common with the available test forms. Also, in this design, it is assumed that the groups of subjects are independent of each other. The goal of the equalisation procedure is to discern whether the differences in the test scores obtained by the groups of subjects are due to intrinsic differences in the subjects, differences related to the tests, or both.

The anchor items are chosen to represent, in proportion, the entire test so that they can be considered a reduced version of the full test. Anchor items can be placed anywhere in the test and may or may not contribute to the subjects' final score [12]. Regarding the number of common items necessary to carry out the equalisation in an effective manner, as a rule of thumb, approximately 30% of the items are used [12].

3.2 *IRT Parameter Linking Methods*

Let $T1$ and $T2$ two test form to be equalised, where $T2$ is the test form that has to be equated to test form $T1$.

In concurrent calibration, the parameters of the two test forms to be equalised are estimated in a single run of the model [5]. Items not in common between the two tests are considered missing. A key feature of this method is that it does not require the association of IRT scores, as the parameters are linked on a common scale during the estimation process, and, therefore, item parameter and ability estimates are placed on the same scale [3].

On the other hand, IRT parameter linking methods require carrying out a transformation that associates the IRT parameter estimates from separate calibrations of two test forms on a common scale. Generally, a linear transformation is adopted:

$$\theta_i^{T2} = a\theta_i^{T1} + b$$

where a and b are the *linking constants* (or *equating constants*) that have to be estimated, whereas θ_i^{T1} and θ_i^{T2} are the ability level of subject i on test forms $T1$ and $T2$, respectively.

For example, in the mean-mean method, the linking constants are estimated on the set of common items as follows:

$$a = \frac{\overline{\hat{\alpha}}_{T1}}{\overline{\hat{\alpha}}_{T2}} \qquad b = \overline{\hat{\beta}}_{T2} - a\overline{\hat{\beta}}_{T1}$$

where $\overline{\hat{\alpha}}_{T1}$ and $\overline{\hat{\alpha}}_{T2}$ are the means of the estimated discrimination parameters, while $\overline{\hat{\beta}}_{T1}$ and $\overline{\hat{\beta}}_{T2}$ are the means of the estimated difficulty parameters. Note that the Rasch model implies $a = 1$, so that $b = \overline{\hat{\beta}}_{T2} - \overline{\hat{\beta}}_{T1}$.

4 Simulation Study

In the simulation study, we consider two test forms $T1$ and $T2$ administered at the first and second occasion, respectively. Each test form has $J = 50$ binary items

Table 1 Configurations: fraction π of subjects taking the test twice and numbers of subjects

Configuration	C1	C2	C3	C4	C5
π	0.5	0.6	0.7	0.8	1.0
N. subjects taking both $T1$ and $T2$	5000	6000	7000	8000	10000
N. subjects taking only $T1$	5000	4000	3000	2000	0
N. subjects taking only $T2$	5000	4000	3000	2000	0
Total number of ranked subjects	15000	14000	13000	12000	10000

generated according to the Rasch model, i.e. model (1) with $\alpha_j = 1, \forall j$. We assume the two test forms have 15 items in common, thus constraining the first 15 difficulty parameters to be equal in the two test forms: $\beta_j^{T1} = \beta_j^{T2}$ for $j = 1, \ldots, 15$. The difficulty parameters are sampled from a standard normal distribution.

We assume the test is taken by N subjects on each occasion with a fraction π repeating the test, namely, attending both occasions. We denote with R the set of $N\pi$ subjects repeating the test and with $\overline{R}$ the set of $N(1-\pi)$ subjects taking only the second test.

To simulate data for the test form administered on the first occasion, the latent abilities are sampled from a standard normal distribution, i.e. $\theta_i^{T1} \sim N(0, 1)$. For the subjects repeating the test, we assume that the latent ability on the second occasion is proportional to that on the first one: $\theta_{i\in R}^{T2} = \delta_i \times \theta_{i\in R}^{T1}$ with change factor $\delta_i \sim N(1.2, 0.2)$. Note that $Pr(\delta_i > 1) = 0.84$; thus, a large majority of subjects have an increasing ability. For the remaining fraction $(1-\pi)$ of *new* subjects, the latent ability is drawn from a standard normal distribution, i.e. $\theta_{i\in\overline{R}}^{T2} \sim N(0, 1)$.

For each test form, the response patterns for $N = 10{,}000$ subjects were simulated using the `sim.rasch` function of the `eRm` package of R [8]. Parameter estimation is obtained through the approximate marginal maximum likelihood algorithm implemented in the R package `ltm` [11].

We repeat the simulation experiment, varying the fraction π of subjects who take the test twice, as described in Table 1. In configuration C5, all the subjects attend both tests. Note that the number of ranked subjects depend on the configuration.

5 Results

To investigate the research questions, we considered the merit rankings obtained by equalising the two test forms in the five scenarios of Table 1 using two distinct approaches: the concurrent calibration and the mean-mean linking method. Under the assumptions of our simulation study, all IRT linking parameter methods yielded similar results; thus, we present only the finding of the mean-mean method. The merit ranking resulting from each strategy was compared to the true (simulated) ranking by the Spearman correlation coefficient (Fig. 1).

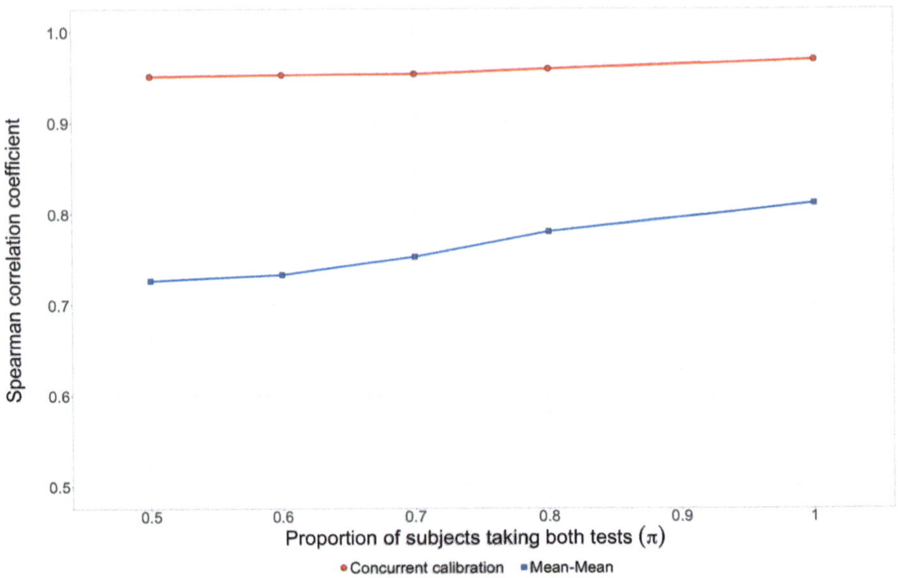

Fig. 1 Spearman correlation between true and estimated rankings

The plot shows for both methods a higher correlation in configurations with a higher proportion of subjects taking both tests. In particular, the correlation coefficient ranges from 0.950 to 0.969 for the concurrent calibration, while it ranges from 0.726 to 0.811 for the mean-mean method. These results highlight the superiority of concurrent calibration with respect to the mean-mean method. Similar results are obtained with other popular IRT linking equalisation methods (not reported here).

In order to evaluate the classification performances of the two methods, we compare the quantiles of the true (simulated) and predicted ranking distributions. In particular, we considered the top 20%, in line with the motivating example presented in Sect. 1; indeed, based on the admission test scores for the faculty of medicine, only the top students are allowed to enrol on the courses, so classifying them correctly is crucial in the assessment process.

As an example, Table 2 reports the classification matrices for both equalisation methods in configuration C1 (50% of subjects taking the test twice). In this configuration, 15,000 subjects are ranked (5,000 assessed only in the first occasion, 5,000 only in the second one and 5,000 in both). The results reported in the classification matrices show that the concurrent calibration correctly classifies 93.1% of subjects. However, among the actual top 20%, only 2489 out of (2489+511), corresponding to 82.9% of the subjects, are correctly classified. For the mean-mean method, the percentages of correctly classified are lower: 85.1% overall and 62.7% among the actual top 20%. For the other configurations, the proportion of subjects correctly

Table 2 Classification matrices for top 20% in configuration C1 for the two equalisation methods

(a) Concurrent calibration				(b) Mean-mean			
	Predicted				*Predicted*		
	Top 20%	Others	Total		Top 20%	Others	Total
Actual				*Actual*			
Top 20%	2489	511	3000	top 20%	1883	1117	3000
Others	511	11489	12000	others	1117	10883	12000
Total	3000	12000	15000	total	3000	12000	15000

classified (not reported here) increases in both equalisation methods for higher values of π as defined in Table 1.

Finally, we investigate if there are differences in the misclassification between the subjects who repeated the test and those who did not. As regards the top 20% of the ranking distribution, we do not observe any difference in the two groups of subjects under all configurations for both the equalisation methods. The results do not change in concurrent calibration when considering the top 10% of the ranking. However, we observe differences when applying the mean-mean method. Indeed, in this case, the percentage of misclassified subjects is about 7% in all configurations for those who did not repeat the test, while this percentage ranges from 9% (configuration C4) to 11% (configuration C1) for those who repeated the test. In this case, the chi-square test is always significant, showing that there may be a dependency between repeating the test and being misclassified in the top 10%.

6 Conclusions

In contexts where multiple test forms are administered, and the possibility to repeat the test is given to the individuals, building a unique final merit ranking could raise a classification problem related to the position of a subject in the ranking reflecting the true subject's ability. This issue is crucial when the ranking is used to select subjects, for example, to enrol students in specific degree courses.

We conducted a simulation study focusing on non-equivalent groups with anchor items, comparing various popular equalisation methods within the IRT paradigm. Our simulation results indicate that concurrent calibration outperforms other traditional IRT linking parameter methods. Both approaches perform better when the proportion of subjects from the same group is higher. However, when the primary objective of test assessment involves ranking based on equalizing different test forms, such as in the Italian admission test for the faculty of medicine, even concurrent calibration needs enhancements to ensure a fairer selection of the top-performing subjects. Additionally, concurrent calibration can only be utilized when the ranking is provided solely during the final test session.

As part of future developments, we intend to expand the simulation study to identify conditions that can enhance the performance of equalisation methods, such as increasing the number of anchor items. Furthermore, we plan to consider more complex frameworks, such as multiple test forms on the same occasion or a test composed of testlets to assess multiple traits.

References

1. Bacci, S., Bertaccini, B., Galluccio, C., Grilli, L., Rampichini, C.: Test equating with evolving latent ability. In: Coretto, P., Giordano, G., La Rocca, M., Parrella, M.L., Rampichini, C. (eds.) Cladag 2023. Book of Abstracts and Short Papers, pp. 54–57. Pearson, Italy (2023)
2. Birnbaum, A.: Some latent trait models and their use in inferring an examinee's ability. In: Lord, F.M., Novick, M.R. (eds.) Statistical Theories of Mental Test Scores, pp. 395–479. Addison-Wesley, Boston (1968)
3. Bjermo, J., Miller, F.: Efficient estimation of mean ability growth using vertical scaling. Appl. Psychol. Meas. **34**, 163–178 (2021)
4. Haebara, T.: Equating logistic ability scales by a weighted least squares method. Jpn. Psychol. Res. **22**, 144–149 (1980)
5. Hanson, B., Béguin, A.: Obtaining a common scale for item response theory item parameters using separate versus concurrent estimation in the common-item equating design. Appl. Psychol. Meas. **1**, 3–24 (2002)
6. Kolen, M.J., Brennan, R.L.: Test Equating, Scaling, and Linking: Methods and Practices. Springer, New York (2014)
7. Loyd, B. H., Hoover, H.D.: Vertical equating using the Rasch model. J. Educ. Meas. **35**, 179–193 (1980)
8. Mair, P., Hatzinger, R.: Extended Rasch modeling: The eRm package for the application of IRT models in R. J. Stat. Softw. **20**(9), 1–20 (2007)
9. Marco, G.L.: Item characteristic curve solutions to three intractable testing problems. J. Educ. Meas. **14**, 139–160 (1977)
10. Rasch, G.: Probabilistic models for some intelligence and attainment tests. Nielsen & Lydiche, Oxford (1960)
11. Rizopoulos, D.: ltm: An R Package for Latent Variable Modeling and Item Response Analysis. J. Stat. Softw. **17**(5), 1–25 (2006)
12. Ryan, J., Brockmann, F.: A Practitioner's introduction to equating with primers on classical test theory and item response theory. In: Council of Chief State School Officers (2009) Available via ERIC. https://files.eric.ed.gov/fulltext/ED544690.pdf
13. Sansivieri, V., Wiberg, M., Matteucci, M.: A review of test equating methods with a special focus on IRT-based approaches. Statistica **77**, 329–352 (2017)
14. Stocking, M., Lord, F.: Developing a common metric in item response theory. Appl. Psychol. Meas. **7**, 201–210 (1983)
15. van der Linden, W.J., Adema, J.J.: Simultaneous assembly of multiple test forms. J. Educ. Meas. **35**, 185–198 (1998)

Hidden Markov and Related Discrete Latent Variable Models: An Application to Compositional Data

Francesco Bartolucci, Michael Greenacre, Silvia Pandolfi, and Fulvia Pennoni

Abstract We review the class of discrete latent variable models, and we propose a new formulation of the hidden Markov model for compositional data. We illustrate some results of the analysis of the expenditures of the Spanish regions over several decades, showing that the approach is promising to cluster regions with different patterns linked to the composition of parts in the system over time. We give particular emphasis to the possible developments of discrete latent variable models that take inspiration from common problems of these models, such as the multimodality of the likelihood function and issues related to the choice of the number of support points of the latent variables.

1 Introduction

This chapter is focused on discrete latent variable (DLV) models, which include variables not directly observable and are assumed to follow a discrete distribution tailored to explain the association between observable variables [5]. Among these models, we mention in particular latent class [12], hidden Markov (HM) [3], and stochastic block models [15].

The inclusion in a statistical model of latent variables implies several advantages: (i) the capability of clustering units in different latent groups, also named components, subpopulations, classes, or states, when the latent variables have a discrete distribution; (ii) the high degree of flexibility, so that complex dependence structures

F. Bartolucci · S. Pandolfi
Department of Economics, University of Perugia, Perugia, Italy

M. Greenacre
Department of Economics & Business, and Barcelona School of Management, Universitat Pompeu Fabra, Barcelona, Spain

F. Pennoni (✉)
Department of Statistics and Quantitative Methods, University of Milano-Bicocca, Milan, Italy
e-mail: fulvia.pennoni@unimib.it

G. Giordano et al. (eds.), *Statistical Models and Learning Methods for Complex Data*, Studies in Classification, Data Analysis, and Knowledge Organization,
https://doi.org/10.1007/978-3-031-84702-8_2

may be dealt with also in the presence of variables of a different nature; (iii) the straightforward interpretation when the latent variables correspond to explanatory factors that cannot be observed; and (iv) the possibility, under certain conditions, to interpret some parameters causally, as the model can consider certain forms of confounding. Specifically, we can account for the effect of unobservable variables affecting both the treatment and the response by latent variables having a suitable distribution. Parameter estimation is based on the maximum likelihood approach through the expectation-maximization (EM) algorithm [10].

Our first aim is to illustrate new possible applications of these models for compositional data [13], which are nonnegative multivariate observations where relative rather than absolute information is relevant; thus, they represent a quantitative measurement of the parts of some total, expressed as proportions summing to 1. Secondly, we aim to discuss general issues, suggest possible solutions, and highlight new research frontiers regarding these models.

The chapter is organized as follows. First, we introduce a new development of the HM model for compositional data. Second, we show an application to analyze the annual capital stock wealth amount of different sectors in Spain. Finally, we conclude with some remarks on future directions of this research.

2 Hidden Markov Model for Compositional Data

HM models find application in the analysis of time-series [23] and longitudinal data [3]. These models may be employed for clustering units in a dynamic fashion, where the same individual may move between clusters across time. In the context of longitudinal data, for a sample of n individuals observed over T time occasions, we denote the vectors of response variables by $\boldsymbol{Y}_{it}$ with elements Y_{ijt}, $i = 1, \dots, n$, $j = 1, \dots, r$, $t = 1, \dots, T$, where r is the number of such variables. For every i, the sequence of discrete latent variables U_{it} collected in the vector $\boldsymbol{U}_i$ is introduced; these variables are assumed to follow a Markov chain of first order with k states.

The HM model assumes that each time-specific vector of response variables $\boldsymbol{y}_{it}$, corresponding to parts of the composition that are proportions summing to 1, follows a Dirichlet distribution, although other choices are also possible [21]. The parameters of this distribution depend on the underlying discrete latent variable; in symbols, we have $\boldsymbol{Y}_{it}|U_{it} = u \sim \text{Dir}(\boldsymbol{\alpha}_u)$, where $\boldsymbol{\alpha}_u$ is the state-specific vector of parameters with $u = 1, \dots, k$.

As usual, each sequence of latent variables $U_{i1}, \dots, U_{iT}$ follows a Markov chain with initial and transition probabilities that, without covariates, are denoted by $\lambda_u = p(U_{i1} = u)$ and $\pi_{uv} = p(U_{it} = v|U_{i,t-1} = u)$, $t = 2, \dots, T$, $u, v = 1, \dots, k$, respectively. With unit-specific covariates, these probabilities are formulated by suitable logit parametrizations based on regression coefficients to account for the effect of such covariates. This formulation is based on the usual assumption that the response variables are conditionally independent given the latent variables, which is a form of *local independence* [3].

Regarding the parametrization of the Dirichlet distribution, we follow an approach that distinguishes the effects of the latent states on the expected value and on the variance (see also [16]). This amounts to normalizing the parameters of the Dirichlet distribution so that their sum is equal to 1. Additionally, we introduce a new scale parameter corresponding to the sum of the parameters used in the traditional formulation of this distribution.

Assuming independence between subjects, the log-likelihood referred to the observed data can be written as $\ell(\boldsymbol{\theta}) = \sum_{i=1}^{n} \log f(\boldsymbol{y}_i)$, where $\boldsymbol{\theta}$ is the vector of all model parameters and $f(\boldsymbol{y}_i)$ is the manifest distribution of the observed compositional parts. The latter may be expressed as

$$f(\boldsymbol{y}_i) = \sum_{u_1=1}^{k} \sum_{u_2=1}^{k} \cdots \sum_{u_T=1}^{k} \lambda_{u_1} \left(\prod_{t=2}^{T} \pi_{u_{t-1}u_t} \right) \left[\prod_{t=1}^{T} f(\boldsymbol{y}_{it} | U_{it} = u_t) \right].$$

In order to maximize $\ell(\boldsymbol{\theta})$, the EM algorithm relies on the complete-data log-likelihood expressed as $\ell^*(\boldsymbol{\theta})$; at the *E-step*, the algorithm computes the posterior expected value of $\ell^*(\boldsymbol{\theta})$ using certain forward-backward recursions [6, 22], and at the *M-step*, it updates the model parameters. Initialization of the algorithm is a crucial aspect and requires careful consideration of different starting values. Model selection is related to the choice of the number of states when it is not known a priori, and it is usually performed according to the information criteria such as the Bayesian information criterion (BIC) [19]; see [1] among others. Suitable functions are implemented for the R environment extending those of the package `LMest` [4] to perform the estimation of the proposed HM model.

3 Analysis of Spatiotemporal Compositional Data

We consider data provided by the BBVA Foundation from Madrid[1] concerning the composition of capital stock wealth in different sectors of the Spanish economy. Data are recorded for an extended period from 1964 to 2019 [11]. In the present work, we mainly concentrate on the problem of the national data on the temporal scale, and we briefly mention how to broaden this to the more detailed spatiotemporal scale across the different autonomous regions of Spain.

For the spatiotemporal framework, the data are collected in vectors $\boldsymbol{y}_{it}$, $i = 1, \ldots, n$, $t = 1, \ldots, T$, where n is the number of regions. The changing total amount across the years is, of course, important to analyze. Still, here it is the changing composition of the capital stock wealth that is of interest, namely, the amounts of each year relative to their respective totals. Hence, compositional data are such that the sum of the elements of each compositional response vector is

[1] https://www.fbbva.es/en/.

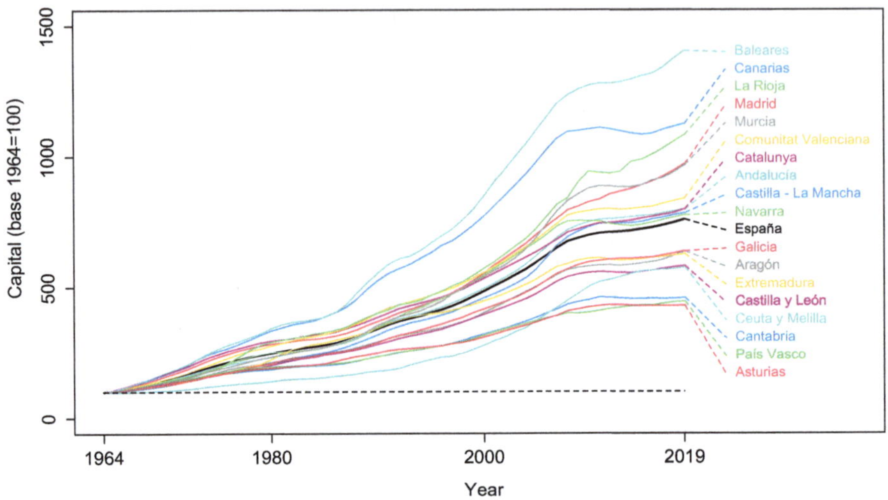

Fig. 1 Trajectories of capital stock wealth, from 1964 to 2019, across the 18 autonomous communities of Spain, along with the trajectory for Spain (España) itself. To facilitate comparison, all amounts are set to 100 at the initial time point. Dashed lines are for indicating labels

fixed at 1 or 100%. This has crucial implications in terms of data analysis. The $r = 15$ different sectors are considered for $T = 56$ years referred to the 17 autonomous communities and to the 2 autonomous cities (included together) of Spain. Trajectories of overall capital stock wealth by Spanish regions are reported in Fig. 1, while those of capital stock wealth by sector are depicted in Fig. 2.

The nature of the data suggests basing the analysis on region-year compositions across sectors [13]; therefore, capital stock wealth amounts across sectors are divided by their total so as to fix the sum to 1. For this analysis, we adopt the model illustrated in Sect. 2. To perform model selection, this model is estimated for $k = 1, \ldots, 6$, where 6 is the maximum number of states that we consider reasonable in order to avoid an excessive model complexity. Results in terms of maximum log-likelihood, number of parameters, and corresponding values of BIC are shown in Table 1. A model with $k = 5$ latent states is adopted for interpretability (despite not being optimal in terms of BIC); this approach is usually adopted in applications to complex and high-dimensional data, where this index may continue to decrease as additional states are added until a very large value of k. In such a situation, it is advisable to choose a suitable value of k also for the interpretability of the latent states.

Estimated means of the compositions given the latent states are visualized through the heat map depicted in Fig. 3. The latent states are ordered increasingly based on the average proportion of capital stock wealth allocated to real estate, with regions in the first subpopulation characterized by the lowest part of capital stock wealth devoted to this sector. On the other hand, it is not possible to detect a clear ordering of the latent states with respect to the composition of capital invested in

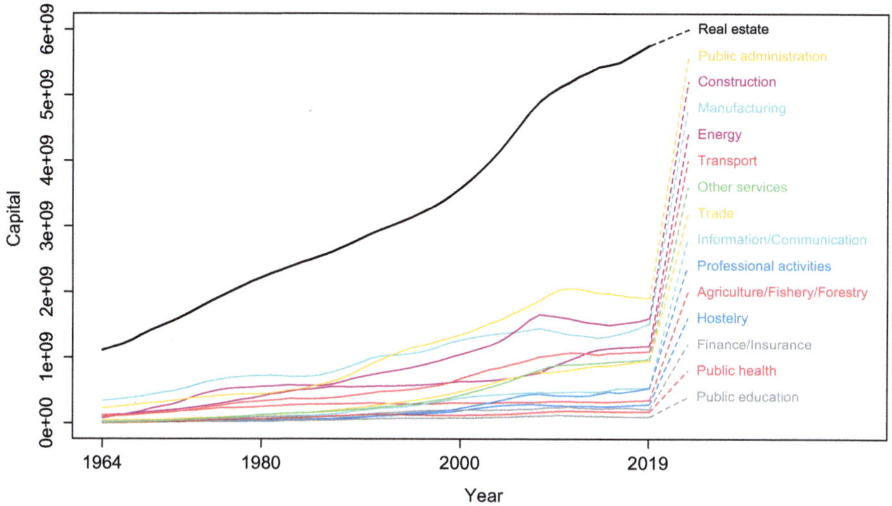

Fig. 2 Trajectories of capital stock wealth, measured in thousands of euros, from 1964 to 2019, for the whole of Spain

Table 1 Number of states, maximum likelihood, number of parameters, and BIC index for proposed HM models

k	$\ell(\hat{\boldsymbol{\theta}})$	#par	BIC
1	36,430.97	15	$-72,818.59$
2	38,229.40	33	$-76,363.43$
3	39,835.10	53	$-79,517.02$
4	40,533.36	75	$-80,849.94$
5	41,294.41	99	$-82,302.67$
6	42,068.90	125	$-83,776.51$

the other sectors, and then, a specific interpretation is in order for each of these subpopulations. Estimates of the initial and transition probabilities are shown in Table 2. It is noteworthy that the majority of regions belong to the last class, characterized by the highest proportion of capital stock wealth in real estate, while no regions are initially assigned to the first three classes. Furthermore, regarding transition probabilities, there is a very high persistence in the same subpopulation. However, it can be observed that the lower triangular elements of the transition matrix in Table 2 are generally larger than the upper triangular elements, which suggests a pattern where some regions tend to transition from the last two classes to the first four classes.

As already mentioned, the approach also allows assigning every region to a specific latent cluster over time, permitting the creation of maps to analyze the spatial pattern of the phenomenon under study. Just to give an idea, in Fig. 4, we report these maps for the first and last available years.

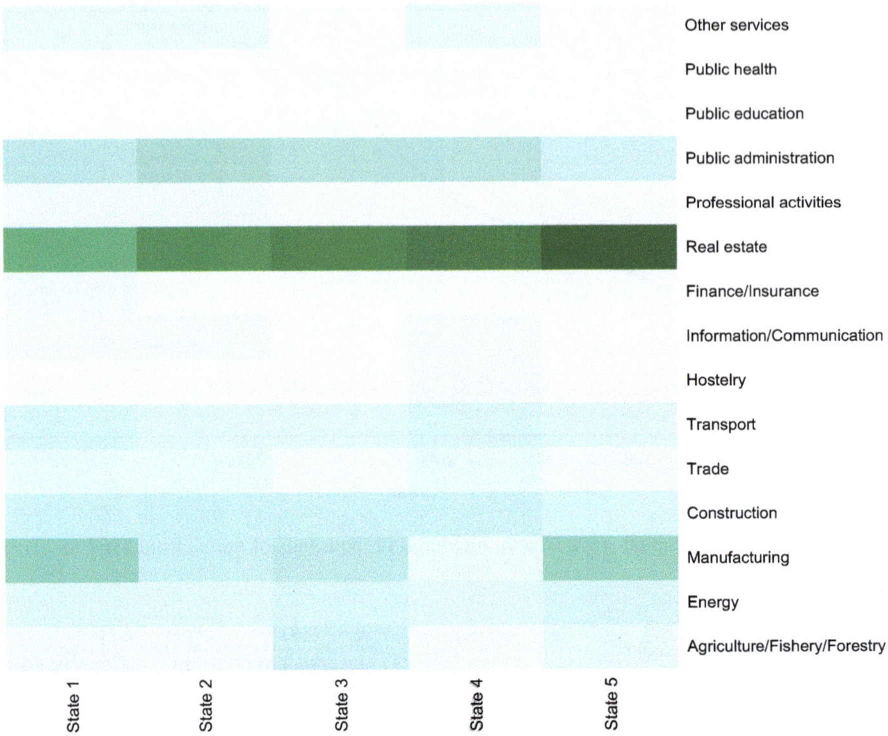

Fig. 3 Heatmap of scaled cluster means under the HM model with five hidden states

Table 2 Estimated initial and transition probabilities under the HM model

Initial probabilities				
$u = 1$	$u = 2$	$u = 3$	$u = 4$	$u = 5$
0.000	0.000	0.000	0.221	0.779

Transition probabilities					
u	$v = 1$	$v = 2$	$v = 3$	$v = 4$	$v = 5$
1	0.966	0.028	0.000	0.007	0.000
2	0.000	1.000	0.000	0.000	0.000
3	0.007	0.035	0.958	0.000	0.000
4	0.000	0.000	0.000	1.000	0.000
5	0.027	0.004	0.027	0.005	0.937

4 Conclusions and Further Developments

In this final section, we mention some issues that are important to overcome the actual limitations of the discrete latent variable (DLV) models. First, even for relatively simple DLV models, the likelihood is typically multimodal, with a potentially huge number of modes. As a possible approach, we can consider

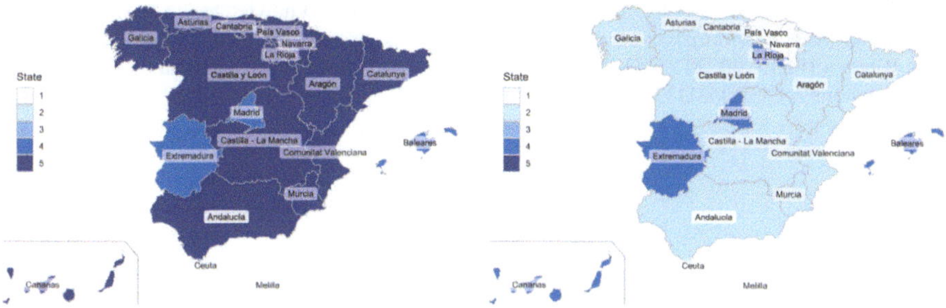

Fig. 4 Regional maps showing the estimated distribution of the latent clusters built by the selected HM model for years 1964 and 2019

choosing appropriate starting values for the estimation algorithm on the basis of deterministic and stochastic random rules [17] and then selecting the best solution among different attempts. Alternatively, we can use new tempered versions of the expectation-maximization (EM) algorithm, which consist in rescaling the objective function depending on a variable, known as temperature, controlling in such a way the prominence of global and local maxima [8]; more recent advanced versions of the EM are also promising at this aim, such as those based on evolutionary algorithms [9].

Second, it is uncommon that in applications clearly separated clusters of units indeed exist. Therefore, the adopted DLV model may identify "spurious" clusters, and model selection criteria may suggest many components with problematic interpretations. As a solution, a latent structure comprising a mixture of continuous distributions can be used, allowing the resulting model to serve as a compromise between discrete and continuous latent variable models [20]. Moreover, we can adopt model selection criteria that avoid a huge number of clusters and focus on the quality of clustering, such as the normalized entropy criterion [7].

Furthermore, accounting for informative missing data and dropout, especially in longitudinal data analysis, may be of primary interest. In this regard, we can consider implementing a joint or shared parameter model where the same latent variables govern the longitudinal process and time to dropout [2]. We can also consider an extra absorbing state and joint modeling missing data as proposed in [18].

Finally, specifically regarding the proposed application illustrated above, it is worth mentioning that as an alternative to the Dirichlet distribution, the HM models can be applied to the compositional data transformed to logratios [14], which are then treated as interval-scale variables with a multivariate normal distribution. These results can be compared with those obtained in the present study. Furthermore, it is also possible to extend the model by considering a spatiotemporal structure, where the latent state of a region in a certain year may depend not only on the previous state but also on the state of neighboring regions.

Acknowledgments F. Bartolucci, S. Pandolfi, and F. Pennoni acknowledge the financial support from the grant *Hidden Markov Models for Early Warning Systems* of Ministero dell'Università e della Ricerca (PRIN 2022TZEXKF) funded by European Union—Next Generation EU, Mission 4, Component 2, CUP J53D23004990006.

References

1. Bacci, S., Pandolfi, S., Pennoni, F.: A comparison of some criteria for states selection in the latent Markov model for longitudinal data. Adv. Data Anal. Classif. **8**, 125–145 (2014)
2. Bartolucci, F., Farcomeni, A.: A discrete time event-history approach to informative drop-out in mixed latent Markov models with covariates. Biometrics **71**, 80–89 (2015)
3. Bartolucci, F., Farcomeni, A., Pennoni, F.: Latent Markov Models for Longitudinal Data. Chapman & Hall/CRC Press, Boca Raton, FL (2013)
4. Bartolucci, F., Pandolfi, S., Pennoni, F.: LMest: An R package for latent Markov models for longitudinal categorical data. J. Stat. Softw. **81**, 1–38 (2017)
5. Bartolucci, F., Pandolfi, S., Pennoni, F.: Discrete latent variable models. Annu. Rev. Stat. Appl. **9**, 425–452 (2022)
6. Baum, L.E., Petrie, T., Soules, G., Weiss, N.: A maximization technique occurring in the statistical analysis of probabilistic functions of Markov chains. Ann. Math. Stat. **41**, 164–171 (1970)
7. Biernacki, C., Celeux, G., Govaert, G.: An improvement of the NEC criterion for assessing the number of clusters in a mixture model. Pattern Recogn. Lett. **20**, 267–272 (1999)
8. Brusa, L., Bartolucci, F., Pennoni, F.: Tempered expectation-maximization algorithm for the estimation of discrete latent variable models. Comput. Stat. **38**, 1391–1424 (2023)
9. Brusa, L., Pennoni, F., Bartolucci, F.: Maximum likelihood estimation for discrete latent variable models via evolutionary algorithms. Stat. Comput. **34**, 1–15 (2024)
10. Dempster, A., Laird, N., Rubin, D.: Maximum likelihood from incomplete data via the EM algorithm. J. R. Stat. Soc. Ser. B **39**, 1–22 (1977)
11. García, F., Ivars, M., Radoselovics, J., Candau, E., Domínguez, J.: El stock de capital en España y sus comunidades autónomas: Análisis de los cambios en la composición de la inversión y las dotaciones de capital entre 1995 y 2022. In: *Documentos De Trabajo, Fundación BBVA* (2023)
12. Goodman L.: Exploratory latent structure analysis using both identifiable and unidentifiable models. Biometrika **61**, 215–231 (1974)
13. Greenacre M.: Compositional Data Analysis in Practice. Chapman & Hall/CRC Press, Boca Raton, FL (2019)
14. Greenacre, M.: Compositional data analysis. Annu. Rev. Stat. Appl. **8**, 271–299 (2021)
15. Holland, P.W., Laskey, K.B., Leinhardt, S.: Stochastic blockmodels: First steps. Soc. Networks **5**, 109–137 (1983)
16. Maier, M.: DirichletReg: Dirichlet regression for compositional data in R. In: Research Report Series / Department of Statistics and Mathematics No. 125 (2014)
17. Maruotti, A., Punzo, A.: Initialization of hidden Markov and semi-Markov models: a critical evaluation of several strategies. Int. Stat. Rev. **89**, 447–480 (2021)
18. Pandolfi, S., Bartolucci, F., Pennoni, F.: A hidden Markov model for continuous longitudinal data with missing responses and dropout. Biom. J. **65**, 1–25 (2023)
19. Schwarz, G.: Estimating the dimension of a model. Ann. Stat. **6**, 461–464 (1978)
20. Skrondal, A., Rabe-Hesketh, S.: Latent variable modelling: A survey. Scand. J. Stat. **34**, 712–745 (2007)
21. Smithson, M., Shou, Y.: Flexible CDF-quantile distributions on the closed unit interval, with software and applications. Commun. Stat. Theory Methods **11**, 3876–3898 (1978)
22. Welch, L.R.: Hidden Markov models and the Baum-Welch algorithm. IEEE Information Theory Society Newsletter **53**, 10–13 (2003)
23. Zucchini, W., MacDonald, I.L., Langrock, R.: Hidden Markov Models for Time Series: An Introduction Using R. CRC Press, Boca Raton, FL (2017)

Modelling Football Players' Field Position via Mixtures of Gaussians with Flexible Weights

Marco Berrettini, Giuliano Galimberti, Thomas Brendan Murphy, and Saverio Ranciati

Abstract This contribution provides an empirical analysis on player position on the field throughout a football match. To identify potential phases of the game, a Bayesian mixture of experts model is defined where the relationship between the location of a player and their teammates is specified using a mixture via a flexible formulation of the component weights, which are modelled as smooth functions of concomitant covariates. The analysis shows that player positions can be decomposed into phases of play that are related to the other teammates' locations on the field.

1 Introduction

Remote sensing analysis is surging as a tool for sport statistics, with telemetry data recording characteristics of players and game objects in competition events. Pettersen et al. [1] present a dataset of body-sensor traces and corresponding videos from three professional football games captured in late 2013 at the Alfheim Stadium in Tromsø, Norway. The dataset used in this contribution consists of bidimensional locations of players on the field recorded during the full match, sampled at 20 Hz using the ZXY Sport Tracking system. Due to privacy reasons, each player is identified only by a random numeric tag, in order to avoid reverse identification. Players' spatial movements throughout the game are complex and dependent on many more factors, i.e., a player's role and its corresponding assigned field area. Berrettini et al. [2] presented an analysis based on mixture models to study how a player's position is affected by another teammate's location. The goal was to

M. Berrettini (✉) · G. Galimberti · S. Ranciati
Department of Statistical Sciences, University of Bologna, Bologna, Italy
e-mail: marco.berrettini2@unibo.it; giuliano.galimberti@unibo.it; saverio.ranciati2@unibo.it

T. B. Murphy
School of Mathematics and Statistics, University College Dublin, Belfield, Ireland
e-mail: brendan.murphy@ucd.ie

G. Giordano et al. (eds.), *Statistical Models and Learning Methods for Complex Data*, Studies in Classification, Data Analysis, and Knowledge Organization,
https://doi.org/10.1007/978-3-031-84702-8_3

identify a finite number of different phases of the game, capturing the local spatial behavior of the players. Mixture modeling is particularly suited for this task, being an unsupervised learning approach. In particular, it allows for establishing the relationships between positions and game phases within the model formulation. In this contribution, the analysis is extended by considering the position of all the other teammates: the aim is to understand to what extent conditioning on the information of more players, instead of a single one, improves the detection of the phases of the game. In addition, in order to shed light on the potential symmetry of the gameplay with respect to the football pitch, the results obtained on the player covering the right fullback position, identified with tag 9, are compared with those from the left fullback (tag 8), operating on the opposite side of the field. The adopted framework is given in detail in the next section, while the results are described in Sect. 3.

2 Model Definition and Inference

To carry out the empirical analysis, further assumptions and preprocessing steps are needed. First, observations are considered to be independent across time: data are thinned out to $i = 1, 2, \ldots, 501$ observations over more than 90 minutes of play, leading to a duration of approximately 10 seconds between consecutive observations, making the assumption more realistic. Secondly, play direction changes mid-game so a 180° rotational transformation is performed on locations from the second half of the football match. The study is carried out separately for the two selected players. In each analysis, the information on the fullback is considered to be dependent on the remaining teammates. In particular, it is assumed that this player's coordinates (y_1, y_2) are approximately distributed as a bivariate Gaussian, with the centroid of the bidimensional location coordinates (x_1, x_2) of the remaining nine outfield teammates (excluding the goalkeeper) taken as concomitant covariates [3]. Let $\mathbf{c}$ be a vector of latent variables representing the potential phases of the game; for each time point i of the game, $c_i = g$ if i belongs to cluster g. Conditioning on $\mathbf{c}_i$ and $\mathbf{x}_i$ then $\mathbf{y}_i$ follows a Gaussian distribution with vector of means $\boldsymbol{\mu}_{c_i}$ and positive definite covariance matrix Σ_{c_i}. Hence, the conditional density can be written as the following mixture:

$$f(\mathbf{y}_i|\mathbf{x}_i) = \sum_{g=1}^{G} \pi_g(\mathbf{x}_i)\phi(\mathbf{y}_i|\mu_g, \Sigma_g),$$

with $\phi(\cdot|\boldsymbol{\mu}_g, \Sigma_g)$ being the density of a bivariate Gaussian distribution with mean μ_g and covariance Σ_g, and component weights $\pi_g(\mathbf{x}_i) = \Pr(c_i = g|\mathbf{x}_i) > 0$, so that $\sum_{g=1}^{G} \pi_g(\mathbf{x}_i) = 1$. In this formulation, at each sampled time point i, the location of the teammates' centroid affects explicitly the probability that the player covering the fullback position is allocated to one of the phases of the game $g = 1, 2, \ldots, G$. Furthermore, the bidimensional location x_1 and x_2 are assumed to have an additive effect on the log-odds of the component weights. To allow for flexible specification of such probabilities, a similar methodology to that proposed by Berrettini et al. [4] for latent class models is adapted to the continuous case; this model is a flexible mixture of experts model [5, 6]. More specifically, log-odds of prior probabilities are expressed as sums of smooth functions of the covariates:

$$\log\left(\frac{\pi_g(\mathbf{x}_i)}{\pi_1(\mathbf{x}_i)}\right) = \eta_{gi} = \beta_{g0} + s_{g1}(x_{i1}) + s_{g2}(x_{i2}), \qquad \text{for } g = 2, \ldots, G, \tag{1}$$

where β_{g0} is the intercept and $s_{g1}(\cdot)$ and $s_{g2}(\cdot)$ are centered smooth functions of longitude and latitude, respectively, represented through Bayesian P-splines [7]

$$s_{gj}(x_{ij}) = \sum_{k=1}^{K} \beta_{gjk} B_{jk}(x_{ij}), \qquad \text{for } j = 1, 2,$$

with $B_{jk}(\cdot)$ being the k-th B-spline basis function for a cubic spline.

Estimation is carried out following the data augmentation scheme suggested by Frühwirth-Schnatter et al. [8]. A Gaussian Markov random field [9] is considered as prior for the β_{gjk} coefficients associated to each $s_{gj}(\cdot)$, with $g = 2, \ldots, G$ and $j = 1, 2$. Furthermore, Gaussian and inverse Wishart priors are assigned, respectively, to $\boldsymbol{\mu}_g$ and Σ_g, for $g = 1, 2, \ldots, G$, as in [10]. The resulting MCMC algorithm does not require any Metropolis-Hastings step and a sequential Gibbs sampling scheme from full conditionals distribution is adopted.

3 Applications to Player Position Data

The proposed model with centroid locations as covariates and an additive structure of the predictors is fitted with $G = 1, 2, \ldots, 7$, and the optimal value for G is selected using the AICM criterion [11]. In order to highlight the advantages of this specification, other additional models are fitted for comparison. Namely, positions of the closest center-back (as in [2]) are also considered, instead of centroid locations. Moreover, models characterized by a linear predictor $\eta_{gi} = \beta_{g0} + \beta_{g1}x_{i1} + \beta_{g2}x_{i2}$ ($g = 2, \ldots, G$) and mixture models without concomitant covariates are examined too. For all these models, observations are allocated into the G components (phases of the game) using the maximum-a-posteriori (MAP) rule.

3.1 Right Fullback (Player 9)

Table 1 summarizes the results obtained on the right fullback. According to AICM, the best model is characterized by $G = 6$ components with additive predictors based on the teammates' centroid. It is worth mentioning that this model is to be

Table 1 Optimal number of components G and corresponding AICMs of each selected model fitted on Player 9 data. Values of the best model overall are highlighted in bold

Covariate	Centroid		Center-back (Player 13)		None
Predictor structure	Additive	Linear	Additive	Linear	–
G	**6**	2	3	2	1
AICM	**6666.563**	7585.646	7229.84	7565.036	7933.032

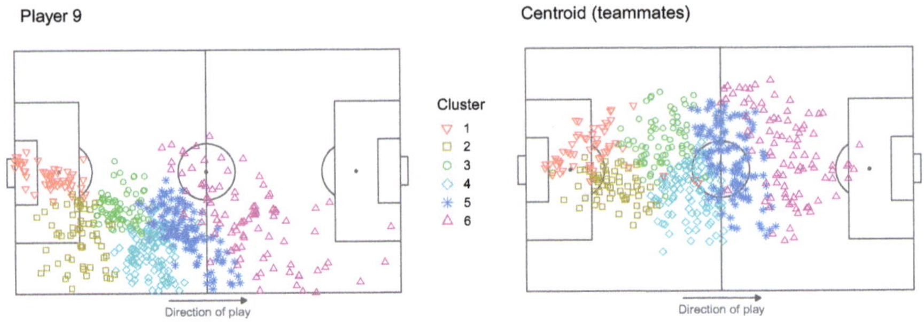

Fig. 1 Locations of Player 9 (left plot) and centroid of the other teammates. Different colors and dot symbols correspond to different clusters

preferred to the model described in [2] for the same player. In particular, a larger number of components allows a richer characterization of the flow of the game. For this model, the partitioned data points are visualized in Fig. 1 which shows the player locations during the match. There is no sharp separation between clusters, and, indeed, a finite mixture of Gaussians without covariates would suggest a single component solution. The identified clusters may still be interpreted as six phases of the game: Clusters 1 and 2 seem to refer to the defensive phase; Clusters 3 to 5 correspond to intermediate phases; Cluster 6 appears to capture the offensive phase. Figure 2 depicts through heatmaps the log-odds of each prior probability relative to its complement, computed as

$$\log\left(\frac{\pi_g(\mathbf{x}_i)}{1-\pi_g(\mathbf{x}_i)}\right) = \log\left(\frac{\pi_g(\mathbf{x}_i)}{\pi_1(\mathbf{x}_i)}\right) - \log\left(\sum_{h\neq g}\frac{\pi_h(\mathbf{x}_i)}{\pi_1(\mathbf{x}_i)}\right), \quad g = 1, \ldots, G, \tag{2}$$

with $\pi_g(\mathbf{x}_i)/\pi_1(\mathbf{x}_i)$ defined as in Eq. (1); to highlight the impact of teammates' centroid location on the position of Player 9, the estimated component densities $\phi(\cdot|\hat{\boldsymbol{\mu}}_g, \hat{\Sigma}_g)$, $g = 1, \ldots, 6$ have been added to the plots. As far as the mixture weights are concerned, it is interesting to note that the main difference between Cluster 6 (offensive phase) and the remaining clusters is due to the effect of the long side direction (x_1) of the field. Similar conclusions can be drawn for Cluster 5 (forward-intermediate phase). On the contrary, the short side seems to contribute especially in characterizing portions of the defensive and the backward-intermediate phases.

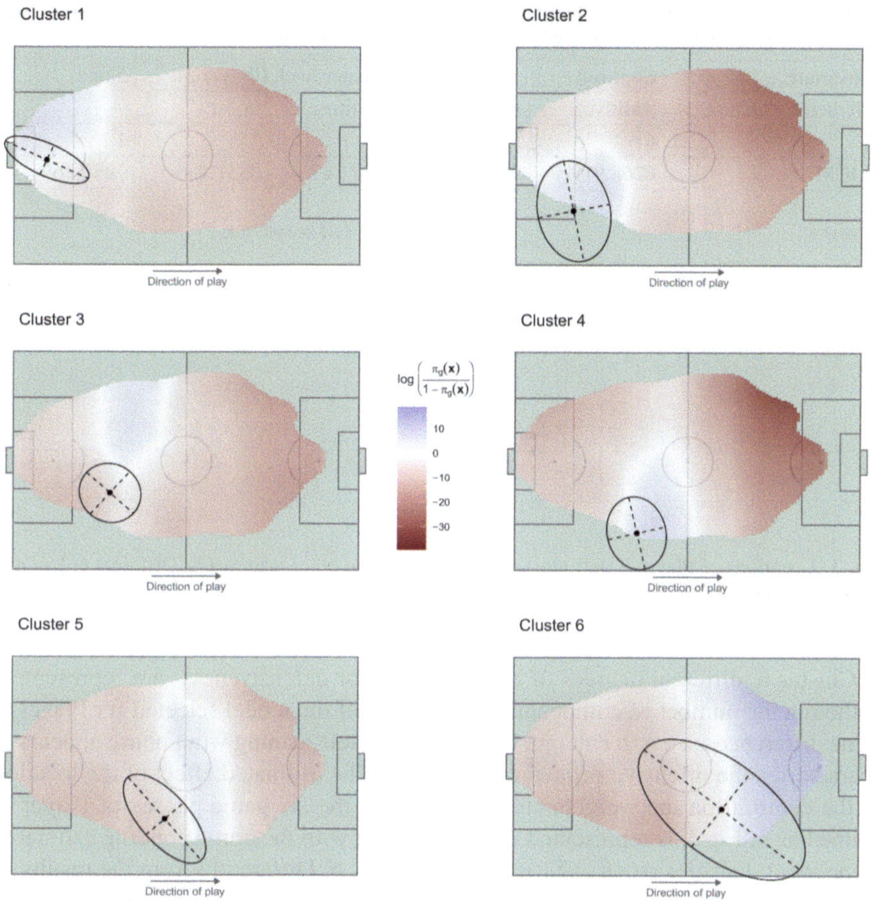

Fig. 2 Estimated log-odds of mixture weights along with component densities for Players 9. The ellipse corresponding to the g-th cluster is centered in $\hat{\boldsymbol{\mu}}_g$, $g \in \{1, \ldots, 6\}$, while the size corresponds to a 0.9 confidence level

3.2 *Left Fullback (Player 8)*

On the opposite side, AICM leads to the selection of a model with the same features: $G = 6$ components and an additive predictor involving the centroid locations (see Table 2). Figure 3 shows the allocations of the positions of Player 8 (left plot) and of the teammates' centroid to the six clusters. Some overlap is present also among these clusters, and, in fact, a finite mixture of Gaussians with constant component weights is still not able to identify more than one cluster, according to AICM. Similarly to the previous subsection, an interpretation of these clusters as game phases can be provided: moving along the long side of the field, from Cluster 1

Table 2 Optimal number of components G and corresponding AICMs of each selected model fitted on Player 8 data. Values of the best model overall are highlighted in bold

Covariate	Centroid		Center-back (Player 2)		None
Predictor structure	additive	linear	additive	linear	–
G	**6**	3	5	3	1
AICM	**6768.919**	7915.021	6871.47	7351.136	7950.369

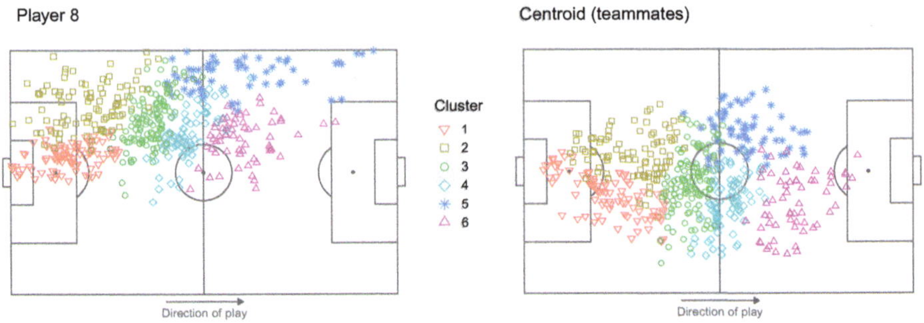

Fig. 3 Locations of Player 8 (left plot) and centroid of the other teammates. Different colors and dot symbols correspond to different clusters

to Cluster 6, different degrees of offensiveness (or defensiveness) are represented. Although the number of components is the same of the model selected for Player 9, some differences seem to emerge. In particular, a left-leaning wing phase appears to characterize the offensive behavior of Player 8. The estimated effect of the location of the centroid on the weights corresponding to the six game phases of Player 8, defined as in Eq. 2, is represented in Fig. 4, along with the corresponding estimated component densities $\phi(\cdot|\hat{\boldsymbol{\mu}}_g, \hat{\boldsymbol{\Sigma}}_g)$, for $g = 1, \ldots, 6$. Differently from the results of the previous subsection, now both dimensions of the field present a notable effect on the offensive phases. In particular, when the centroid moves to the left side of the field, the odds of belonging to Cluster 5 increase. On the contrary, the backward-intermediate phase is associated to a unique cluster (Cluster 3) whose odds vary only along the long side dimension of the field. The interpretation of the remaining estimated effects is in line with the one given in Sect. 3.1.

4 Conclusions

The results seems to indicate that the inclusion of covariates in the specification of the prior probabilities not only leads to models with lower AICM but also helps in the identification of different game phases. This is particularly evident when considering additive predictors. Also, the choice of the covariates plays a fundamental role, as the improvements obtained when using the teammates' centroid location suggest. From a practical point of view, it is possible to conclude

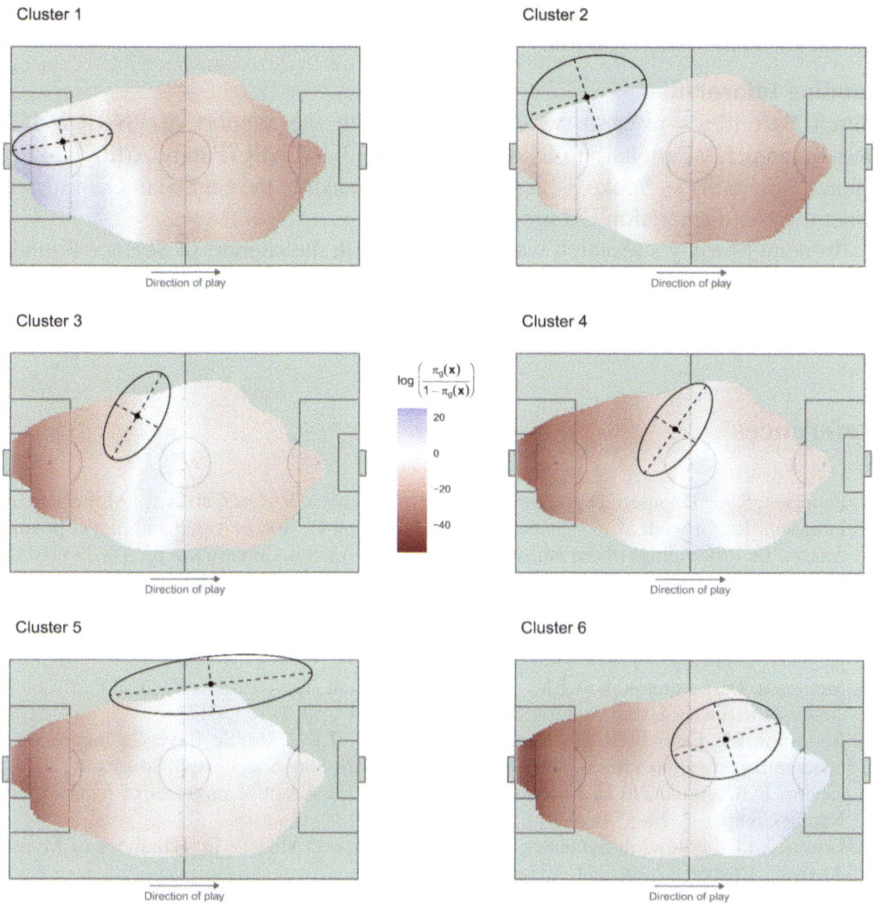

Fig. 4 Estimated log-odds of mixture weights along with component densities for Player 8. The ellipse corresponding to the g-th cluster is centered in $\hat{\boldsymbol{\mu}}_g$, $g \in \{1, \ldots, 6\}$, while the size corresponds to a 0.9 confidence level

that the two fullbacks, although playing similar positions, seem to react differently to the moves made by the teammates. In particular, the second analysis highlights an offensive phase which takes place on the left wing for Player 8, which has no analog on the right for Player 9. This may be the result of some specific tactical choices made by the coach combined with individual characteristics of the players. In spite of the promising results, however, it is worth mentioning that some aspects of the analysis may require some improvements. First of all, it can be seen that some ellipses representing the fitted bivariate Gaussian components tend to exceed the limits of the field (see Figs. 2 and 4), hinting at the possibility of using asymmetric or truncated alternative distribution to handle these positional data. In addition, the independence assumption could be questionable. In this perspective, studying the spatial patterns instead of single positions (as suggested by Beernaerts et al. [12]) represents a promising solution.

Funding Information
Marco Berrettini's research was conducted with the support of PNRR-M4C2-Investimento 1.3, Partenariato Esteso PE00000013—"FAIR (Future Artificial Intelligence Research)"—Spoke 8 "Pervasive AI," funded by the European Commission under the NextGeneration EU programme.

Brendan Murphy's research was conducted with the support of Science Foundation Ireland (12/RC/2289_P2) at Insight the SFI Research Centre for Data Analytics at University College Dublin.

References

1. Pettersen, S.A., Johansen, D., Johansen, H., Berg-Johansen, V., Gaddam, V.R., Mortensen, A., Langseth, R., Griwodz, C., Stensland, H.K., Halvorsen, P.: Soccer video and player position dataset. In: Proceedings of the 5th ACM Multimedia Systems Conference, pp. 18–23 (2014)
2. Berrettini, M., Galimberti, G., Murphy, T.B., Ranciati, S.: Modelling soccer players field position via mixture of Gaussians with flexible weights. In: Coretto, P., Giordano, G., La Rocca, M., Parrella, M.L., Rampichini, C. (eds.) CLADAG 2023 Book of abstracts and short papers, pp. 351–354. Pearson, London (2023)
3. Memmert, D., Lemmink, K.A.P.M., Sampaio, J.: Current Approaches to Tactical Performance Analyses in Soccer Using Position Data. Sports Med 47, pp. 1–10 (2017)
4. Berrettini, M., Galimberti, G., Ranciati, S., Murphy, T.B.: Flexible Bayesian modelling of concomitant covariate effects in mixture models. J. Roy. Stat. Soc. C-App. To appear (2024)
5. Jacobs, R.A., Jordan, M.I., Nowlan, S.J., Hinton, G.E.: Adaptive mixtures of local experts. Neural Comput. **3**, 79–87 (1991)
6. Gormley, I.C., Frühwirth-Schnatter, S.: Mixture of Experts Models. In: Handbook of Mixture Analysis, Chapter 12, pp. 271–307 (2018)
7. Lang, S., Brezger, A.: Bayesian P-splines. J. Comput. Graph. Stat. **13**, 183–212 (2004)
8. Frühwirth-Schnatter, S., Pamminger, C., Weber, A., Winter-Ebmer, R.: Labor market entry and earnings dynamics: Bayesian inference using mixtures-of-experts Markov chain clustering. J. App. Economet. **27**, 1116–1137 (2012)
9. Rue, H., Held, L.: Gaussian Markov Random Fields: Theory and Applications. CRC Press, New York (2005)
10. Frühwirth-Schnatter, S.: Finite Mixture and Markov Switching Models. Springer, Berlin (2006)
11. Raftery, A.E., Newton, M.A., Satagopan, J.M., Krivitsky, P.N.: Estimating the integrated likelihood via posterior simulation using the harmonic mean identity. In: Bernardo, J.M., Bayarri, M.J., Berger, J.O., Dawid, A.P., Heckerman, D., Smith, A.F.M., West, M. (eds.) Bayesian Statistics, vol. 8, pp. 371–416. Oxford University, Oxford (2007)
12. Beernaerts, J., De Baets, B., Lenoir, M., Van de Weghe, N.: Spatial movement pattern recognition in soccer based on relative player movements. PloS One **15**, e0227746 (2020)

Estimation Issues in Multivariate Panel Data

Silvia Bianconcini and Silvia Cagnone

Abstract Latent variable models are a powerful tool in various research fields when the constructs of interest are not directly observable. However, the likelihood-based model estimation can be problematic when dealing with many latent variables and/or random effects since the integrals involved in the likelihood function do not have analytical solutions. In the literature, several approaches have been proposed to overcome this issue. Among them, the pairwise likelihood method and the dimension-wise quadrature have emerged as effective solutions that produce estimators with desirable properties. In this study, using a simulation study, we compare a weighted version of the pairwise likelihood method with the dimension-wise quadrature for a latent variable model for binary longitudinal data.

1 Latent Variable Models for Longitudinal Binary Data

Let $\mathbf{y}_1, \mathbf{y}_2, \ldots, \mathbf{y}_p$ be vectors of p binary observed variables each of them observed at T different occasions, such that $\mathbf{y}'_j = (y_{1j}, y_{2j}, \ldots, y_{Tj})$. $\mathbf{z}' = (z_1, z_2, \ldots, z_T)$ is a T-dimensional vector of latent variables that account for the associations among the p items at each time point, whereas $\mathbf{u}' = (u_1, u_2, \ldots, u_p)$ represent p-dimensional vector of random effects accounting for the associations of the same item at different time points. The joint density of the observed variables can be defined as [1]

$$f(\mathbf{y}) = \int_{R^q} g(\mathbf{y} \mid \mathbf{z}, \mathbf{u}) h(\mathbf{z}, \mathbf{u}) d\mathbf{z} d\mathbf{u}$$

where $g(\mathbf{y} \mid \mathbf{z}, \mathbf{u})$ is referred to as measurement part of the model, being

S. Bianconcini (✉) · S. Cagnone
Department of Statistical Sciences, University of Bologna, Bologna, Italy
e-mail: silvia.bianconcini@unibo.it; silvia.cagnone@unibo.it

G. Giordano et al. (eds.), *Statistical Models and Learning Methods for Complex Data*, Studies in Classification, Data Analysis, and Knowledge Organization,
https://doi.org/10.1007/978-3-031-84702-8_4

$\mathbf{y} = (\mathbf{y}_1, \mathbf{y}_2, \ldots, \mathbf{y}_p)$, and $h(\mathbf{z}, \mathbf{u})$ as structural part of the model. The dimension of the integral is $q = p + T$.

The measurement part of the model is defined as a generalized linear model with the random component given by

$$g(\mathbf{y}|\mathbf{z}, \mathbf{u}) = \prod_{t=1}^{T}\prod_{j=1}^{p} g(y_{tj}|z_t, u_j) = \prod_{t=1}^{T}\prod_{j=1}^{p} \pi_{tj}(z_t, u_j)^{y_{tj}}(1 - \pi_{tj}(z_t, u_j))^{(1-y_{tj})},$$

where the first equality comes from the conditional independence assumption between items and over time. Each $g(y_{tj}|z_t, u_j)$ follows a Bernoulli distribution of parameter $\pi_{tj}(z_t, u_j)$, that is, the probability of success of item j at time t. The systematic component defines the linear predictor:

$$\eta_{tj} = \alpha_{0tj} + \alpha_{tj} z_t + u_j$$

where α_{0tj}'s are item and time-dependent intercepts and α_{tj}'s are item and time-dependent factor loadings. We consider the logit as the link function between the systematic component and the conditional means of the random component.

As for the structural part of the model, we assume that the latent variables follow a nonstationary autoregressive process of the first order [5] as follows:

$$z_t = \phi z_{t-1} + \delta_t \tag{1}$$

where ϕ is the autoregressive coefficient, $\delta_t \sim N(0, 1)$ and $z_1 \sim N(0, \sigma_1^2)$.

Moreover, the joint density $h(\mathbf{z}, \mathbf{u})$ is a multivariate normal with zero mean vector and block diagonal covariance matrix $\boldsymbol{\Psi}$ that contains the matrices $\boldsymbol{\Omega} = diag_{j=1,\ldots,p}\{\sigma_{uj}^2\}$ and the autocovariance matrix $\boldsymbol{\Gamma}$ of the latent variables whose inverse has the very well-known special pattern:

$$\boldsymbol{\Gamma}^{-1} = \begin{bmatrix} \frac{1}{\sigma_1^2} + \phi^2 & -\phi & 0 & \ldots & 0 & 0 & 0 \\ -\phi & 1+\phi^2 & -\phi & \ldots & 0 & 0 & 0 \\ 0 & -\phi & 1+\phi^2 & \ldots & 0 & 0 & 0 \\ \vdots & \vdots & \vdots & \ddots & \vdots & \vdots & \\ 0 & 0 & 0 & \ldots & -\phi & 1+\phi^2 & -\phi \\ 0 & 0 & 0 & \ldots & 0 & -\phi & 1 \end{bmatrix}$$

2 Model Estimation

Model estimation is usually performed by using a full maximum likelihood method. Given a sample of size n, the log-likelihood results

$$L(\boldsymbol{\theta}) = \sum_{i=1}^{n} \log f(\mathbf{y}_i, \boldsymbol{\theta}) = \sum_{i=1}^{n} \log \int_{R^q} g(\mathbf{y}_i \mid \mathbf{z}_i, \mathbf{u}_i) h(\mathbf{z}_i, \mathbf{u}_l) d\mathbf{z}_i d\mathbf{u}_i \tag{2}$$

where $\boldsymbol{\theta}$ is the vector of parameters to be estimated. Problems related to the maximization of the log-likelihood arise due to the presence of the latent variables. Maximation techniques, such as expectation-maximization (EM) or direct maximization algorithms, are commonly applied.

Furthermore, in the presence of discrete observations, the multidimensional integral in (2) is not solvable analytically. Recent solutions proposed in the literature to address this problem include the pairwise likelihood (PL) approach [6] and the dimension-wise (DW) quadrature method [4]. An initial comparison between these two methods has shown that DW, with a reduced number of integrals (small or equal to 2), outperforms the classical PL method [2]. This study compares DW with a weighted version of PL [7], aiming to evaluate whether or under which conditions one of the two methods can be preferred over the other. The weighted version does not include all the log-pairwise components to avoid including irrelevant information that could impact the accuracy of the estimators.

2.1 Pairwise Likelihood of Order d

The PL estimator is obtained by maximizing bivariate likelihood products containing the most model parameter information. In the latent variable model for longitudinal binary data described in Sect. 2, the bivariate density for a pair of responses is

$$f(y_{ijt}, y_{ij't'}; \theta) =$$
$$\int_{\mathbb{R}^4} g(y_{ijt}|z_{it}, u_{ij}) g(y_{ij't'}|z_{it'}, u_{ij'}) h(z_t, z_{t'}, u_j, u_{j'}) dz_t dz_{t'} du_j du_{j'}.$$

The dimension of the integrals involved in the expression of $f(y_{ijt}, y_{ij't'}; \theta)$ is four, and if $j = j'$ or $t = t'$, it reduces to three. Thus, they can be easily approximated using the Gauss Hermite (GH) quadrature method. Due to the autoregressive pattern assumed for the latent variable process, closer pairs are expected to be more informative. Hence, we use a PL likelihood constructed from marginal probabilities of observed pairs distant less than $d \geq 0$ time points. This produces a weighted log PL likelihood of order d defined as

$$pl^{(d)}(\theta; \mathbf{y}) = \sum_{i} \sum_{j,j',t<t'} \log f(y_{ijt}, y_{ij't'}; \theta) \mathcal{I}_{[0,d]}(t' - t). \tag{3}$$

$\mathcal{I}_{[0,d]}$ is the indicator function, equal to 1 if $(t' - t) \in [0, d]$ and 0 otherwise.

2.2 The Dimension-Wise Method

The DW method is based on the following representation of the marginal density function:

$$f(\mathbf{y};\theta) = |\mathbf{C}_{mo}| \int_{R^q} \frac{\prod_{j=1}^{p} g(y_j|\mathbf{C}_{mo}\mathbf{b}^* + \mathbf{b}_{mo})h(\mathbf{C}_{mo}\mathbf{b}^* + \mathbf{b}_{mo})}{\phi(\mathbf{b}^*;\mathbf{0},\mathbf{I})}\phi(\mathbf{b}^*;\mathbf{0},\mathbf{I})d\mathbf{b}^* =$$

$$= |\mathbf{C}_{mo}| \int_{R^q} m(\mathbf{b}^*)\phi(\mathbf{b}^*;\mathbf{0},\mathbf{I})d\mathbf{b}^* = |\mathbf{C}_{mo}|E_{\phi}[m(\mathbf{b}^*)] \quad (4)$$

where $\mathbf{b} = (\mathbf{z}, \mathbf{u})$, $\Sigma_{mo} = \mathbf{C}_{mo}\mathbf{C}'_{mo}$ and $\phi(\cdot)$ is the standard normal density function.

DW consists in approximating the function $m(\mathbf{b}^*)$ as follows [4]:

$$\hat{m}(\mathbf{b}^*) = \sum_{l=0}^{s}(-1)^l \binom{q-s+l-1}{l} m_{s-l}(\mathbf{b}^*) = \sum_{l=0}^{s} A_l m_{s-l}(\mathbf{b}^*) \quad (5)$$

where $m_{s-l}(\mathbf{b}^*) = m(0,\cdots,b^*_{k_1},0\cdots,b^*_{k_{s-l}},\cdots,0)$ and $A_l = (-1)^l \binom{q-s+l-1}{l}$. Replacing (5) in (4), we obtain the approximate density function:

$$f_a(\mathbf{y};\theta) = f_L + |\mathbf{C}_{mo}| \times \left[\sum_{l=0}^{s-1} A_l \int_{R^{s-l}} \sum_{k_1<\ldots<k_{s-l}} m_{s-l}(\mathbf{b}^*)\phi(b^*_{k_1})\cdots\phi(b^*_{k_{s-i}})db^*_{k_1}..db^*_{k_{s-l}}\right] \quad (6)$$

where f_L denotes the classical Laplace approximation of the integral when $s = 0$. The dimension of the integrals in expression (6) depends on the choice of s. With small values of s, the integrals can be easily approximated using the GH quadrature. The limiting cases obtained by placing $s = 0$ and $s = q$ are the classical Laplace and the adaptive GH quadrature method, respectively.

3 Simulation Study

We perform a simulation study with three items ($p = 3$) observed on T occasions equal to 6,15,30 on 200 and 1000 individuals ($n = 200, 1000$). The autoregressive parameter ϕ is set equal to 0.5 and 0.9.

We consider the UnWeighted (UW) PL method where all the pairs are involved and the PL of order $d = 1, 2, 3, 4$. As for DW, we set $s = 0, 1, 2, 3$. For both methods, the number of quadrature points is fixed at eight. Five hundred replications are generated for each condition of the study.

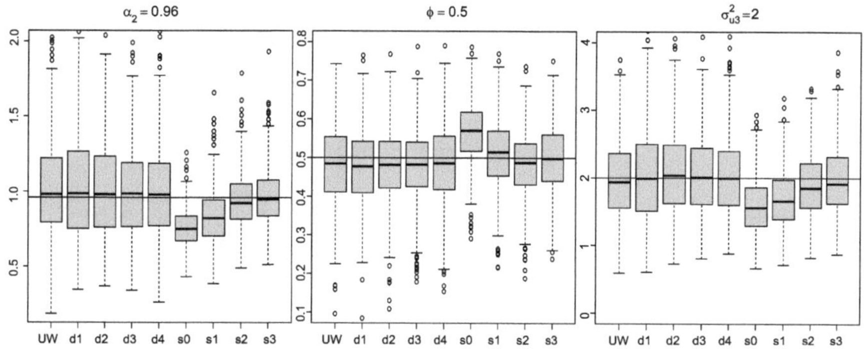

Fig. 1 Boxplots of parameter estimates, $T = 6$, $\phi = 0.5$, $n = 200$

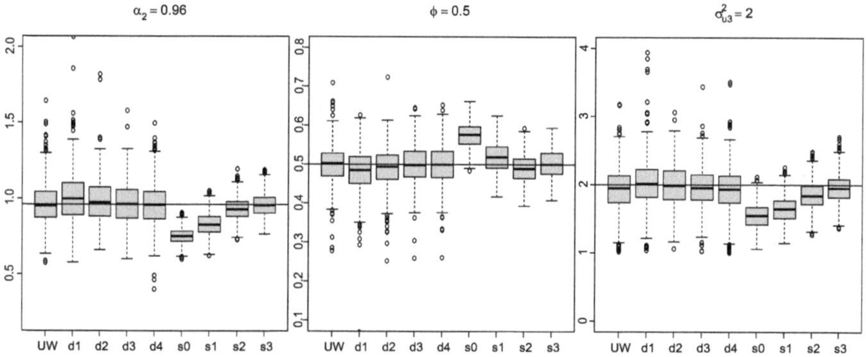

Fig. 2 Boxplots of parameter estimates, $T = 6$, $\phi = 0.5$, $n = 1000$

The boxplots of parameter estimates for α_2, ϕ, and $\sigma^2_{u_3}$ are presented in the following Figs. 1 and 2. The behavior observed for these estimators has also been observed for the remaining parameter estimates, but they are not reported due to space constraints.

Figure 1 illustrates the results for the scenario with six occasions ($T = 6$) observed for 200 individuals ($n = 200$) and with the autoregressive parameter set equal to 0.5 ($\phi = 0.5$). It can be noticed that the performance of the DW method improves as s increases. The estimator obtained with $s = 0$, which corresponds to the Laplace estimator, exhibits the highest bias, while the estimator obtained with $s = 3$ demonstrates the best accuracy. Moreover, the UW pairwise and all versions of pairwise of order d perform less efficiently than DW with $s = 3$.

With a larger sample size, as shown in Fig. 2, the efficiency of all the considered estimators improves, and their behavior remains consistent with the previous scenario. Notably, for the estimator of the random effect variance $\sigma^2_{u_3}$, the UW PL method outperforms PL of any order d.

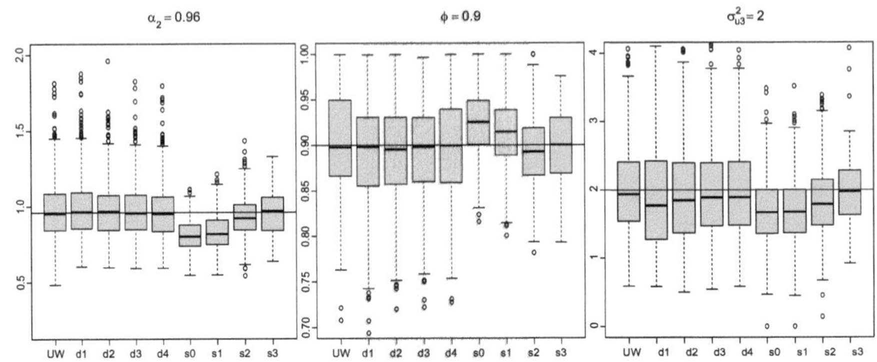

Fig. 3 Boxplots of parameter estimates, $T = 6, \phi = 0.9, n = 200$

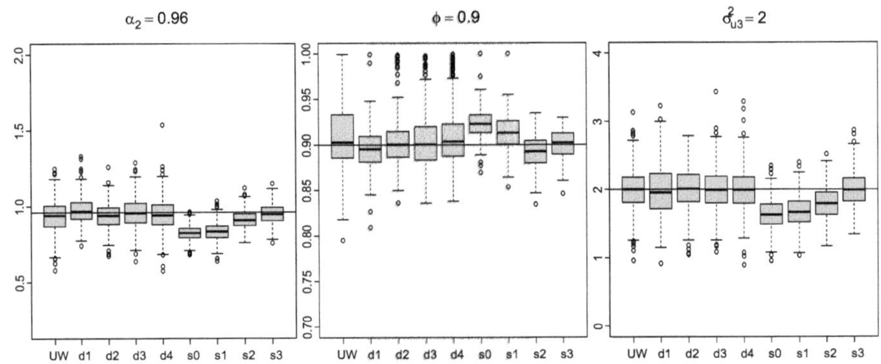

Fig. 4 Boxplots of parameter estimates, $T = 6, \phi = 0.9, n = 1000$

In Figs. 3 and 4, the results for $T = 6$ and $\phi = 0.9$ are reported for both small and large sample sizes, respectively. The behavior of the estimators is consistent with the patterns observed in the previous cases. For $n = 1000$, even DW with $s = 2$ exhibits a good performance.

Regarding the computational time (Table 1), it becomes evident that the PL method is considerably faster than the DW method. In particular, the UW PL method exhibits computational times comparable to DW with $s = 2$ for most scenarios, except for the case with $n = 200$ and $\phi = 0.5$ where its computational time is similar to that of DW with $s = 0$ and $s = 1$. The weighted PL methods up to order three also demonstrate faster computational times than the UW PL. It is worth noting that DW with $s = 3$ requires more than 1 hour for small sample sizes and more than 3 hours for large sample sizes, but it provides estimators with much better properties with respect to all the other competitors.

The case of $T = 15$ is shown in Fig. 5. It is worth noticing that, even with the small sample size, DW with $s = 2$ has remarkable accuracy and outperforms all the PL estimators. Moreover, the UW PL performs similarly, or even better, than

Table 1 Average computational time (in seconds)

		$\phi = 0.5$		$\phi = 0.9$	
		$n = 200$	$n = 1000$	$n = 200$	$n = 1000$
DW					
	$s0$	223.81	582.53	323.14	761.19
	$s1$	233.64	605.65	338.67	799.03
	$s2$	415.23	1042.38	692.90	1682.62
	$s3$	3629.66	10015.45	3998.11	11204.19
PL					
	UW	274.81	1260.98	533.59	2179.94
	$d1$	106.74	299.43	113.92	960.64
	$d2$	201.75	769.09	122.95	1238.82
	$d3$	201.28	1077.24	182.34	1681.39
	$d4$	261.52	1385.26	182.72	2685.29

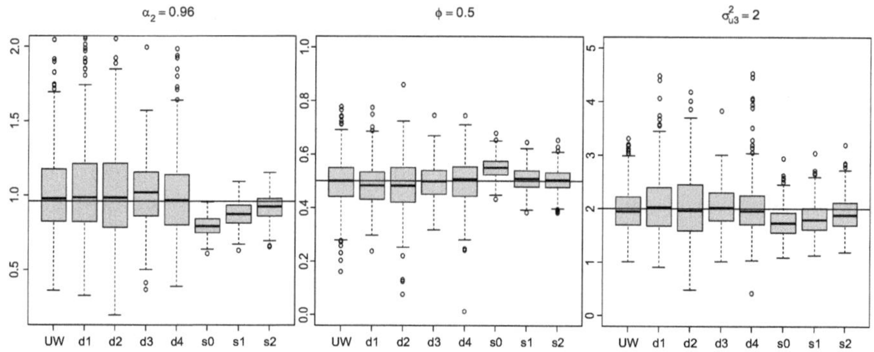

Fig. 5 Boxplots of parameter estimates, $T = 15$, $\phi = 0.5$, $n = 200$

the weighted PL estimators in this scenario. The only case in which the PL of order $d = 2$ and $d = 3$ outperforms the UW PL is when a larger number of waves are considered, that is, when $T = 30$, as illustrated in Fig. 6. Still, in this scenario, DW with $s = 2$ is superior to all the other estimators.

4 Conclusions

In this work, we conducted a comparative analysis of two limited information methods to address computational challenges in generalized linear latent variable models for longitudinal binary data. Unlike previous investigations within a time series context, where the performance of the two methods has been found to depend on the variability in the data [3], our findings for multivariate panel data reveal that DW, particularly with $s = 3$, outperforms both the unweighed and weighted PL. For

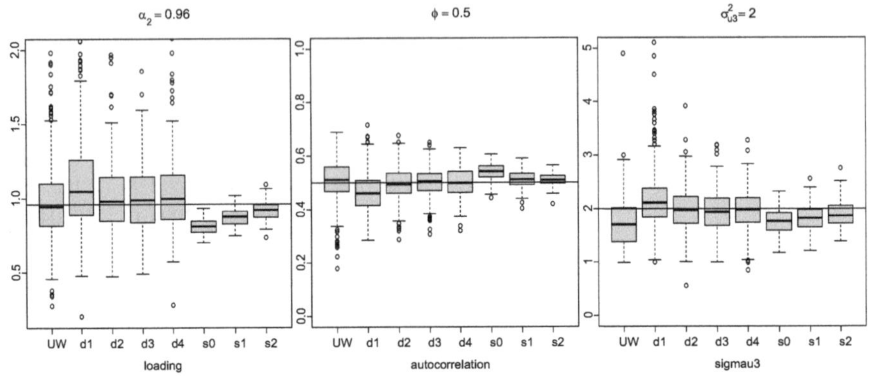

Fig. 6 Boxplots of parameter estimates, $T = 30$, $\phi = 0.5$, $n = 200$

large sample sizes and a number of occasions greater than or equal to six, DW with $s = 2$ also performs well.

For a small number of occasions T, we have observed no relevant differences between the UW PL and the PL of order d (where d ranged from 1 to 4). It is worth noting that only when $T = 30$ the PL of order $d = 2$ and $d = 3$ outperform the UW PL. However, it is important to emphasize that such a scenario is unrealistic for panel data.

Future investigations of the dimension-wise method should study the asymptotic properties of estimators in longitudinal settings and compare its performance with other approaches, such as variational approximation. In this initial study, we varied T rather than p due to its expected greater impact on pairwise order d estimators, but future research should also consider varying p. Additionally, it would be interesting to explore the method's potential in longer panel data characterized by more time points.

References

1. Bartholomew, D., Knott, M., Moustaki, I.: Latent Variable Models and Factor Analysis: A Unified Approach. Wiley Series in Probability and Statistics, New York (2011)
2. Bianconcini, S., Cagnone, S.: Comparison between different estimation methods of factor models for longitudinal ordinal data. In: Quantitative psychology. Springer Proceedings in Mathematics and Statistics, Berlin, pp. 9–21 (2021)
3. Bianconcini, S., Cagnone, S.: The dimension-wise quadrature estimation of dynamic latent variable models for count data. Comput. Stat. Data Anal. **177**, 107585 (2023)
4. Bianconcini, S., Cagnone, S., Rizopoulos, D.: Approximate likelihood inference in generalized linear latent variable models based on the dimension-wise quadrature. Electron. J. Stat. **11**, 4404–4423 (2017)

5. Cagnone, S., Moustaki, I., Vasdekis, V.: Latent variable models for multivariate longitudinal ordinal responses. Br. J. Math. Stat. Psychol. **62**, 401–415 (2009)
6. Lindsay, B.: Composite likelihood methods. In: Prabhu, N.U. (ed.). Statistical Inference from Stochastic Processes, pp. 221–239. American Mathematical Society, Providence (1988)
7. Varin, C., Czado, C.: A mixed autoregressive probit model for ordinal longitudinal data. Biostatistics **11**, 127–138 (2010)

An Application of Natural Language Processing Analysis on TripAdvisor Reviews

Barzizza Elena, Biasetton Nicolò, Disegna Marta, and Prayag Girish

Abstract Film tourism and the motives driving film tourists are important topics in tourism studies; nonetheless, only recently social media comments have been employed to understand and analyse film tourism. We analyse TripAdvisor reviews spanning from 2015 to 2022 that pertain to the Lord of the Rings and Hobbit movie trilogies, employing various known natural language processing methods to explore the connection between tourists' motivations and emotions. Such user-generated content is analysed to identify both stable trends and significant changes in tourist behaviours and preferences. We find a prevalent motive among visitors to be the allure of the movie sets, while emotions reported range from surprise and anticipation to occasional moments of fear. Theoretical insights and practical implications for destination management are discussed.

1 Introduction

Film-induced tourism, characterised by travellers visiting destinations after being inspired by films [1, 2], is a highly personal experience and leads to different emotions and behaviours from tourists [3]. This phenomenon, often driven by a desire to immerse oneself in the settings and experiences portrayed on screen, has led to a deeper exploration of the motivations and behaviours of tourists who engage in this form of travel. While the appeal of film locations is evident, encompassing aspects of romance, nostalgia, and fantasy, understanding the intricacies of

B. Elena · B. Nicolò (✉) · D. Marta
Università degli studi di Padova, Deparment of Management Engineering, Vicenza, Italy
e-mail: elena.barzizza@unipd.it; nicolo.biasetton@unipd.it; marta.disegna@unipd.it

P. Girish
Department of Management, Marketing and Tourism, University of Canterbury Business School, Christchurch, New Zealand

University of Johannesburg, Johannesburg, South Africa
e-mail: girish.prayag@canterbury.ac.nz

G. Giordano et al. (eds.), *Statistical Models and Learning Methods for Complex Data*, Studies in Classification, Data Analysis, and Knowledge Organization,
https://doi.org/10.1007/978-3-031-84702-8_5

tourist experiences and sentiments towards these destinations remains a complex endeavour. Traditionally, research in this field has relied on various methods to dissect the drivers behind film-induced tourism, including surveys, economic analyses, and destination image assessments. However, with the proliferation of social media platforms and the wealth of User Generated Content (UGC hereafter) they host, leveraging these resources for deeper insights into tourist behaviours and preferences can be of great value [4, 5]. TripAdvisor, a prominent travel platform, has emerged as a valuable repository of reviews and opinions, offering researchers a rich source of data to explore the nuances of film tourism. This paper aims to contribute to the evolving discourse on film-induced tourism by employing Natural Language Processing (NLP) techniques to analyse the impact of the Lord of the Rings (LOTR) trilogy and the Hobbit trilogy on creating a strong destination image for New Zealand. Through an exploration of TripAdvisor reviews, we seek to address fundamental questions regarding tourists' perceptions, interests, and emotional connections to film destinations. Specifically, we aim to uncover what aspects of film locations resonate most with tourists, the underlying motivations driving their visits, and the sentiments they express towards these destinations. This paper is structured as follows: Sect. 2 outlines through a brief description the analytical methodology employed. Section 3 provides insight into the data on which the case study delves into, considering both the collection step and a data description. In Sect. 4, we present the findings of our analysis of UGC from TripAdvisor, focusing on reviews related to the LOTR trilogy and the Hobbit trilogy shot in New Zealand. Finally, Sect. 5 discusses the results and concludes the paper, summarising key insights and implications for both theoretical understanding and managerial practice in the context of film-induced tourism.

2 Method

2.1 Text Mining

Text mining involves the extraction of structured and valuable information from textual data. In this study, we preprocess the textual reviews to analyse them in terms of term frequencies, bigram frequencies, term co-occurrence, and term frequencies through time, searching for specific words, combinations of words, or bigrams defined by experts. Before conducting such analysis, the following preprocessing steps were performed: (1) Text lowering (2) Punctuation removal (3) Numbers removal (4) English stopwords removal—eliminating common but non-informative words like determiners and prepositions (5) White space stripping (6) Tokenisation—dividing the document into unigrams (each single element from a string) and bigrams (each sequence of two adjacent elements from a string) (7) Stemming—reducing each token to its root form

Once the reviews are cleaned, a Document Term Matrix (DTM) is constructed, wherein each row represents a document (i.e. tourist review) and each column represents a token (i.e. preprocessed word). Each cell of the matrix contains the frequency with which a token appears in a document, enabling the identification of frequent words, expert-identified words, and associations among words.

2.2 *Topic Modelling*

Latent Dirichlet Allocation (LDA) [6], belonging to the family of topic models, is a very well-known model utilised to extract topics from the textual data. LDA is an unsupervised learning topic model that assumes each document consists of a mixture of topics, and each topic is a probability distribution over words. The model outputs two distributions: one for each document over topics (document-topic distribution θ) and one for each topic over words (topic-word distribution φ). These distributions are estimated using algorithms such as variational inference or Gibbs sampling (see [7]). In order to select the optimal number of topics to extract from the LDA model, multiple procedures have been proposed in the literature [7–9]. In this work, to validate the number of topics, the density-based methodology proposed by Cao et al. is adopted. The estimated topic-word distribution φ will be used to label the topics. In fact, it represents the predictive probability or distribution of sampling a new instance of vocabulary word from a generic topic, and it gives a list of words under each topic ranked according to their probability values.

2.3 *Sentiment and Emotion Analysis*

Sentiment Analysis (SA) [10] involves computationally identifying and categorising opinions expressed in text as positive, neutral, or negative. In this study, a lexicon-based sentiment analysis approach [11, 12] is adopted, wherein sentiment values are assigned to tokens based on matches with words in a predefined sentiment lexicon. The final sentiment orientation of a document is determined by aggregating the sentiment values of its constituent tokens. Emotion Analysis (EA) extends SA by identifying emotions expressed in text. In this study, a lexicon-based approach to EA is employed, utilising the NRC Word-Emotion Association Lexicon (EmoLex),[1,2] which associates words with sentiments (negative and positive) and eight basic emotions [13]. This approach allows for the extraction of both sentiments and

---

[1] https://nrc.canada.ca/en/research-development/products-services/technical-advisory-services/sentiment-emotion-lexicons.

[2] https://saifmohammad.com/WebPages/NRC-Emotion-Lexicon.html.

emotions from the reviews, providing deeper insights into consumer feelings and emotions expressed in the text.

3 Data

The data utilised in this study comprise user-generated reviews and opinions sourced from TripAdvisor, a prominent travel platform, focusing on experiences related to the Hobbit movies in New Zealand. This section provides a detailed description of the data collection process, including the scraping methodology and characteristics of the extracted dataset.

The collection of user reviews and opinions involved web scraping, a technique utilised to extract content and data from web pages (in this case TripAdvisor) efficiently. Specifically, the scraping process commenced with gathering information from a web page cataloguing activities available in Hinuera, New Zealand, the primary setting for the Hobbit movies. All pages featuring different activities were scraped, and reviews authored by tourists who engaged in experiences, tours, or locations associated with the Hobbit movies were included. An automated Python script facilitated the collection of 3285 textual reviews written between 2015 and 2022.

Each review in the dataset contains several key attributes: activity, the name of the experienced activity, corresponding to the TripAdvisor listing; URL, the link to the activity page on TripAdvisor; reviewer (1), the nickname of the reviewer; Type (4), the classification of the reviewer's travel type (e.g. solo, couple, friends, family, business); review title (3), the title provided by the reviewer for the review; review text (5), the detailed textual content expressing the reviewer's opinion on the experience; overall rating (2), the reviewer's rating of the experience on a five-point Likert-type scale; and date (6), the date when the review was written.

The temporal distribution of the reviews indicates that the majority were written in 2018 (40%), followed by 2019 (24.4%) and 2020 (17.4%), with fewer reviews from 2015 to 2017. Over the years, there has been an increase in the number of reviewers omitting information regarding their travel party and type. However, among those who reported such information, the majority travelled in couples or with family members.

This comprehensive dataset serves as the foundation for the application of the analyses described at Sect. 2 aiming to delve into the motivations and emotions of tourists engaging in film-induced tourism experiences in New Zealand.

4 Results

In the textual analysis based on frequency, we drew upon insights from prior literature [14, 15] to identify eight fundamental motivations for film tourism. These motivations encompassed various aspects such as landscapes, movie learning, technological aspects, immersive experiences, destination appeal, film sets, enjoyment, and social interaction. Consequently, we conducted a search within the reviews for words associated with these motives, including terms like "landscapes", "scenery", "learning", "movies", "film", "shooting", "locations", "technology", "fantasy", "world", "destination", "set", "fun", "excitement", "socialise", "interests", "experience" and "memorable". Using the Document Term Matrix (DTM), we were able to discern the most frequently used words in the reviews. Notably, certain terms like "technology" and "socialise" were absent in the visitor reviews. The results were presented in Fig. 1, showcasing the overall relative frequency of all identified words and the relative frequency of the top six words over the years, respectively. Among these, "set", "movies", and "experience" emerged as the top three most frequently used words, with "experience" gaining prominence since 2020, surpassing the others. Additionally, the analysis of bigrams, sequences of two adjacent elements from the tokenised text, revealed the appearance of 20 bigrams containing the previously identified 18 words. Notably, phrases like "movie-set" and "hobbiton-movie" stood out as the most frequent over time (see Fig. 1).

Moving on to topic modelling, we employed the Latent Dirichlet Allocation (LDA) model on preprocessed reviews. Utilising various indices proposed in literature, we determined the optimal number of topics to be four, based on [8, 9]. These topics were labelled as follows: "On-site experience", "packaged tour activities", "other regional activities", and "movie set details" based on the most relevant words identified for each topic (see Fig. 2).

Furthermore, sentiment and emotional analysis were conducted using a lexicon-based sentiment analysis (SA) approach. Each review was associated with a single topic, representing the most likely topic the reviewer wished to express. The average sentiment and emotion scores were computed by topic, revealing a predominantly positive sentiment across all topics, with "anticipation" being the primary emotion observed (Fig. 3). Overall, the results underscored the significance of on-site experiences and highlighted the positive sentiment prevalent in the reviews, offering valuable insights into the motivations and emotions of film tourists.

5 Discussion and Conclusions

Considering that film tourism and the motives driving film tourists are prominent topics in tourism studies [2, 15] in this work, we draw upon UGC investigations to explore the evolution of film tourists' motivations and emotions in Hinuera, New Zealand. From a theoretical standpoint, we establish a connection between motivation and emotions, drawing upon the framework of goal-directed behaviour theory [16] and applying it within the context of film tourism. While this link has primarily been explored in heritage tourism contexts [17, 18], our analysis reveals how motives, embedded in both cognitive (landscape, scenery, destination) and emotive cues (fun, excitement), serve as energising forces driving experiences at

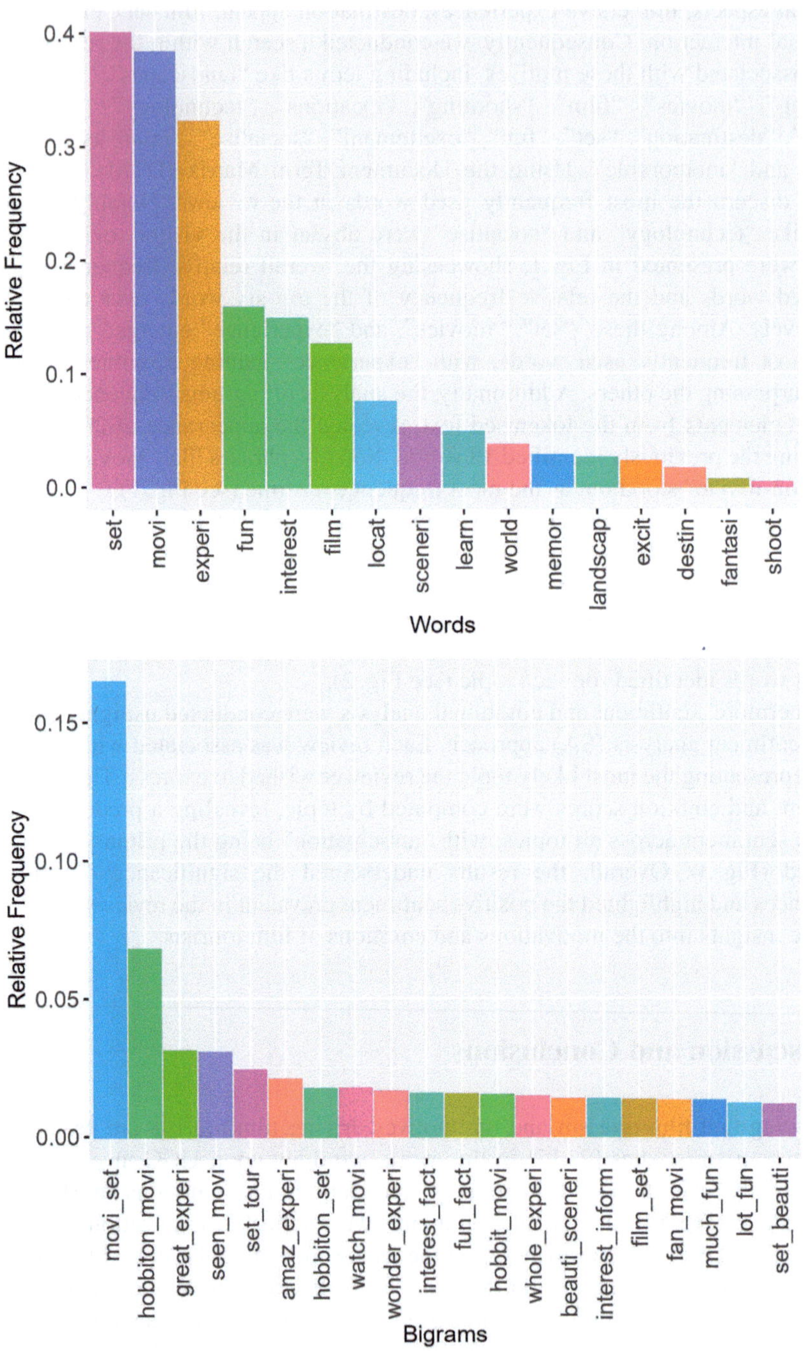

Fig. 1 Word frequency and bigram frequency

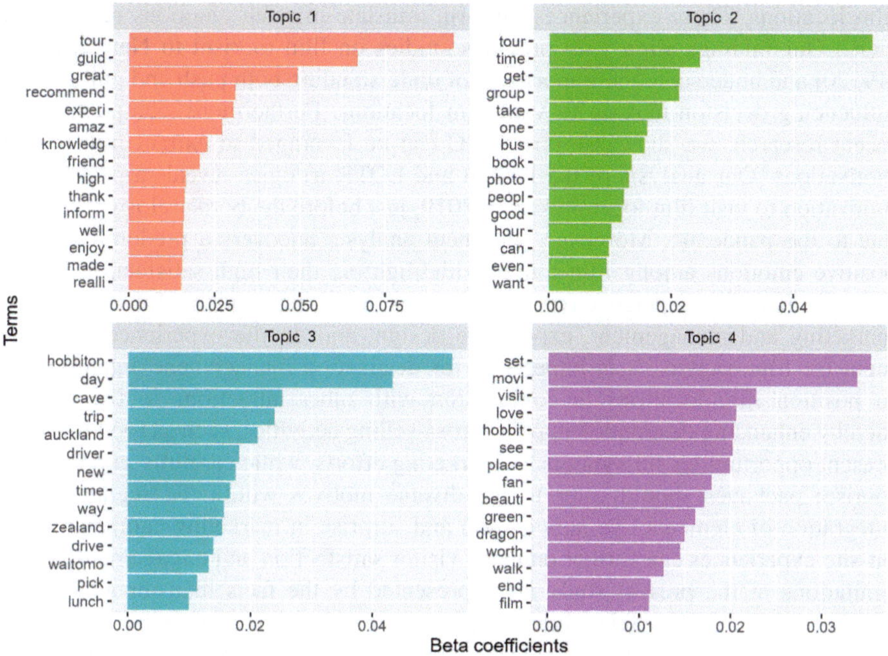

Fig. 2 LDA

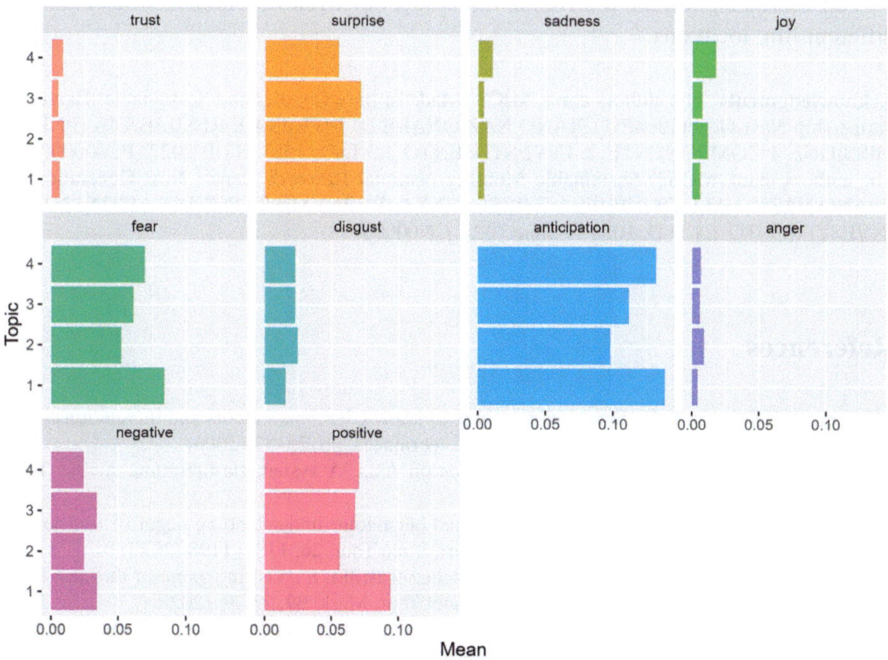

Fig. 3 Reviews' sentiment and emotions by topic

film locations. These experiences, in turn, translate into felt emotions reported in UGC. Our findings align with previous studies on film tourism to New Zealand [19, 20] and underscore the importance of understanding both push and pull factors influencing decision-making to visit film locations. Furthermore, complementing recent research, we showcase how UGC can be leveraged to extract film tourist motivations. Our analysis of the Hobbit and LOTR trilogies reveals a peak in the motivation to visit film location sets in 2019, just before the border closure in 2020 due to the pandemic. Moreover, sentiment analysis uncovers a predominance of positive emotions among film tourists, highlighting their high satisfaction levels. From a managerial perspective, our findings have implications for destination marketing and management, experience design, and on-site experience management for film tourists. It is imperative for destination management organisations to position film locations in conjunction with other attractions in the region, thereby enhancing destination appeal. Cross-selling activities, such as food tourism, present opportunities for synergistic marketing efforts, while carefully curated film tourism packages should cater to the diverse motives within the segment. The integration of elements like anticipation and surprise in marketing campaigns and on-site experiences can further enhance visitor satisfaction and engagement. Some limitations of the present study are represented by the facts that our analysis is restricted to UGC from TripAdvisor, representing only a subset of online comments about the destination and that the UGC analysed pertains specifically to attractions directly connected to the Hobbit and LOTR trilogies in the primary region where Hobbiton is located, potentially overlooking other film locations. Moreover, while we considered eight main motives in our study, other relevant motives may exist for different film locations.

Acknowledgments The authors thank MICS (Made in Italy Circular and Sustainable) Extended Partnership Next-GenerationEU (PIANO NAZIONALE DI RIPRESA E RESILIENZA (PNRR)-MISSIONE 4 COMPONENTE 2, INVESTIMENTO 1.3-D.D. 1551.11-10-2022, PE00000004). The authors thank MOST, Sustainable Mobility National Research Centre Next-GenerationEU (PIANO NAZIONALE DI RIPRESA E RESILIENZA (PNRR)-MISSIONE 4 COMPONENTE 2, INVESTIMENTO 1.4-D.D. 1033 17 June 2022, CN00000023).

References

1. Beeton, S., Bowen, H.E., Santos, C.A.: State of knowledge: mass media and its relationship to perceptions of quality. In: Quality Tourism Experiences, pp. 25–37 (2006)
2. Nakayama, C.: Film-induced tourism studies on Asia: A systematic literature review. Tour. Rev. Int. **25**, 63–78 (2021)
3. Jørgensen, T., Reichenberger, I.: Breaking bad behaviour: understanding negative film tourist behaviour through moral disengagement. Curr. Issue Tour. **26**, 1183–1198 (2023)
4. Gómez-Morales, B., Nieto-Ferrando, J., Sánchez-Castillo, S.: Visiting game of thrones: film-induced tourism and television fiction. J. Travel Tour. Mark. **39**, 73–86 (2022)
5. Du Y., Li J., Pan B., Zhang Y.: Lost in Thailand: A case study on the impact of a film on tourist behavior. J. Vacat. Mark. **26**, 365–377 (2020)

6. Blei D.M., Ng A.Y., Jordan M.I.,: Latent dirichlet allocation. J. Mach. Learn. Res. **3**, 993–1022 (2003)
7. Griffiths T.L., Steyvers M.: Finding scientific topics. In: Proceedings of the National academy of Sciences, vol. 101(suppl_1), pp. 5228–5235 (2004)
8. Deveaud R., SanJuan E., Bellot P.: Accurate and effective latent concept modeling for ad hoc information retrieval. Document numérique **17**(1), 61–84 (2014)
9. Cao J., Xia T., Li J., Zhang Y., Tang S.: A density-based method for adaptive LDA model selection. Neurocomputing **72**(7–9), 1775–1781 (2009)
10. Birjali M., Kasri M., Beni-Hssane, A.: A comprehensive survey on sentiment analysis: approaches, challenges and trends. Knowl.-Based Syst. **226**, 107–134 (2021). Num Pages: 134
11. Jurek A., Mulvenna M.D., Bi Y.: Improved lexicon-based sentiment analysis for social media analytics. Security Informatics **4**, 1–13 (2015)
12. Hu M., Liu B.: Mining and summarizing customer reviews. In: Proceedings of the Tenth ACM SIGKDD International Conference on Knowledge Discovery and Data Mining, pp. 168–177 (2004)
13. Plutchik R.: The Emotions: Facts, Theories, and a New Model. Random House, New York (1962)
14. Oviedo-Garcıa, M. n., Castellanos-Verdugo, M., Trujillo-Garcıa, M. A., Mallya T.: Film-induced tourist motivations: the case of Seville (Spain). Curr. Issue Tour. **19**, 713–733 (2016)
15. Moritz P., Hrivnák, M., Mazúchová, L., Sándorová, Z.: Who take a part on film tourism? The analysis of determinants of visiting film locations. Int. J. Tour. Res. **26**(1), e2573 (2023)
16. Perugini M., Bagozzi R.P.: The distinction between desires and intentions. Eur. J. Soc. Psychol. **34**, 69–84 (2004)
17. Alrawadieh Z., Prayag G., Alrawadieh Z.: A cognitive appraisal perspective of emotional accessibility at heritage sites: empirical evidence from the UNESCO World Heritage Site of Petra. J. Herit. Tour. **18**, 145–163 (2023)
18. Prayag G., Pung J.M., Lee C., Chiappa G.D.: The self-concept and psychological antecedents of intention to recommend a heritage site: the moderating effects of visitor type. Tour. Manag. Perspect. **42**, 100962–100962 (2022)
19. Carl D., Kindon S., Smith K.: Tourists' experiences of film locations: New Zealand as 'Middle-Earth'. Tour. Geogr. **9**, 49–63 (2007)
20. Larsen G., George V.: The social construction of destination image: A New Zealand film example. In: Creating Images and the Psychology of Marketing Communication, pp. 117–139 (2006)

Testing Linearity in the Single Functional Index Model for Dependent Data

Lax Chan and Aldo Goia

Abstract Motivated by checking the linearity assumption in the regression model with functional covariate and real response in the case of dependent data, a new test for the single functional index model is introduced. The critical region of the test is derived by a bootstrap procedure. The finite sample performances of the test are evaluated by a simulation study, and an application in the framework of functional time series forecasting is performed.

1 Introduction

Functional data analysis (FDA) is a relatively recent and well-developed branch of statistics which studies data having infinite dimensionality nature (functions, curves, images, etc.). The interest of the scientific community toward this discipline has grown over the years: to have an idea about the variety of covered topics, one can, for instance, refer to the monographs [8, 9], or [11] and the papers published in some recent special issues (see, e.g., [1–3]).

In such a framework, many statistical problems deal with the prediction of a real random variable Y via a regression model involving a functional covariate X, and a challenging issue is the specification of the link between these objects. In particular, testing the simplest and interpretable specification of linearity has attracted lots of interests, as attested by recent contributions in the literature (see, for instance, [6]).

One of the main problems in checking linearity in FDA is the difficulty in visualizing the link: in this context, a useful help comes from the Single Functional Index Model (SFIM) that, thanks to a projective strategy, allows to visualize a link function. In particular, it postulates that given X, a random element valued in a Hilbert space $\mathcal{H}$ of real functions defined over a compact interval $\mathcal{T}$ and

L. Chan (✉) · A. Goia
Dipartimento di Studi per l'Economia e l'Impresa, Università del Piemonte Orientale, Novara, Italy
e-mail: lax.chan@uniupo.it; aldo.goia@uniupo.it

G. Giordano et al. (eds.), *Statistical Models and Learning Methods for Complex Data*, Studies in Classification, Data Analysis, and Knowledge Organization,
https://doi.org/10.1007/978-3-031-84702-8_6

equipped with an inner product $\langle \cdot, \cdot \rangle$ and an associated norm $\|\cdot\|$, the relationship between X and Y is modeled through an unknown real link function g acting on a projection of the functional regressor along an unknown direction θ, constrained for identifiability:

$$Y = g\left(\langle \theta, X \rangle\right) + \mathcal{E} \qquad \|\theta\|^2 = 1 \text{ and } \theta\left(t\right) > 0 \text{ for a fixed } t \in \mathcal{T}$$

where $\mathcal{E}$ is a centered random error uncorrelated with X. Hence, once an estimate of g is obtained from a sample, if its plot exhibits a straightness shape, the linearity of the regression model could be investigated. For more insights on the model, its estimators, and some asymptotics, one could refer to [7] and [10].

The focus of this paper is to introduce a test of linearity in the SFIM framework, when the observed data $\{(X_i, Y_i), i = 1, \dots, n\}$ are identically distributed as (X, Y) and dependent. This situation could happen, for instance, in the functional time series framework where a real valued random process is divided in n functional random objects X_i and the aim is to predict $Y_i = \phi\left[X_{i+1}\right]$, where $\phi\left[\cdot\right]$ is a real functional modelizing a characteristic of X_{i+1} as, for instance, the maximum, the value observed at a specific time, a variation, and so on. The test procedure adopted is based on the conditional moment test approach introduced and analyzed in the i.i.d. case in [4] and [5]. Since the null distribution of the test statistic is not available for dependent data, a bootstrap procedure to determine the critical region is adopted. To assess the performances of the test, a numerical study on synthetic data is provided showing how the estimated power of the test behaves when one moves away from the null hypothesis of linearity. An application of the proposed test to the problem of modeling the prediction of peaks load in a teleheating system based on the daily load curves of the previous day is illustrated. Note that no comparisons are done with other concurrent tests of linearity since the available ones are introduced only in the i.i.d. case.

This work is organized as follows: the general background, the test statistic, and the bootstrap procedure are introduced in Sect. 2, the performances of the test are analyzed in Sect. 3, and the obtained results on the real world dataset are presented in Sect. 4.

2 The Test Procedure

Consider the sequence $\{(X_i, Y_i), i = 1, \dots, n\}$ of random objects identically distributed as (X, Y) and dependent. The SFIM writes

$$Y_i = g\left(\langle \theta, X_i \rangle\right) + \mathcal{E}_i \qquad \|\theta\|^2 = 1 \text{ and } \theta\left(t\right) > 0 \text{ for a fixed } t \in \mathcal{T}$$

where $\mathcal{E}_i$ are i.i.d. centered random errors with each one uncorrelated with X_i.

Define the set of affine functions:

$$\mathcal{G}_0 = \{g_0^{\beta}(u) = \beta_0 + \beta_1(u), \beta_0 \in \mathbb{R}, \beta_1 \in \mathbb{R}\setminus\{0\}\}$$

and a set $\mathcal{G}_1$ of real functions g_1^{β} such that $\mathcal{G}_0 \cap \mathcal{G}_1 = \emptyset$.

If one would like to test the following hypothesis

$$H_0 : g \in \mathcal{G}_0 \qquad \text{vs.} \qquad H_1 : g \in \mathcal{G}_1$$

one can use the following test statistic that has been introduced in [4]:

$$Q_n\left(\widehat{\theta}\right) = \frac{1}{n(n-1)h} \sum_{i=1}^{n} \sum_{j=1, j\neq i}^{n} \widehat{\mathcal{E}}_i \widehat{\mathcal{E}}_j K_{ij}^{\widehat{\theta}},$$

where:

- $\widehat{\theta}$ is an estimate of θ that can be calculated by using the first step of the procedure proposed in [7] which combines a spline approximation of the functional coefficient θ and the one-dimensional Nadaraya-Watson approach.
- $\widehat{\beta}$ is an estimate for $\beta = (\beta_0, \beta_1)$ obtained, under the null hypohesis, via a standard OLS approach.
- $\widehat{\mathcal{E}}_i = Y_i - g_0^{\widehat{\beta}}\left(\left\langle X_i, \widehat{\theta}\right\rangle\right)$ are the estimated errors under the null hypothesis.
- $K_{ij}^{\widehat{\theta}} = K\left(\left|\left\langle X_i - X_j, \widehat{\theta}\right\rangle\right| / h\right)$, where $K(\cdot)$ is a kernel function and h is a bandwidth.

The introduced test statistic rests on the conditional moment test principle for which the quantity $\mathbb{E}[\mathcal{E}\mathbb{E}[\mathcal{E}|X]w(X)]$, where $w(X)$ is a suitable positive weight function, is zero under the null hypothesis and positive otherwise. For more details, refer to [4].

The standardised version of the test statistic is

$$T_n = n\sqrt{h}\, Q_n\left(\widehat{\theta}\right) / \nu_n\left(\widehat{\theta}\right),$$

where

$$\nu_n^2\left(\widehat{\theta}\right) = \frac{2}{n(n-1)h} \sum_{i=1}^{n} \sum_{j=1, j\neq i}^{n} \widehat{\mathcal{E}}_i^2 \widehat{\mathcal{E}}_j^2 \left(K_{ij}^{\widehat{\theta}}\right)^2.$$

Since an asymptotic theory for this test statistic has been developed only in the i.i.d. case but not in the dependent one, to derive the critical region of the test, one can profit from the following bootstrap approach:

1. Estimate β and θ and then compute the errors $\widehat{\mathcal{E}}_i$ under the null hypothesis and the value of the test statistic T_n.

2. Compute the bootstrap version $T_n^\star$ of the test statistics by using the bootstrapped errors $\widehat{\mathcal{E}}_i^\star$ obtained from direct resampling with replacement of the estimated errors $\widehat{\mathcal{E}}_i, i = 1, \ldots, n$.
3. Repeat step 2. a large number B of times and compute the $(1 - \alpha)$-quantile $\tau_\alpha^\star$ of the distribution of $T_n^\star$.
4. Compare the value of T_n with $\tau_\alpha^\star$ and reject the null hypothesis if the former is larger than the latter.

3 A Simulation Study

In the experiments conducted in this section, the data are generated according to the model:

$$Y_i = g\left(\langle \theta, X_i \rangle\right) + \sigma \mathcal{E}_i \qquad i = 1, \ldots, n$$

where:

- $n = 50, 100, 200$.
- $X_i(t) = 2a_i t^2 + b_i \sin(2\pi t) + c_i \cos(4\pi t), t \in [0, 1]$, where a_i, b_i, c_i are mutually independent AR(1) processes with uniform innovation over $(-1, 1)$.
- $\theta(t) = \sqrt{2}\sin(2\pi t), t \in [0, 1]$.
- All the curves are discretized over a grid of 100 equispaced points.
- $\mathcal{E}_i$ are i.i.d. standard Gaussian.
- $\sigma^2 = \rho^2 Var\left(g\left(\langle \theta, X \rangle\right)\right)$ with $\rho^2 = 0.2$, so that the theoretical coefficient of determination R^2 is about 0.83.

The aim of the simulation study is to evaluate the level and the power of the test when the null hypothesis of a linear affine link is considered:

$$H_0 : g(u) = \beta_0 + \beta_1 u$$

against the following alternatives

$$H_1^{(1)}(\gamma) : g(u) = \beta_0 + \beta_1 u + \gamma\left(u\sqrt{2}\pi / (\pi + 2)\right)^2$$

$$H_1^{(2)}(\gamma) : g(u) = \beta_0 + \beta_1 u + \gamma \sin\left(u\sqrt{2}\pi^2 / (\pi + 2)\right)$$

$$H_1^{(3)}(\gamma) : g(u) = \beta_0 + \beta_1 u + \gamma \cos\left(u\sqrt{2}\pi^2 / (\pi + 2)\right)$$

where γ controls the departure from the null hypothesis (here $\gamma = 0.2, 0.4, 0.6, 0.8, 1$). To generate the models under the null as well as alternative hypotheses, the parameters $\beta_0 = 0$, $\beta_1 = 1$ are used. To appreciate how the different perturbations

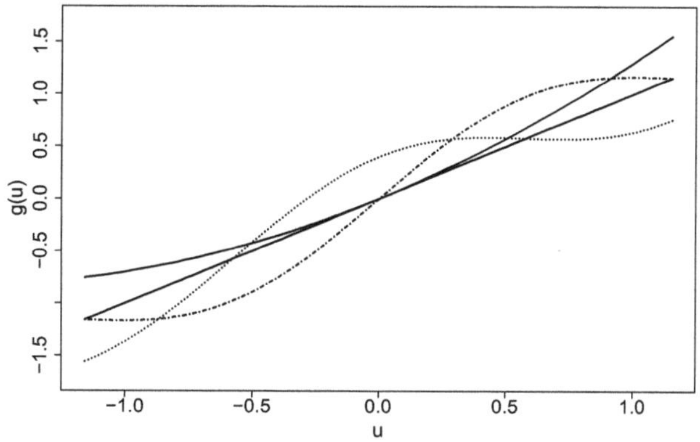

Fig. 1 Shapes of the link functions g under H_0 (solid), $H_1^{(1)}(0.4)$(dashed), $H_1^{(2)}(0.4)$(dotdash), and $H_1^{(3)}(0.4)$(dotted)

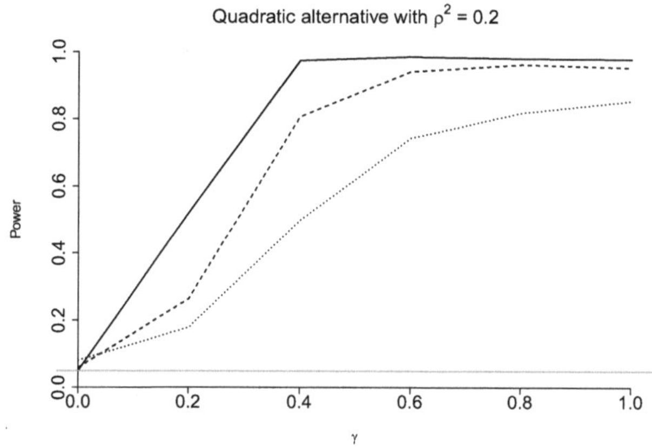

Fig. 2 Power curves for the alternatives $H_1^{(1)}$ varying γ. The gray line represents the nominal level, the dotted line $n = 50$, the dashed line $n = 100$, and the solid line $n = 200$

impact on the departure from the null hypothesis, the graphs of the introduced functions g (with $\gamma = 0.4$) are depicted in Fig. 1.

For every experimental setting, the empirical level (for $\gamma = 0$) and the power (when $\gamma > 0$) of the test are calculated as the proportion of times in which the null hypothesis is rejected at the nominal level $\alpha = 5\%$ over 1000 Monte Carlo replications.

The panels in Figs. 2, 3, and 4 display the behavior of the estimated powers varying γ for the different considered scenarios. By looking at the plots, it emerges that the test performs relatively well for any sample size n both in terms of level and power: the latter increases with γ, as one would naturally expect. More in detail, the family indexed by $H_1^{(3)}$, where the cosinus perturbation acts both on the central part and the tails of the interval (see the dotted line in Fig. 1), produces the best result also for small sample sizes and γ rather small. On the other hand, the detection of the families indexed by $H_1^{(1)}$ and $H_1^{(2)}$ seem to be more complex, at least when γ and

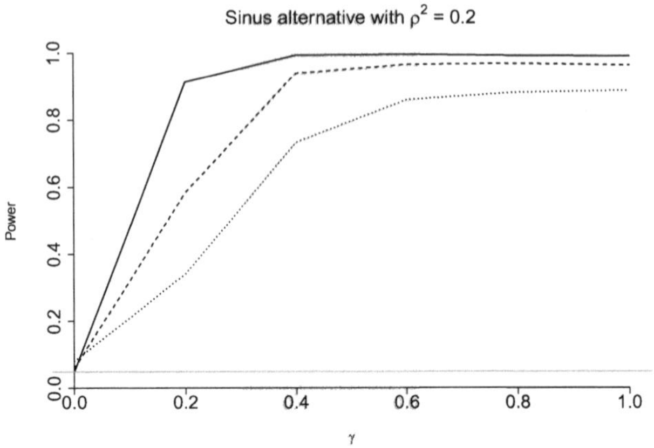

Fig. 3 Power curves for the alternatives $H_1^{(2)}$ varying γ. The gray line represents the nominal level, the dotted line $n = 50$, the dashed line $n = 100$, and the solid line $n = 200$

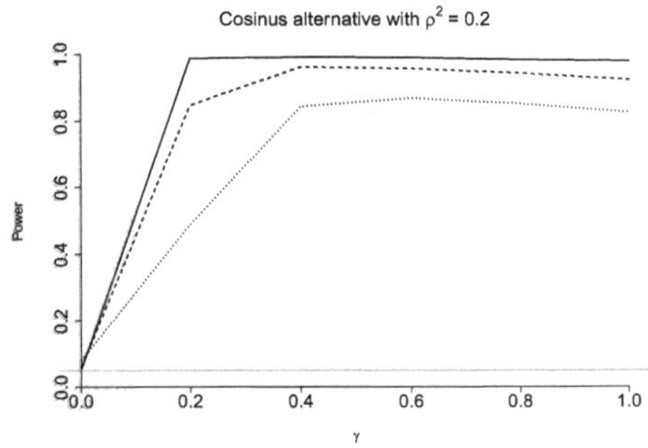

Fig. 4 Power curves for the alternatives $H_1^{(3)}$ varying γ. The gray line represents the nominal level, the dotted line $n = 50$, the dashed line $n = 100$, and the solid line $n = 200$

the sample size are small: this can be explained by the natures of the perturbations that interest mainly the tails of the interval for $H_1^{(1)}$ and the central parts of the two halves of the interval for $H_1^{(2)}$ (see the dashed and dot dash lines, respectively, in Fig. 1).

4 Application to Real World Data

The dataset used in this section consists of hourly measurements of heat consumption in a district heating system in Turin (a northern Italian town), during the periods from the 15th of October to the 20th of April, in the years 2001–2002, 2002–2003, 2003–2004, and 2004–2005. Taking advantage of the functional nature of the data, the observed series of each period is divided in 186 functional observations, each one coinciding with a specific daily load curve: in this way, the resulting dataset consists in a matrix with 744 rows (one for each curve) and 24 columns (one of each of discretisation points). The plot in Fig. 5 displays that functional dataset.

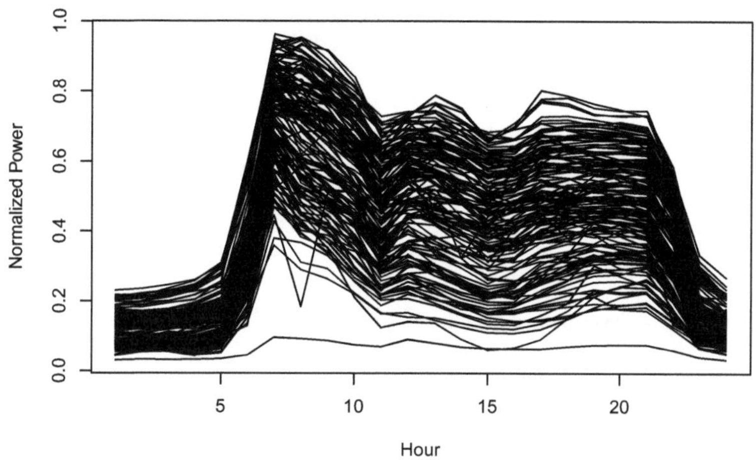

Fig. 5 The sample of daily load curves

The goal of this section is to specify a model to predict the observed peak load of a specific day by using the load curve from the previous day. The main question is if a linear specification could be reasonable.

First of all, note that due to the presence of a trend in the time series of the peaks in each period, the differenced series are considered instead of the original data. Moreover, as one can see from Fig. 5, the load curves exhibit a vertical shift, and then, it is more useful to take as covariate the first derivative of the load curves instead of the original data. Finally, the resulting SFIM consists of:

- Y_i differenced series of daily peaks
- X_i first derivative of the daily load curve of the day before
- $\mathcal{E}_i$ i.i.d. centered errors

In order to evaluate the performances of the estimated model in a cross-validation perpsective, the initial sample is split into a training set corresponding to the data observed in the first three periods and a test set containing the data of the last one. The plots in Fig. 6 show the estimated direction $\widehat{\theta}$ and the link function $\widehat{g}$ obtained using the estimation technique illustrated in [7]. The predictive abilities of the model are evaluated by using the relative mean square error of prediction (RMSE) and the mean absolute percentage errors (MAPE) calculated from the predicted peaks: the forecasting results are very good with RMSE equals to 0.0780 in the training set and 0.0849 in the test set, whereas the MAPE is 0.0495 in the training set and 0.0538 in the test set. Moreover, the percentage of days in which the absolute percentage errors do not exceed the threshold of 5% is about 65% both on the training set and the test set.

Since the graphic of the estimated link function $\widehat{g}$ exhibits a linear behavior, one can suspect that a linear specification could be adopted. Then, the linearity test illustrated in Sect. 2 is performed with the bootstrap step based on 1000 iteration.

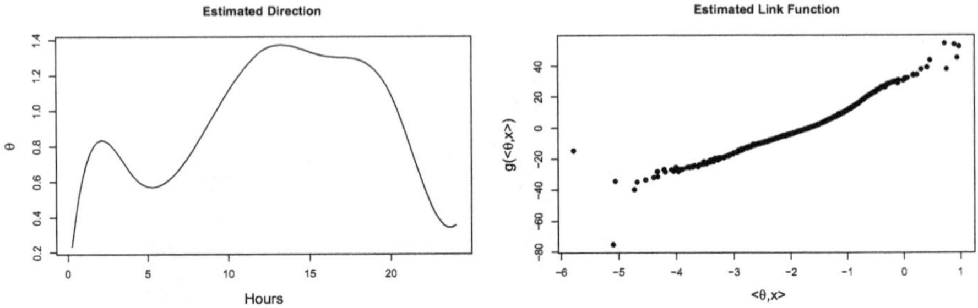

Fig. 6 Left panel—The estimated direction $\widehat{\theta}$. Right panel—The estimated link function $\widehat{g}$

The estimated p–value is 0.373, and so, it is possible to conclude in favor of the compatibility of the null hypothesis of linearity with the empirical evidence: to forecast daily peaks by using the (derivatives) of the load curves of the day before, a linear specification can be a good choice.

5 Concluding Remarks

A linearity test is proposed and illustrated here for the SFIM in the context of dependent data, a bootstrap approach is used to derive the critical region of the test. We show that the test achieves good performances in various simulation scenarios. It would be of interest in a future work to develop a full asymptotic theory for the test statistics in the dependent context.

Acknowledgments Lax Chan holds a career grant supported by the European Commission-FSE REACT-EU, PON Ricerca e Innovazione 2014–2020. A. Goia and L. Chan are members of the Gruppo Nazionale per l'Analisi Matematica, la Probabilità e le loro Applicazioni (GNAMPA) of the Istituto Nazionale di Alta Matematica (INdAM).

References

1. Aneiros, G., Horová, I., Hušková, M., Vieu, P.: On functional data analysis and related topics. J. Multivar. Anal. **189**, 104861 (2022)
2. Aneiros, G., Cao, R., Fraiman, R., Vieu, P.: Editorial for the special issue on functional data analysis and related topics. J. Multivar. Anal. **170**, 1–2 (2019)
3. Aneiros, G., Cao, R., Vieu, P.: Editorial on the special issue on Functional Data Analysis and Related Topics. Comput. Stat. **34**, 447–450 (2019)
4. Chan, L., Delsol, L., Goia, A.: A link function specification test in the single functional index model. Adv. Data Anal. Classif. (2023). https://doi.org/10.1007/s11634-023-00545-7
5. Chan, L., Delsol, L., Goia, A.: Goodness–of–fit test for single functional index model. In: Book of Abstracts and Short Papers. CLADAG (2023)

6. Cuesta-Albertos, J.A., Garcıa-Portugués, E., Febrero-Bande, M., Gonzalez-Manteiga, W.: Goodness-of-fit tests for the functional linear model based on randomly projected empirical processes. Ann. Stat. **47**(1), 439–467 (2019)
7. Ferraty, F., Goia, A., Salinelli, E., Vieu, P.: Functional projection pursuit regression. Test **22**, 293–320 (2013)
8. Ferraty, F., Vieu, P.: Nonparametric Functional Data Analysis. Springer, New York (2006)
9. Horvath L., Kokoszka P.: Inference for Functional Data with Applications. Springer, New York (2012)
10. Novo, S., Aneiros-Perez, G., Vieu, P.: Automatic and location-adaptive estimation in functional single-index regression. J. Nonparametr. Stat. **31**, 364–392 (2019)
11. Ramsay J.O., Silverman B.W.: Functional Data Analysis, 2nd edn. Springer, New York (2005)

A Multistep Approach for Streamflow Classification

Marcella Corduas and Domenico Piccolo

Abstract The article presents a strategy for classifying streamflows into groups based on their temporal dynamics. It compares dynamic patterns of river discharges over time, capturing seasonality and short-term components using linear time series models. The final classification offers a means of identifying hydrological regimes and gives a preliminary understanding of the potential impacts of climate change on streamflows. The approach is illustrated by analyzing streamflow data from 221 stations in the United States in two nonoverlapping periods.

1 Introduction

The temporal dynamics of streamflows are affected by various factors related to climate, watershed features, and land use. Weather (precipitation and temperature) interacts with topography, geological properties, soil, and land cover characteristics to influence the flow generation process. The classification of rivers into groups that are similar with respect to the determinants or the signature of their flows is a topic of renewed interest in the last two decades because of possible variations in hydrological regimes related to climate change [22]. This may alter, in fact, the balance between snow and rainfall, the frequency and intensity of extreme hydrological events, the variability and persistence of river flows and cause relevant consequences on natural ecosystems [8] and water-resources management [1, 9].

Two main approaches for regime classification have been discussed in the literature. The first one classifies locations according to a number of attributes describing the surrounding environment. The second one bases the classification on a set of indexes that measure various components of the flow regime (i.e., the seasonal pattern, timing of extreme flows, frequency and duration of floods, droughts, etc.). Both approaches produce a static picture with information on the

M. Corduas (✉) · D. Piccolo
Department of Political Sciences, University of Naples Federico II, Naples, Italy
e-mail: marcella.corduas@unina.it; domenico.piccolo@unina.it

G. Giordano et al. (eds.), *Statistical Models and Learning Methods for Complex Data*, Studies in Classification, Data Analysis, and Knowledge Organization,
https://doi.org/10.1007/978-3-031-84702-8_7

hydrological regimes in a certain region at a given moment and do not account for the river flows dynamics over long periods of time.

In this article, the problem is placed in a different perspective. The classification is derived from the comparison of the dynamic patterns of river discharges observed along the years. The effect of factors driving the hydrological response is, in fact, embedded in the temporal dynamics of the streamflow series [5, 6, 21]. In particular, the seasonality and short-term components of a time series are captured by means of a linear model. The related parameters are used to establish the similarity between streamflow series and, then, to produce the final classification. Furthermore, the parametric representation of a time series in nonoverlapping temporal intervals allows to verify whether changes in streamflow dynamics have affected regime class membership over time. Our aim is to explore how statistical tools can provide a useful strategy for the identification of river regimes in a changing environment. The proposed approach is illustrated by analyzing streamflow daily data from 221 stations in the contiguous United States during two reference periods [15].

2 Methods

Streamflow time series are typically characterized by a marked seasonal pattern, due to the alternating of wet and dry periods, and a persistent or long-term component. Seasonality often explains much of the time series variability and appears as a deterministic component in the spectrum. This makes the observed series unsuitable for stochastic modelling, because the marked seasonal pattern obscures the remaining dynamic components, and, thus, it must be preliminarily treated. Numerous techniques, such as nonlinear models [12], neural networks [23], or machine learning algorithms [19], have been applied to model and predict this type of data. Here, for the aim of the present work—the characterization of the dynamics of river discharges—we will refer to a linear model because this is flexible enough to approximate a stationary time series even in the presence of nonlinear or long memory components. At this stage, we assume that the effects of data skewness, calendar effects, outliers, and missing value have already been removed by preliminary analysis and transformations and that the time series W_t has zero mean.

Thus, W_t is described by the harmonic regression model:

$$W_t = \sum_{j=1}^{[s/2]} \left[\alpha_{wj} \sin(2\pi jt/s) + \beta_{wj} \cos(2\pi jt/s)\right] + Z_t \tag{1}$$

where s denotes the seasonal period and Z_t follows a stationary autoregressive model, *AR(p)*:

$$\varphi(B) Z_t = a_t, \tag{2}$$

where a_t is a Gaussian white noise (WN) process with constant variance σ^2_{aw} and $\varphi(B) = 1 - \varphi_1 B - ... - \varphi_p B^p$ and $B^k Z_t = Z_{t-k}$, for all $k = 0, 1, ...$ It is well known that any process with an absolutely continuous spectrum can be adequately approximated by an autoregressive model [3, p.130–133]. Then (2) describes both short and long memory stationary components. The order p can be selected by BIC criterion, so that parsimonious models are preferred. Thus, the time series W_t is characterized by the coefficients estimated by GLS: $\widehat{\boldsymbol{\delta}}_{\boldsymbol{w}} = (\hat{\alpha}_{w1}, ..., \hat{\alpha}_{wk}, \hat{\beta}_{w1}, ..., \hat{\beta}_{wk})'$ and $\widehat{\boldsymbol{\varphi}}_{\boldsymbol{w}} = (\hat{\varphi}_{w1}, ..., \hat{\varphi}_{wp})'$. Given two independent time series W_t and Y_t, the dissimilarity will be measured by comparing the seasonal and nonseasonal coefficients separately because, as already mentioned, the two components (deterministic seasonality and stochastic inertia) have a very different weight in the generating process of hydrological series, and in addition, these components are uncorrelated according the Wold theorem for stationary processes. Specifically, the procedure is articulated as follows:

- *STEP 1: Comparison of seasonal components*
  Seasonal components are compared by evaluating the Mahalanobis distance [6]: $M_{wy} = (\widehat{\boldsymbol{\delta}}_w - \widehat{\boldsymbol{\delta}}_y)'(\sigma^2_{aw}\boldsymbol{\Omega}_w + \sigma^2_{ay}\boldsymbol{\Omega}_y)^{-1}(\widehat{\boldsymbol{\delta}}_w - \widehat{\boldsymbol{\delta}}_y)$, where $\sigma^2_{a*}\boldsymbol{\Omega}_*$ is the covariance matrix of $\widehat{\delta}_*$.
- *STEP 2: Comparison of stochastic components*
  The dissimilarity between the residual components is measured by means of the *AR* metric [7, 17, 18]: $D_{wy} = \sqrt{\sum_{j=1}^{\infty} \left(\varphi_{wj} - \varphi_{yj}\right)^2}$.
- *STEP 3: Clustering*
  The corresponding distance matrices $\mathcal{M}$ and $\mathcal{D}$ are analyzed separately. They become the input of a clustering algorithm in order to identify groups of time series having similar seasonal pattern and different level of inertia. Here, we use the complete linkage method because it does not require the preliminary specification of the cluster number and produces compact clusters. The number of clusters K^* is selected by maximizing the Dunn's index: $I(k) = \frac{\min_{i \neq j}(\delta(C_i, C_j))}{\max_{0 \leq h \leq k}(\Delta(C_h))}$, where $\delta(C_i, C_j)$ denotes the distance between two clusters and $\Delta(C_h)$ the cluster diameter.

At the end of the procedure, each time series is characterized by two labels specifying the seasonal regime and the level of inertia. These features can be summarized in a two-way table that provides the typical hydrological regimes for the considered region. The proposed approach allows further analyses. Firstly, it helps to account for watersheds where streamflow data are not available or are available for a too short period. In such a case, the assignment of the river flow to a certain regime has to be performed using variables describing landscape features that may influence the flow [16]. For this aim, the classification of a set of streamflow series in similar groups according to the seasonal components (as obtained in STEP 1) becomes a useful starting point. Random forest algorithm [2] can be applied to predict the observed regime classes by means of a small set of environmental variables. These can be selected from a larger set according to the variable importance measure [10]. An additional river to be classified would be attributed to the class for which the Euclidean distance between the standardized variables describing the additional site and the class centroid is minimum. This problem will not be further explored in the empirical analysis due to space constraints. Secondly, by analyzing the classifications of streamflows from data observed in two periods of time, one can gain an insight into the possible changes of the hydrological regime that a given time series may have suffered during that time. This can be done by assessing the stability of the regime classification in terms of seasonal pattern and the changes in the class membership.

3 Results

The analysis has been conducted on a subset of time series of mean daily discharge (feet^3/sec) of unregulated streamflows in the contiguous United States, belonging to the CAMELS dataset [15] and available from the US Geological Survey at https://waterdata.usgs.gov/nwis/. We have selected 221 sites where more than 70 years of data are available. Two nonoverlapping time intervals have been considered: from 1930.10.01 (or later, depending on data availability) to 1974.09.30 and from 2000.10.01 to 2021.09.30. The first period is used as a reference for detecting changes in the membership class in recent years, which may be due to the effect of climate change. This problem has been addressed by other studies at global and local scale [4, 11, 14].

Firstly, we concentrate our attention on the study of data observed in the second period of time, from 2000.10.01 to 2021.09.30. The complete link clustering of the Mahalanobis distance matrices, $\mathcal{M}$, leads to the identification of six clusters. This partition will be denoted with C_2. In order to interpret the results and identify the hydrological regimes, we have evaluated the (standardized) mean daily discharge at each station. Figure 1 illustrates the shape of the average seasonal pattern and the periods of major/minor flows of the series belonging to each group. Specifically, the clusters describe:

- G1: *strong fall/spring regime* characterized by high discharge in fall and spring and low discharge in winter and summer with high variability. Most of the sites are located in the North Atlantic and Pacific Northwest Coast.
- G2: *intermittent winter/spring regime* with flashy peaks in winter and spring and a low flow in fall and summer. Locations are in the mid-Atlantic and central valleys and in the Pacific Coast.
- G3: *intermittent regime* identified by a very weak seasonality, dominated by the occurrence of short precipitation events over the year, flashy peaks in summer and early fall with no distinct low-flow period. Locations are in the Gulf Coast.
- G4: *weak winter regime* such that the wettest time of the year occurs in the late spring and early summer, whereas winter is the driest season. Locations are in the upper Great Lakes and Northern Great Plains.

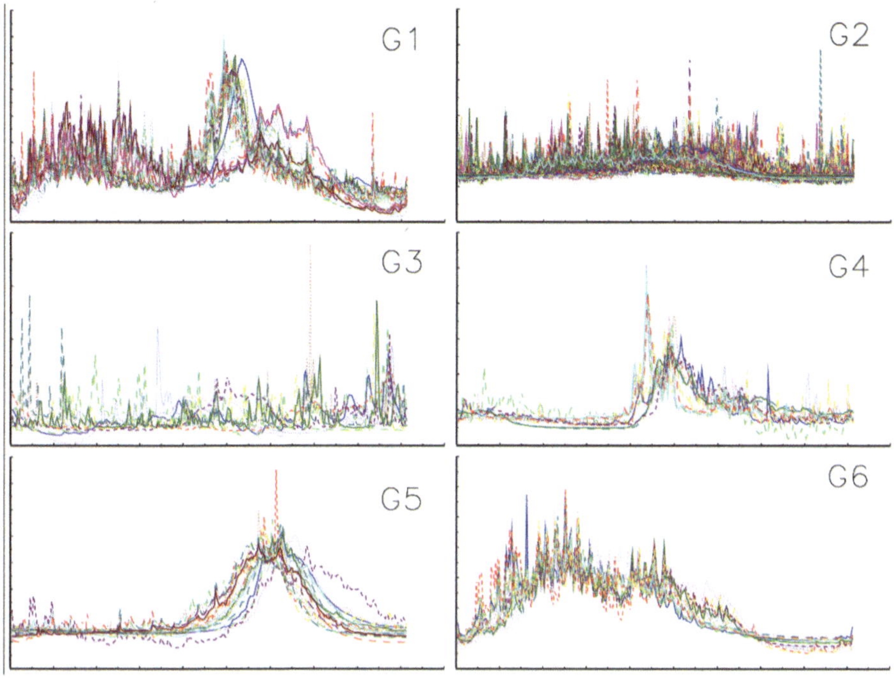

Fig. 1 Average daily discharge of clustered time series (2nd period; water year: 1st October/30th September)

- G5: *snowmelt regimes* with seasonal pattern is dominated by snowmelt, with high discharge in spring and summer and very low discharge in winter and fall. Most sites are located in the Rocky Mountain and Northern Great Plains.
- G6: *strong winter regime* with precipitations in fall and winter. The flow trend is characterized by high discharge in those seasons against a rapid decline in spring and summer. Most sites are located in the Northwest Coast.

Clusters are also well categorized in terms of catchment features. In particular, Fig. 2 shows the box plots of some characteristics: mean elevation, mean slope, mean daily precipitation, fraction of precipitation falling as snow, frequency of high precipitation days, and slope of the flow duration curve (which is an indicator of streamflow variability [20]). The dynamics of the series can be further discriminated by measuring the AR distance between the residual components of each series. Each group G_i can be split in subclusters according to different level of inertia (see, STEP 2). Specifically, applying the clustering method to the AR distance matrix, D, we recognize three clusters of series. These are described by increasing level of inertia (low, moderate, high) as shown by the parametric spectral densities of the series representing the cluster centroids (Fig. 3). In the end, in the considered period (2000–2021), there are 13 clusters with different seasonal patterns and level of inertia (Table 1).

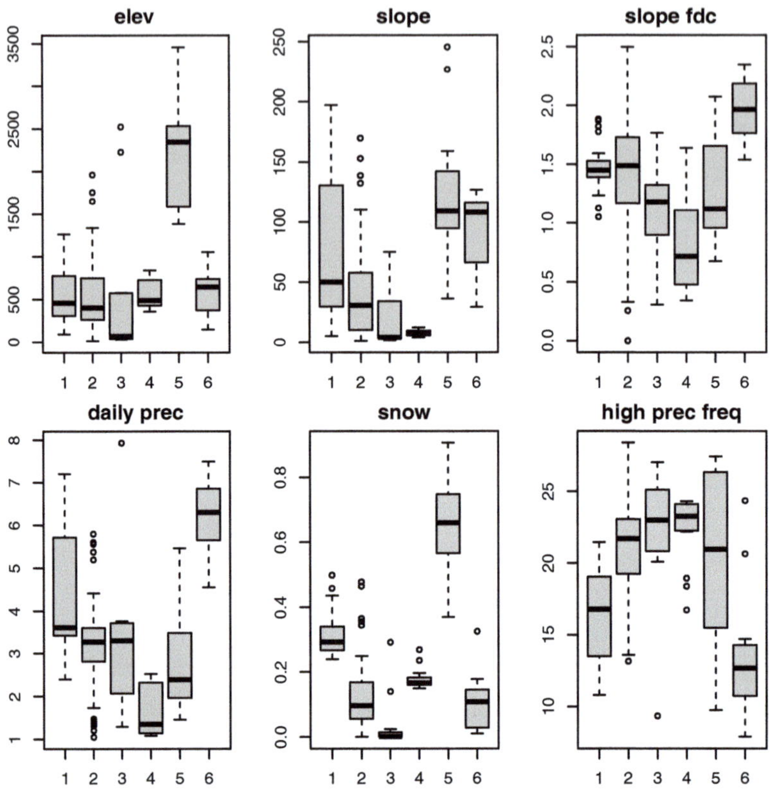

Fig. 2 Climatic and physiographic features of the clusters

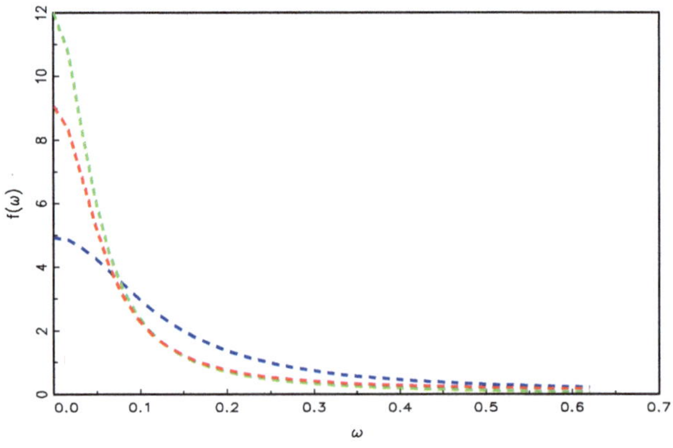

Fig. 3 Spectral densities of cluster centroids over the band of frequency $[0, \pi/5]$ (blue line=low level of inertia, red=moderate level of inertia, green=high level of inertia)

Table 1 Final classification for the dataset observed the period 2000–2021

	Inertia		
Seasonality	High	Medium	Low
Strong fall/spring	2	17	6
Intermittent winter/spring	0	72	61
Intermittent	0	6	6
Weak winter	0	10	5
Snowmelt	0	15	5
Strong winter	0	15	1

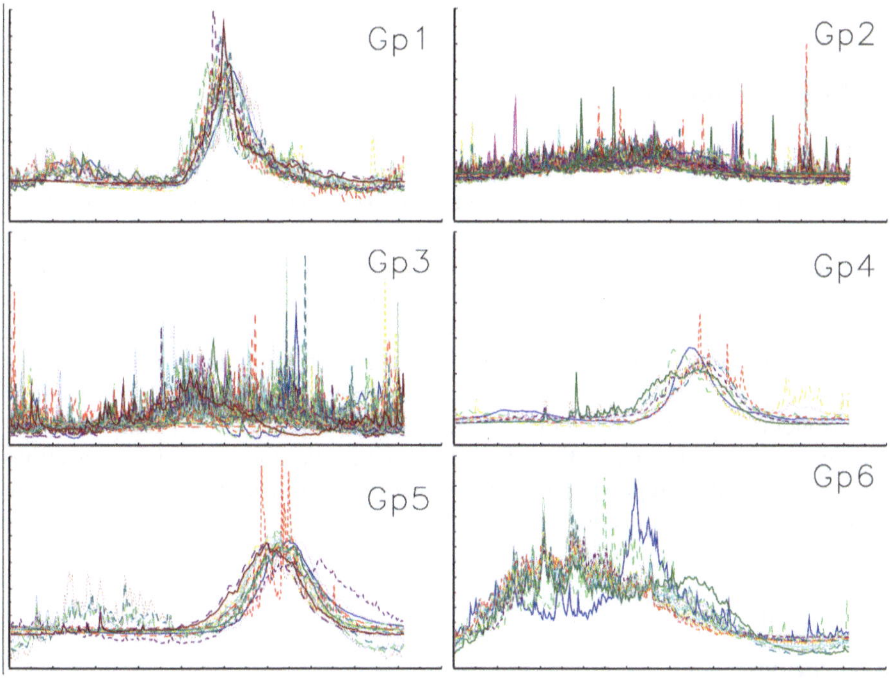

Fig. 4 Average daily discharge of clustered time series (1st period, water year)

In order to assess changes to regime class memberships due to modifications of the river flow trend in the last 20 years, the proposed procedure has been applied to the streamflow series observed up to 1974 (from 1930.10.01 or later to 1974.09.30). The corresponding regime classification, denoted with C_1, has been compared with the most recent classification C_2. The seasonal patterns of the clusters in C_1 (Fig. 4) show a remarkable similarity with those identified from C_2 in the recent years (Fig. 1). This is confirmed by the value (0.78) of the total unmatched probability mass in the confusion matrix (recall that the index takes value 1 for identical clusterings [13]). However, a large number of series (about 33%) have changed their class memberships. The examination of the mean pattern and seasonal variability of these series has revealed that variations are caused by a number of factors, including anticipation of seasonal peak due to early snowmelt, increased winter rainfall, increase in the dry periods, and number of flashy peaks. Moreover, using the asymptotic test based on the AR metric [7], we have verified the stability of the AR generating process of the residuals Z_t in the two considered time periods, for the series with changed hydrological regime. Specifically, the inertia varies in 70% of the series, and of these, 65% exhibit increased contribution to the total variance of the spectral components with period greater than 1 year. This raises the possibility that the basic properties of the temporal dynamics of those rivers have been substantially impacted by climate change.

4 Final Remarks

This study classified flow regimes using statistical tools that are widely available. The approach relies on using the parameters of the underlying generating processes to compare the temporal dynamics of streamflows. This gets beyond the traditional approach's need for a large number of indices describing the flow properties and the hydrological basin. The effectiveness of the recommended classification strategy is supported by the empirical case study. Six regime classes have been identified from the streamflow time series of 221 stations in the contiguous United States. These classes are well characterized by the annual mean pattern of the analyzed streamflow series together with the environmental and climate variables of the catchments. Our findings offer an informative basis for additional discussion about the connections between streamflow regimes and climate factors as well as for projecting future regime changes based on simulated scenarios generated by models driven by climate data.

References

1. Ali, G., Tetzlaff, D., Soulsby, C., McDonnell, J.J., Capell, R.: A comparison of similarity indices for catchment classification using a cross-regional dataset. Adv. Water Resour. **40**, 11–22 (2012)
2. Breiman, L.: Random forests. Mach. Learn. **45**, 5–32 (2001)
3. Brockwell, P.J., Davis, R.A.: Time Series: Theory and Methods. Springer, New York (1991)
4. Brunner, M.I., Melsen, L.A., Newman, A.J., Wood, A.W., Clark, M.P.: Future streamflow regime changes in the United States: assessment using functional classification. Hydrol. Earth Syst. Sci. **24**(8), 3951–3966 (2020)
5. Chiang, S.M., Tsay, T.K., Nix, S.J.: Hydrologic regionalization of watershed. Part I. Methodology development. J. Water Resour. **128**, 1–11 (2002)
6. Corduas, M.: Clustering streamflow time series for regional classification. J. Hydrol. **407**, 73-80 (2011)
7. Corduas, M., Piccolo, D.: Time series clustering and classification by the autoregressive metric. Comput. Stat. Data Anal. **52**(4), 1860–1872 (2008)

8. Dhungel, S., Tarboton, D.G., Jin, J., Hawkins, C.P.: Potential effects of climate change on ecologically relevant streamflow regimes. River Res. Appl. **32**(9), 1827–1840 (2016)
9. Eum, H.I., Vasan, A., Simonovic, S.P.: Integrated reservoir management system for flood risk assessment under climate change. Water Resour. Manage. **26**, 3785–3802 (2012)
10. Genuer, R., Poggi, J., Tuleau-Malot, C.: Variable selection using random forest. Pattern Recognit. Lett., **31**(14), 2225–2236 (2010)
11. Gupta, A., Carroll, R.W., McKenna, S.A.: Changes in streamflow statistical structure across the United States due to recent climate change. J. Hydrol. **620**, 129474 (2023)
12. Fathian, F., Fard, A.F., Ouarda, T.B., Dinpashoh, Y., Nadoushani, S.M.: Modeling streamflow time series using nonlinear SETAR-GARCH models. J. Hydrol. **573**, 82–97 (2019)
13. Meilă, M.: Comparing clusterings—an information based distance. J. Multiv. Anal. **98**(5), 873–895 (2007)
14. Milly, P., Dunne, K., Vecchia, A.: Global pattern of trends in streamflow and water availability in a changing climate. Nature **438**, 347–350 (2005)
15. Newman, A.J., et al.: Development of a large-sample watershed-scale hydrometeorological data set for the contiguous USA: data set characteristics and assessment of regional variability in hydrologic model performance. Hydrol. Earth Syst. Sci. **19**(1), 209–223 (2015)
16. Olden, J.D., Kennard, M.J., Pusey, B.J.: A framework for hydrologic classification with a review of methodologies and applications in ecohydrology. Ecohydrology **5**(4), 503–518 (2012)
17. Piccolo, D.: Una topologia per la classe dei processi ARIMA. Statistica **44**(1), 47–59 (1984)
18. Piccolo, D.: A distance measure for classifying ARIMA models. J. Time Ser. Anal. **11**(2), 153–164 (1990)
19. Rasouli, K., Hsieh, W.W., Cannon, A.J.: Daily streamflow forecasting by machine learning methods with weather and climate inputs. J. Hydrol. **414**, 284–293 (2012)
20. Sawicz, K., Wagener, T., Sivapalan, M., Troch, P.A., Carrillo, G.: Catchment classification: empirical analysis of hydrologic similarity based on catchment function in the eastern USA. Hydrol. Earth Syst. Sci. **15**(9), 2895–2911 (2011)
21. Toth, E.: Catchment classification based on characterisation of streamflow and precipitation time series. Earth Syst. Sci. **17**(3), 1149–1159 (2013)
22. Yang, M., Olivera, F.: Classification of watersheds in the conterminous United States using shape-based time-series clustering and Random Forests. J. Hydrol. **620**, 129409 (2023)
23. Zealand, C.M., Burn, D.H., Simonovic, S.P.: Short term streamflow forecasting using artificial neural networks. J. Hydrol. **214**, 32–48 (1999)

Identification of Misogynistic Accounts on Twitter Through Graph Convolutional Networks

Emiliano del Gobbo, Alex Cucco, and Lara Fontanella

Abstract Misogyny refers to the deeply ingrained bias against women, characterised by feelings of hatred, aversion, and distrust primarily due to their gender. With a large dataset of Italian-written tweets, our research aims to analyse the individuals who generate and disseminate misogynistic information. To achieve this, we first use a Graph Convolutional Network approach to categorise Twitter accounts into a binary misogyny scheme, leveraging both textual and relational data from friend/follower relationships. Then, we compare the retrieved misogynistic and non-misogynistic communities, considering both network centrality measures and linguistic features.

1 Introduction

As highlighted in several reports [e.g. 1, 2], women are consistently one of the primary targets for online abusive content. Across various social media platforms, content directed at women often involves offensive language, frequently including direct allusions to comparisons between women and objects or animals. Additionally, there's a more pronounced tendency to incite violence against women compared to other groups, particularly through sexualised aggression, which stands out as the most common form of online hostility targeting women [2]. These forms of abusive content are closely linked to misogyny, broadly defined as the cultural mindset of directing hatred towards females solely because of their gender

E. del Gobbo (✉) · L. Fontanella
Department of Economics, Management and Territory, University of Foggia, Foggia, Italy
e-mail: emiliano.delgobbo@unifg.it; lara.fontanella@unich.it

A. Cucco
Department of Legal and Social Sciences, G. d'Annunzio University, Chieti-Pescara, Italy

National Heart and Lung Institute, Imperial College, London, UK
e-mail: alex.cucco@unich.it

G. Giordano et al. (eds.), *Statistical Models and Learning Methods for Complex Data*, Studies in Classification, Data Analysis, and Knowledge Organization,
https://doi.org/10.1007/978-3-031-84702-8_8

[3]. More precisely, misogyny encompasses various behaviours or sentiments, including sexual and physical violence against women, their exclusion, advocacy for patriarchy, demeaning actions, and intentional marginalisation [4].

Within the field of hate speech research, there has been a primary emphasis on developing methods to automatically identify abusive content on the Internet [see 5–7, for a review]. However, there has been a recent change in focus towards investigating the behaviour and connections of individuals who propagate abusive remarks on popular online platforms [8, 9]. Only a limited number of studies have utilised a network modelling approach, integrating graph-based centrality measures [10, 11] as well as relational data [12, 13] into classification models alongside textual data to enhance classification performance. The significance of relational data lies in the concept of homophily, wherein people, both offline and online, tend to associate more with those who share similarities. In hate speech classification, beyond linguistic signals, it is possible to leverage accounts' relational networks and the knowledge that members of a particular community are more likely to create abusive content [12]. Still, as far as our current knowledge extends, networked data has not yet been employed to analyse producers of misogynistic online content.

Our study aims to investigate the originators of misogynistic content disseminated in Italian on the Twitter platform (now rebranded as X). Our primary objective is to address an automated classification task, categorising Twitter accounts into a binary misogyny scheme. To achieve this, we employ a Graph Convolutional Network methodology, leveraging relational and textual features extracted from the platform's shared content and the friend/follower network. Subsequently, we conduct a comparative analysis between misogynistic and non-misogynistic communities, assessing network centrality metrics and textual aspects. This comprehensive methodology reveals crucial insights into the distinct linguistic and relational patterns characterising these two categories.

2 Text Graph Convolutional Network

The task of categorising Twitter accounts into a misogyny scheme translates into assigning a binary attribute to each user node. In traditional machine learning classification tasks, it is commonly assumed that data points, consisting of feature-label pairs, are generated independently and identically from a distribution across the domain. Consequently, classification is carried out independently on each unit. As pointed out by London and Getoor [14], this approach does not suit the classification of social network data where neighbouring units have higher chances of sharing the same label and links may be more likely observed between instances of the same class. In such a case, models that independently classify each unit might overlook a wealth of valuable information, potentially resulting in suboptimal performance. In our study, in order to account for potential dependencies between Twitter accounts, we adopt the Text Graph Convolutional Network (TextGCN) approach proposed by Yao et al. [15] which enables us to take into account both the relational structure and the content that each user shares. To apply TextGCN, we construct a heterogeneous graph denoted as $\mathcal{G} = \{\mathcal{V}, \mathcal{E}\}$, where $\mathcal{V}$ represents nodes $\{v_i\}_{i=1,\dots,N_a}$, with N_a being the total number of nodes. These nodes include N_u Twitter accounts and N_w terms extracted from the corpus vocabulary. We consider an undirected graph with no loops, such that $e_{i,j} = e_{j,i}$, $e_{i,i} = 0$, and $e_{i,j} \geq 0$, where equality with zero holds if and only if $v_i, v_j \notin \mathcal{E}$. The graph relational structure can be represented through the $N_a \times N_a$ block adjacency matrix:

$$\mathbf{A} = \begin{bmatrix} \mathbf{U} & \mathbf{M} \\ \mathbf{M}' & \mathbf{W} \end{bmatrix}. \tag{1}$$

The $N_u \times N_u$ matrix $\mathbf{U}$, referring to the Twitter accounts' network, is built by setting $a_{i,j}^{(U)} = 1$ if there is a friend/follower link between account nodes v_i and v_j and 0 otherwise. The $N_w \times N_w$ matrix $\mathbf{W}$ refers to the word co-occurrences and $a_{i,j}^{(W)}$ is equal to the raw counts of co-occurrences between word nodes v_i and v_j in the whole corpus. Finally, the $N_u \times N_w$ matrix $\mathbf{M}$ represents the word occurrences in the accounts' textual content, and the weight of the edge between an account node v_i and a word node v_j $\left(a_{i,j}^{(M)}\right)$ is the term frequency-inverse document frequency (TF-IDF) of the word in the account document. Here, the term frequency is the number of times the j-th word appears in the i-th document, while the inverse document frequency is the logarithmically scaled inverse fraction of the number of documents that contain the j-th word.

In this framework, it is possible to apply Graph Convolutional Networks (GCNs) [16] to classify the nodes based on the entire graph structure. Like in standard artificial neural networks, in GCNs, a hidden layer is an intermediate layer of neurons positioned between the input layer and the output layer. Each neuron in the hidden layer receives inputs from neurons in the previous layer, performs a specific operation, and then applies an activation function to produce an output which represents the input of the following layer. Unlike convolutional neural networks, where each hidden layer performs a convolution on grid data, GCNs operate a convolution on the graph. For more details and graphical intuition, readers are encouraged to refer to Kipf and Welling [16], Bronstein et al. [17], and Yao et al. [15].

To produce node-level output, $\mathbf{Y}$, GCNs takes as input a feature description $\mathbf{x}_v$ for each node $v \in \mathcal{V}$, consolidated in the feature matrix $\mathbf{X}$, and a representation of the graph's structure via the adjacency matrix $\mathbf{A}$. Each neural network hidden layer can be expressed as a nonlinear function [16]:

$$\mathbf{H}^{(l+1)} = f\left(\tilde{\mathbf{D}}^{-\frac{1}{2}}\tilde{\mathbf{A}}\tilde{\mathbf{D}}^{-\frac{1}{2}}\mathbf{H}^{(l)}\boldsymbol{\Omega}_l\right) \tag{2}$$

where $\tilde{\mathbf{A}} = \mathbf{A} + \mathbf{I}_{N_a}$ is the adjacency matrix with added self-connections, $\tilde{\mathbf{D}} = diag(\tilde{\mathbf{A}}\mathbf{1}_{N_a})$ is the diagonal degree matrix, $\boldsymbol{\Omega}_l$ denotes the layer-specific trainable weight matrix, and $f(\cdot)$ is an activation function such as the $\text{ReLU}(\cdot) = \max(0, \cdot)$. $\mathbf{H}^{(l)}$ is the matrix of activations in the l-th layer, with $\mathbf{H}^{(0)} = \mathbf{X}$.

In our analysis, following Yao et al. [15], we adopt a two-layer GCN which enables the exchange of information among nodes that are at most two steps apart, allowing to capture both direct links between the users and their semantic similarities. In addition, we do not take into consideration node properties, so that the feature matrix is set equal to the identity matrix, $\mathbf{X} = \mathbf{I}$.

The implemented GCN model is described by the following formula [15, 16]:

$$\mathbf{Z} = \text{softmax}\left(\tilde{\tilde{\mathbf{A}}}\,\text{ReLU}\left(\tilde{\tilde{\mathbf{A}}}\mathbf{X}\boldsymbol{\Omega}_0\right)\boldsymbol{\Omega}_1\right) = \text{softmax}\left(\tilde{\tilde{\mathbf{A}}}\,\text{ReLU}(\mathbf{B})\boldsymbol{\Omega}_1\right) = \text{softmax}(\mathbf{C}) \quad (3)$$

Here, $\tilde{\tilde{\mathbf{A}}} = \tilde{\mathbf{D}}^{-\frac{1}{2}}\tilde{\mathbf{A}}\tilde{\mathbf{D}}^{-\frac{1}{2}}$. The activation function for the first layer, $\text{ReLU}(b_{ij}) = \max(0, b_{ij})$, is applied. Additionally, $z_{i,j} = \text{softmax}(\boldsymbol{c}_i)_j = \frac{\exp(c_{ij})}{\sum_k \exp(c_{ik})}$, with $\{j, k\} \in \{1, \ldots, K\}$, represents the activation function for the second layer. After the training of the weights, each layer encapsulates document and word embeddings. Our chosen loss function is the cross-entropy error computed over all labelled documents, and we utilise the Adam optimisation algorithm [18].

3 Misogynistic Communities on Twitter

3.1 Materials

We gathered real-time Italian tweets (August–December 2022) using search terms related to about fifty politically active women, feminists, journalists, influencers, and female television celebrities. Initially, we collected 1,002,226 tweets from 204,095 user accounts. By eliminating inactive and low-activity accounts, as well as media sources, we acquired a final collection of 82,807 tweets from 7,371 users. The relationships between the users were based on their friendships and followerships, and we built the textual corpus by compiling all tweets from each account into a single document.

By employing node centrality criteria, we selected a sample of 942 accounts for the purpose of manual annotation. Among the annotated accounts, we identified 44.6% as exhibiting misogynistic material. Additional information on the data collection procedure and the selection of the training set of nodes can be found in Fontanella et al. [19], which also includes the presentation of the preliminary findings.

To construct the adjacency matrix blocks **W** (*word-word edges*) and **M** (*users-word edges*), we performed the screening and cleaning process (tokenization; lowercase conversion; special characters, URL, mentions and stop-word removal) of user documents in order to extract the relevant content and remove any unwanted nuisance terms. Furthermore, the tokens with less than 500 occurrences were

Table 1 Comparison of accuracy and F1 score: performance metrics

Model	Adjacency matrix	Layers	Accuracy	$F1_M$	$F1_{NM}$
$\text{TextGCN}_{2L,A}$	$\mathbf{A}$	2 Layers	0.668	0.682	0.653
GCN_{2L,A_1}	$\mathbf{A}_1$	2 Layers	0.630	0.684	0.552
$\text{TextGCN}_{2L,A_2}$	$\mathbf{A}_2$	2 Layers	0.542	0.337	0.650
GCN_{1L,A_1}	$\mathbf{A}_1$	1 Layer	0.590	0.450	0.673
$\text{TextGCN}_{1L,A_2}$	$\mathbf{A}_2$	1 Layer	0.489	0.459	0.516

$F1_M$—F1 score for the nodes labelled as misogynistic
$F1_{NM}$—F1 score for the nodes labelled as non-misogynistic

removed from the vocabulary. From this initial vocabulary, we selected non-domain-specific terms. Two-by-two association scores, computed through the 'keyness' function of the quanteda package for R [20], were used to compare how words were distributed across the different search domains and find and remove domain-specific terms. Ultimately, we included terms from a lexicon tailored for identifying derogatory language [21], focusing on terms directed towards women, which might have been eliminated in the previous pruning steps.

3.2 *Cross-Validation Results: Evaluating Misogyny Detection in Annotated Accounts*

To evaluate the GCN model's performance, a 20-fold cross-validation procedure on the training set was conducted. In each iteration, one-twentieth of the labelled users was treated as unlabelled, along with the other truly unlabelled nodes.

Table 1 presents the model evaluation metrics for a two-layer model utilising the adjacency matrix $\mathbf{A}$ defined in Eq. (1). We compare these results with the performance of two other models: a one-layer and a two-layer model constructed solely from the relational information contained in the user adjacency matrix, denoted as $\mathbf{A}_1 = [\mathbf{U}]$. Additionally, we compare with the performance of one-layer and two-layer models that disregard friend/follower relationships and focus solely on textual content. In this latter case, the adjacency matrix is constructed as $\mathbf{A}_2 = \begin{bmatrix} \mathbf{0} & \mathbf{M} \\ \mathbf{M}' & \mathbf{W} \end{bmatrix}$.

As evident from the results, the incorporation of both relational and textual features demonstrates an enhancement in classification performance. This underscores the significance of integrating diverse sources of information, wherein the combined strength of relational data, capturing social connections, and textual features, elucidating semantic context, synergistically contribute to improved classification accuracy.

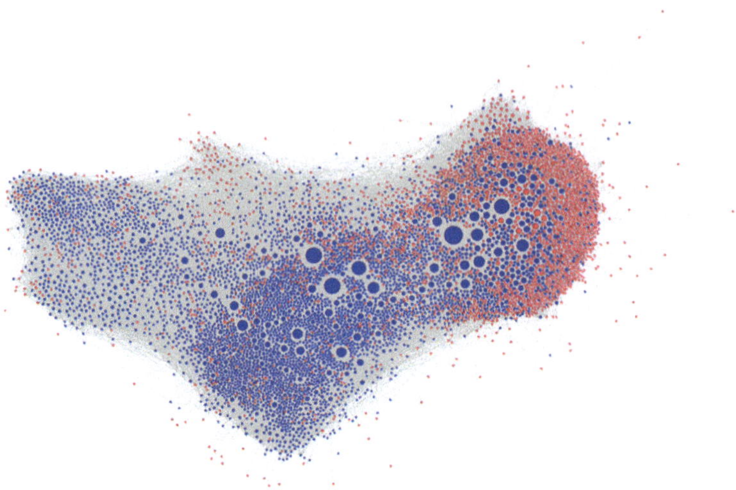

Fig. 1 Classified undirected network–*misogynistic accounts are depicted in red*

3.3 Comparing the Characteristics of Misogynistic and Non-misogynistic Communities

When applying the trained TextGCN$_{2L,A}$ model to predict labels for unlabelled account nodes based on the binary misogyny classification scheme, it reveals that approximately 37% of the total nodes are categorised as misogynistic. The classified network is represented in Fig. 1.

If we consider the directed network based on the friendship link, Fig. 2 illustrates specific network characteristics distinguishing misogynistic accounts from non-misogynistic ones. The analysis reveals that misogynistic accounts demonstrate a lower number of incoming connections (*indegree*) and a higher number of outgoing connections (*outdegree*), indicating a tendency to follow more users while being followed by fewer. Additionally, these accounts exhibit reduced values for *hub* and *authority* centrality indices, implying a diminished significance within the network structure.

Figure 3 illustrates the language differences that may be seen between the misogynistic and non-misogynistic groups. The comparison cloud, which accompanies the traditional word clouds that display the language used in both communities, allows for a comparative examination of word frequency between the two sub-corpora. This comparative cloud specifically contrasts misogynistic and non-misogynistic accounts, aiding in the study. Additionally, this comparison allows for an evaluation of the relative prominence of words, demonstrating variations in word usage among the groups. The information related to misogynistic circumstances prominently exhibits a clearly disparaging vocabulary.

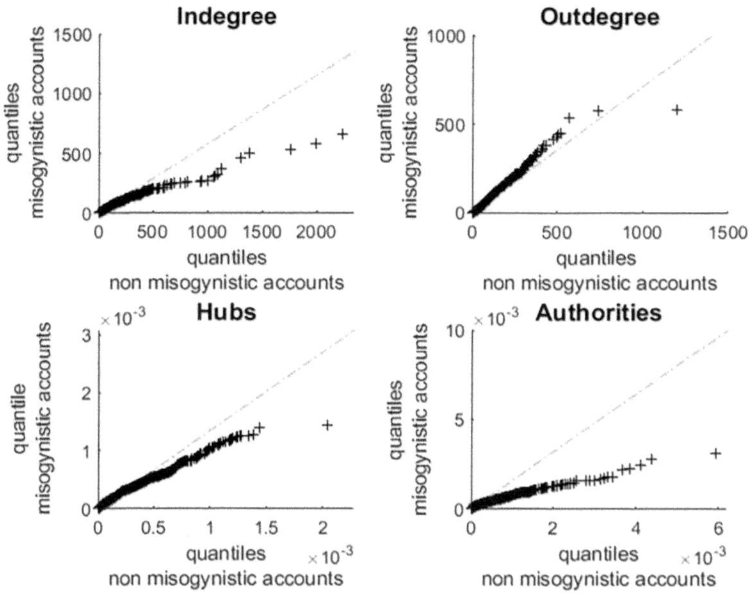

Fig. 2 Centrality measures' qqplots for the directed network based on the friendship link

4 Conclusion and Future Works

Based on our findings, the use of a two-layer GCN that combines relational and textual data shows remarkable potential in identifying misogynistic Twitter accounts. Consistent with previous studies on hater networks, Twitter accounts that promote misogyny tend to form clusters [22, 23]. These accounts typically have fewer followers but follow a larger number of other accounts. Moreover, they tend to hold less influential positions within the network relationships [13]. Furthermore, they demonstrate clear and unique linguistic patterns in their use of words [13].

An inherent limitation of the TextGCN methodology is its handling of document nodes, where it treats all words within a document as a whole, disregarding the sequential arrangement of the text. Competitive strategies are presented and discussed in Liu et al. [24]. For future work, we aim to enhance GCN's efficacy by integrating knowledge of text sequences, drawing inspiration from the approach proposed by Liu et al. [24], which combines Transformer and GCN architectures.

In order to improve the accuracy of classification, our upcoming endeavours will also focus on two main aspects. Firstly, we plan to enlarge the annotated Twitter account dataset. Secondly, we aim to incorporate user-specific factors, such as the total number of friends and followers, post count, retweets, replies, and mentions, as well as linguistic factors, such as sentiment scores, into the feature matrix **X**, which will be used for classification.

Fig. 3 Word clouds depicting sub-corpora of misogynistic and non-misogynistic accounts, alongside a word cloud comparison

Finally, we intend to perform cross-domain research in the future. This involves training the model using data from a specific domain, such as political women's textual data, and then using it to classify information in a wider network.

Acknowledgments This research was undertaken as part of the ICOMIC (Identifying and Counteracting Online Misogyny in Cyberspace) Project funded by EU Next Generation, MUR-Fondo Promozione e Sviluppo-DM 737/2021.

References

1. Amnesty International Italia. Barometro dell'odio: Sessismo da tastiera (2020). https://www.amnesty.it/barometro-dellodio-sessismo-da-tastiera/
2. European Union Agency for Fundamental Rights. Online content moderation—Current challenges in detecting hate speech (2023). https://fra.europa.eu/en/publication/2023/online-content-moderation
3. Johnson, A.G.: The Blackwell Dictionary of Sociology: A User's Guide to Sociological Language. Wiley, New York (2000)
4. Code, L.: Encyclopedia of Feminist Theories. Routledge, London (2000)
5. Fortuna, P., Nunes, S.: A survey on automatic detection of hate speech in text. ACM Comput. Surv. **51**(4), 1–30 (2018)
6. Tontodimamma, A., Nissi, E., Sarra, A., Fontanella, L.: Thirty years of research into hate speech: topics of interest and their evolution. Scientometrics **126**, 157–179 (2021)
7. Fontanella L., Chulvi B., Ignazzi E., Sarra A., Tontodimamma A.: How do we study misogyny in the digital age? A systematic literature review using a computational linguistic approach. Humanit. Soc. Sci. Commun. **11**(478), 1–15 (2024)
8. Siegel A.A.: Online Hate Speech. In Persily, N., Tucker, J.A. (eds.) Social Media and Democracy: The state of the field, prospects for reform, pp. 56–88. Cambridge University Press, Cambridge (2020)
9. Capozzi, A., Lai, M., Basile, V., Poletto, F., Sanguinetti, M., Bosco, C., Patti, V., Ruffo, G., Musto, C., Polignano, M., et al.: "Contro L'Odio": a platform for detecting, monitoring and visualizing hate speech against immigrants in Italian Social Media. IJCoL **6**(6-1), 77–97 (2020)
10. Chatzakou, D., Kourtellis, N., Blackburn, J., De Cristofaro, E., Stringhini, G., Vakali, A.: Mean birds: detecting aggression and bullying on Twitter. In: WebSci '17: Proceedings of the 2017 ACM on Web Science Conference (2017)
11. Unsvåg, E.F., Gambäck, B.: The effects of user features on Twitter hate speech detection. In: Proceedings of the 2nd Workshop on Abusive Language Online (ALW2), pp. 75–85 (2018)
12. Mishra, P., Del Tredici, M., Yannakoudakis, H., Shutova, E.: Author profiling for abuse detection. In: Proceedings of the 27th International Conference on Computational Linguistics (2018)
13. Ribeiro, M., Calais, P., Santos, Y., Almeida, V., Meira, W. Jr.: Characterizing and detecting hateful users on Twitter. In: Proceedings of the International AAAI Conference on Web and Social Media (2018)
14. London, B., Getoor L.: Collective Classification of Network Data. In: Aggarwal, C.C. (ed.) Data Classification: Algorithms and Applications, pp. 399–416. CRC Press, New York (2014)
15. Yao, L., Mao, C., Luo, Y.: Graph convolutional networks for text classification. In: Proceedings of the AAAI Conference on Artificial Intelligence, vol. 33, pp. 7370–7377 (2019)
16. Kipf, T.N., Welling M.: Semi-supervised classification with graph convolutional networks. In: 5th International Conference on Learning Representations, ICLR 2017 (2017)
17. Bronstein, M.M., Bruna, J., LeCun, Y., Szlam, A., Vandergheynst, P.: Geometric deep learning: going beyond euclidean data. IEEE Signal Process. Mag. **34**(4), 18–42 (2017)
18. Kingma, D.P., Ba, J.: Adam: a method for stochastic optimization. In: International Conference on Learning Representations (ICLR), San Diega, CA, USA (2015)
19. Fontanella L., del Gobbo, E., Cucco, A.: Identification of misogynistic accounts on Twitter through Graph Convolutional Networks. In: Coretto, P., Giordano, G., La Rocca, M., Parrella, M.L., Rampichini, C. (eds.) CLADAG 2023 Book of Abstracts and Short Papers, pp. 159–162. Pearson Education Resources, New York (2023)
20. Benoit, K., Watanabe, K., Wang, H., Nulty, P., Obeng, A., Müller, S., Matsuo, A.: quanteda: An R package for the quantitative analysis of textual data. J. Open Source Softw. **3**(30), 774 (2018)
21. Tontodimamma, A., Fontanella, L., Anzani, S., Basile, V.: An Italian lexical resource for incivility detection in online discourses. Qual. Quant. **57**, 3019–3037 (2023)

22. Johnson, N.F., Leahy, R., Johnson Restrepo, N., Velasquez, N., Zheng, M., Manrique, P., Devkota P., Wuchty, S.: Hidden resilience and adaptive dynamics of the global online hate ecology. Nature **573**(7773), 261–265 (2019)
23. Mathew, B., Dutt, R., Goyal, P., Mukherjee, A.: Spread of hate speech in online social media. In: WebSci '19: Proceedings of the 10th ACM Conference on Web Science (2019)
24. Liu, B., Guan, W., Yang, C., Fang, Z., Lu, Z.: Transformer and graph convolutional network for text classification. Int. J. Comput. Intell. Syst. **16**(1), 161 (2023)

2 Research Process and Methodology

To gain a comprehensive overview of the two countries' research output, it is necessary to use a globally consistent dataset that contains a large body of multidisciplinary peer-reviewed literature. This paper provides an analysis of Scopus-registered journal articles published between 2017 and 2022 in the subject areas of economics, business, and finance with at least one Polish or Hungarian author. The retrieved dataset contains 2790 publications for Hungary and 9676 publications for Poland with an overlapping 106 publications indicating Hungarian and Polish research cooperation.

The research question of this study is related to the topics discussed by the analyzed journal articles: How the main topics of the scientific publications can be described? More precisely: What are the main topics in Hungary and Poland, how popular these topics are compared to each other, and how does the topic's importance evolve?

To perform country-level comparisons, the analysis covers both the Hungarian and Polish datasets (excluding the duplications) and uses all the Hungarian and Polish abstracts related to the journal articles published during the given period. The analysis aims to identify the main topics of the scientific publications and to determine how popular these topics are compared to each other in Hungary and Poland.

The research uses the latent Dirichlet allocation Method (LDA), the most popular method of topic analysis proposed by Blei et al. [7] to identify topics mentioned in the abstracts of the selected publications (documents of the corpus). This algorithm works in unsupervised mode (meaning that the learning process uses only a corpus of documents with no labels indicating topics mentioned in them).[1]

Two main steps in the LDA analysis can be distinguished. During the first step, topics occurring in documents are identified. The second step allows the content of each document to be expressed as a mixture of the topics found in the first step [8].

To describe the LDA method more formally, let's assume that we are going to process documents containing words from the dictionary:

$$V = \{V_1, V_2, \ldots, V_{LV}\} \tag{1}$$

with LV unique elements (words, terms) and that the analysis covers a set of documents (corpus)

[1] We applied the following query in the Scopus database for Hungary: AFFILCOUNTRY (HUNGARY) AND (LIMIT-TO (PUBYEAR, 2022) OR LIMIT-TO (PUBYEAR, 2021) OR LIMIT-TO (PUBYEAR, 2020) OR LIMIT-TO (PUBYEAR, 2019) OR LIMIT-TO (PUBYEAR, 2018) OR LIMIT-TO (PUBYEAR, 2017)) AND (LIMIT-TO (SUBJAREA, 'ECON') OR LIMIT-TO (SUBJAREA, 'BUSI')) AND (LIMIT-TO (DOCTYPE, 'AR')), whereas subject area 'BUSI' indicates 'Business, Management, and Accounting' and 'ECON' indicates 'Economics, Econometrics, and Finance'. The query for Poland was similar and the two subsets were downloaded separately.

$$D = \{D_1, D_2, \ldots, D_{LD}\}, \tag{2}$$

which is composed of LD documents where every D_i document can be treated as a sequence of words (terms) having the form

$$D_i = \left(V_1^i, V_2^i, \ldots, V_{n_i}^i\right). \tag{3}$$

The goal of the analysis is the identification of a set of topics:

$$T = \{T_1, T_2, \ldots, T_{LT}\} \tag{4}$$

with LT elements occurring in documents belonging to the corpus D.

Every T_j topic can be defined as a sequence of pairs:

$$T_j = \left(\left(V_1, \phi_{j,1}\right), \left(V_2, \phi_{j,2}\right), \ldots, \left(V_{LV}, \phi_{j,LV}\right)\right), \tag{5}$$

where V_k is a word belonging to the dictionary V and $\phi_{j,k}$ is a probability of occurrence of the word V_k in the topic T_j. Values $\phi_{j,k}$ reflect the significance of the k-th word in the j-th topic. The holistic definition of all topics is provided by the matrix Φ which has a form

$$\Phi = \begin{bmatrix} \phi_{1,1} & \cdots & \phi_{1,LV} \\ \cdots & \cdots & \cdots \\ \phi_{LT,1} & \cdots & \phi_{LT,LV} \end{bmatrix}, \tag{6}$$

where $\phi_{j,k}$ can be interpreted as a probability of the occurrence of the k-th word in the j-th topic.

Topics defined in the matrix Φ allow us to define the content of documents from the corpus D. It is assumed that every D_i document can be expressed as

$$D_i = \left(\left(T_1, \theta_{i,1}\right), \left(T_2, \theta_{i,2}\right), \ldots, \left(T_{LT}, \theta_{i,LT}\right)\right), \tag{7}$$

where T_j represents the given topic and $\theta_{i,l}$ is the probability of occurrence of the l-th topic in the i-th document (which can be interpreted as the contribution of the l-th topic in the i-th document). The content of all documents belonging to the corpus D is presented as the matrix Θ:

$$\Theta = \begin{bmatrix} \theta_{1,1} & \cdots & \theta_{1,LT} \\ \cdots & \cdots & \cdots \\ \theta_{LD,1} & \cdots & \theta_{LD,LT} \end{bmatrix}, \tag{8}$$

where $\theta_{i,l}$ defines as a contribution of the l-th topic in the i-th document.

Estimation of the Φ and Θ matrices is usually carried out using the collapsed Gibbs sampling method [9].

Values from the matrix Φ can be used for topic visualization. Every topic can be presented by the empirical density function having a form of the Dirichlet distribution over words. A modified way of the topics' visualization was proposed in the paper [10]. This method requires the following calculations.

First, the importance of every word in the whole corpus is calculated. Measures of importance form the vector:

$$s = \left[s_1 \ldots s_{LV}\right], \tag{9}$$

where

$$s_i = \sum_{k=1}^{LT} \phi_{k,i}. \tag{10}$$

Next, the matrix Φ is transformed into the matrix Γ using the formula:

$$\gamma_{k,i} = \lambda \phi_{k,i} + (1 - \lambda) \frac{\phi_{k,i}}{s_i}, \tag{11}$$

where λ is a real value belonging to the range [0; 1]. In the study presented here λ equal to 0.75 was used, sorted values of the k-th row of the Γ matrix are used to present main features of the k-th topic. Presentation in the form of the word cloud graph can be very useful.

Let's assume that we have a group of documents chosen from the corpus and indices of documents belonging to this group are members of the set G. Then, the significance of topics for the set G can be expressed by summing up elements of the Θ matrix corresponding to the selected group of documents and is represented by the vector:

$$r = \left[r_1 \ldots r_P\right] \tag{12}$$

with elements

$$r_k = \sum_{j \in G} \theta_{j,k}. \tag{13}$$

Next, the elements of the r vector are normalized using the formula:

$$r_k^{'} = \frac{r_k}{\sum_{l=1}^{P} r_l}. \tag{14}$$

The elements of the $r^{'}$ vector describe the significance of topics for documents belonging to the G set. The process of data analysis to answer the research question is presented in Fig. 1.

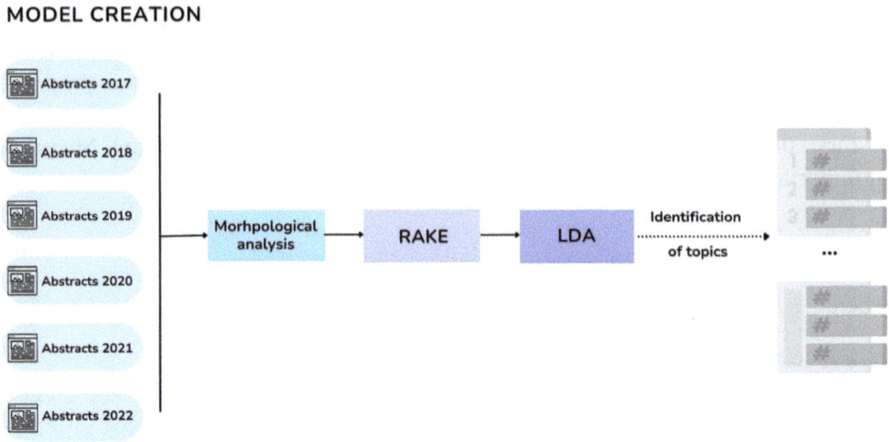

Fig. 1 The process of data analysis in topic modelling using LDA

The subsequent phases of the work included the following steps: The process started with using the merged Hungarian and Polish 2017–2022 dataset. Abstracts retrieved from the Scopus database were subjected to morphological analysis. Next, keywords and key phrases were identified using the RAKE algorithm. This way, only nouns and adjectives were included in the analytics, and key phrases were inserted into the corpus in place of the word sequences that make up these phrases. To get meaningful results, it was necessary to restrict further analysis to nouns and adjectives. Finally, using the LDA method resulted in finding the main topics. Interpretation happened in the form of assigning labels to each of the topics. As a last step, calculating the absolute and relative values of topic importance provided grounds for comparative analytics and investigating the changes in topic popularity over time.

3 Results

In the LDA method, the number of topics is determined by a researcher. In the analysis described here, it was assumed that obtained topics should be consistent with main research areas in economics and management and that the overlap of topics should be minimal. Finally, eight topics were identified: 'quantitative methods'; 'natural resources and the environment'; 'public finance and financial markets'; 'entrepreneurship and innovations'; 'education'; 'company management'; 'the state, law, and security'; and 'regional development'.

The analysis of the topics' appearance had also more subsequent steps. First, the significance of every topic was calculated by summing up assigned values to publications that represented the topics' contribution. The aggregation of measures indicating topics' contribution was performed separately for every year (Fig. 2).

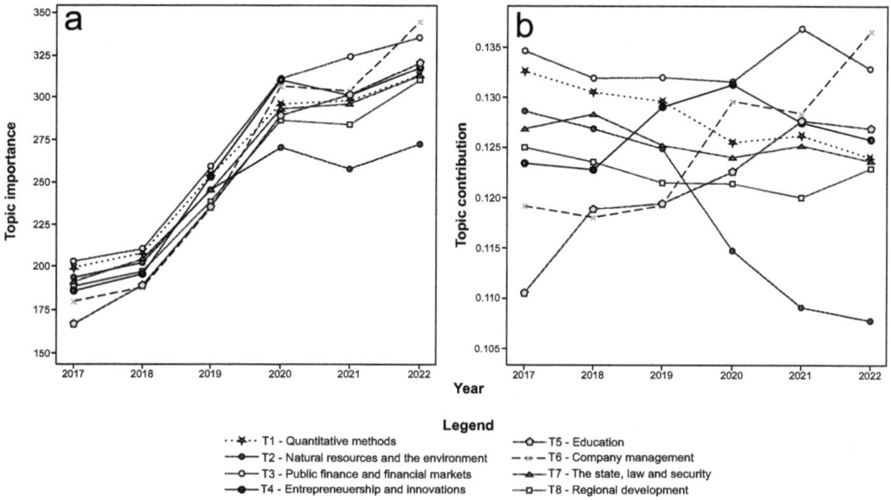

Fig. 2 Importance of topics (**a**) and relevance (**b**) in publications by Hungarian and Polish authors registered in Scopus from 2017 to 2022 in the fields of business, economics, and finance

Figure 2a indicates a general increase in the importance of every topic. It should be noticed that the growing trend is caused by the increasing number of articles published in each subsequent year.

Changes in the contribution of individual topics to the scientific output published in the given year however show a different pattern (Fig. 2b).

The importance of the topic related to company management grew the most, but education is also among the most emerging topics. On the other hand, the issue of natural resources and the environment was losing its importance to the greatest extent. A noticeable decline can also be observed in the relative importance of the area related to quantitative methods. The topic related to public finance and financial markets had the greatest importance for almost the entire period of the analysis (except for the last year covered by the research when it was overtaken by company management issues).

4 Concluding Remarks

The lesson from this research is that the LDA method can indeed find latent topics inside different documents and can provide interpretable, with little overlap in meaning topics. In the case of Hungary and Poland, the most important topics in business, economics, and finance from 2017 to 2022 were quantitative methods; natural resources and the environment; public finance and financial markets; entrepreneurship and innovations; education; company management; the state, law, and security; and regional development. Interestingly, looking closer at the evolution of relative topic importance in the two countries, the research could find that the relevance of topic 2 ('natural resources and environment') expressed relative to all publications from the given year decreased the most from 2017 to 2022 in Hungary and Poland, while topic 6 (company management) and topic 5 (education) increased considerably.

Finally, we have to consider the limitations of this study. Despite our efforts to be methodologically rigorous and make the widest possible use of the available databases, our research is not without limitations.

An aspect that one must consider is the language of the documents. Non-English research papers can also be significant since they can play a crucial role in developing the vocabulary of new disciplines and in the dissemination of the new vocabulary of scientific disciplines, the dissemination of new findings to the wider scientific community, and raising the interest of the new scientific generations. Locally relevant research is also important, even though it is not represented among the high-impact English-language journals [11], like those, that are registered in the Scopus dataset. The solution for overcoming the language limitations can be an extension of the analyses to publications by national journals.

Another opportunity is to combine more data sources, e.g. Web of Science, Crossref, Google Scholar, or Science Direct. However, this research strategy anticipates many challenges, e.g. handling possible overlaps and differences in database structures.

The extension of the analytics can happen in terms of geographical coverage: This research focused specifically on two countries, but comparative analytics may offer particularly interesting results if we extend our scope to Europe or the world. Finally, it is worth noting that quantitative methods might unveil interesting patterns and associations, but to gain a deeper understanding, mixed methods research might be especially fruitful in this field that incorporates qualitative techniques, e.g. an explanatory sequential research design [12].

References

1. Crane, D.: Invisible Colleges; Diffusion of Knowledge in Scientific Communities. University of Chicago Press, Chicago (1972)
2. Aria, M., Cuccurullo, C.: Bibliometrix: An R-tool for comprehensive science mapping analysis. J. Informetrics **11**(4), 959–975 (2017)
3. Yang, S., Yuan, Q., Dong, J.: Are scientometrics, informetrics, and bibliometrics different? Data Science and Informetrics **01**(01), 50 (2020). Number: 01 Publisher: Scientific Research Publishing
4. Hess, D.J.: Science Studies: An Advanced Introduction. NYU Press, New York (1997)
5. Ivancheva, L.: Scientometrics today: A methodological overview. Collnet J. Scientometrics Inf. Manag. **2**(2), 47–56 (2008). ISBN: 0973-7766 Publisher: Taylor & Francis.
6. Greskovits, B., Bohle, D.: Development paths on europe's periphery: Hungary's and poland's return to europe compared. Pol. Sociol. Rev. **33**, 3–27 (2001)
7. Blei, D.M., Ng, A.Y., Jordan, M.I.: Latent dirichlet allocation. J. Mach. Learn. Res. **3**(Jan), 993–1022 (2003)

8. Lula, P.l.: Statystyczne Modelowanie Zawartości Dokumentów Tekstowych. In: Wydawnictwo Uniwersytetu Ekonomicznego (2018)
9. Yao, L., Mimno, D., McCallum, A.: Efficient methods for topic model inference on streaming document collections. In: Proceedings of the 15th ACM SIGKDD International Conference on Knowledge Discovery and Data Mining, pp. 937–946 (2009)
10. Sievert, C., Shirley, K.: LDAvis: a method for visualizing and interpreting topics. In: Proceedings of the Workshop on Interactive Language Learning, Visualization, and Interfaces, pp. 63–70 (2014)
11. Hicks, D., Wouters, P., Waltman, L., De Rijcke, S., Rafols, I.: Bibliometrics: the Leiden Manifesto for research metrics. Nature **520**(7548), 429–431 (2015). ISBN: 0028-0836 Publisher: Nature Publishing Group UK London
12. Creswell, J.W., Clark, V.P.: Mixed Methods Research. SAGE Publications, New York (2011)

Circular Kernel Classification with Errors in Variables

Marco Di Marzio, Stefania Fensore, Chiara Passamonti, and Agnese Panzera

Abstract Classification of data observed with errors is a relevant topic which has still not been fully explored. We address this problem in the directional context and discuss two nonparametric approaches which rely on kernel estimation of circular densities. We provide some asymptotic L_2 properties for the proposed methods, along with some simulation results.

1 Introduction

Circular data are observations having the unit circle as the sample space. After choosing both an origin and a sense of rotation, they can be measured, using a periodic scale, as angles ranging in $[-\pi, \pi)$ radians or $[0, 360)$ degrees. Circular data are typical of many scientific fields, for example, biology (migration paths, directions of animals), meteorology (wind and marine current directions), and geology (orientations of joints and faults, landforms, oriented stones). For a recent, detailed overview of advances in directional statistics, see [9] and [10].

Classification of circular data has been studied by many scientists using both parametric and nonparametric approaches. A chord-length classification rule based on the distance between a new observation and the observations of two known circular populations has been introduced by SenGupta and Roy [12]. Then, [13] proposed a method involving a generalized likelihood ratio test for classifying toroidal and cylindrical data in two populations. Lopez-Cruz et al. [11] extended the naive Bayes classifier to the case where the conditional probability distribution of the predictor is assumed to be von Mises-Fisher. Fernandes and Cardoso [8] adapted

M. Di Marzio (✉) · S. Fensore · C. Passamonti
G. d'Annunzio University, Pescara, Italy
e-mail: marco.dimarzio@unich.it; stefania.fensore@unich.it; agnese.panzera@unifi.it

A. Panzera
University of Florence, Florence, Italy
e-mail: chiara.passamonti@unich.it

G. Giordano et al. (eds.), *Statistical Models and Learning Methods for Complex Data*, Studies in Classification, Data Analysis, and Knowledge Organization,
https://doi.org/10.1007/978-3-031-84702-8_10

the logistic regression to classify directional data. Recently, [3] and [4] introduced some nonparametric classification methods for spherical data which are based on kernel estimation of the population densities and on local logistic regression.

Above contributions refer to classification methods based on data which are directly observable. Here, we extend the classification task to the case where data are affected by measurement errors. Due to the deterioration or imperfections of measurement devices, or the difficulty of directly accessing the variables of interest, measurement contamination arises in a wide variety of data (from meteorology to astronomy, biology, geology). For this reason, the so-called *errors-in-variables* context, which refers to this kind of data, has been explored by many authors. Until now, the problems which have been mainly studied are density and regression estimation. Specifically, kernel-type estimators for these two topics have been derived both in the Euclidean setting (see, e.g., [1] and [2] and the references therein) and in the circular case (see, e.g., [5, 6]).

In this chapter, we discuss this classification task using two nonparametric approaches which rely on kernel estimation of circular densities with errors in variables. The first approach employs a weight function known as circular deconvolution kernel. The second one is based on a *corrected* version of the naive kernel density estimator, where an estimate of the extra bias due to the presence of measurement errors is subtracted. The chapter is organized as follows. In Sect. 2, we present the errors-in-variables context along with the density estimators which will be involved in the classification task, while in Sect. 3, the discriminant rule is studied. Finally, Sect. 4 reports some simulation studies.

2 Errors-in-Variables Problem

Suppose that we aim to estimate nonparametrically the density f_Θ of a random angle Θ, and, instead of observing an i.i.d. random sample $\Theta_1, \ldots, \Theta_n$, the contaminated version $\Phi_1, \ldots, \Phi_n$ is available. Formally, assume that

$$\Phi_i = (\Theta_i + \varepsilon_i)\mathsf{mod}(2\pi), \tag{1}$$

where the ε_is are independent copies of a random angle ε, independent of Θ, whose density f_ε is assumed to be a known circular density, which is symmetric around 0.

Now, since the density f_Φ of Φ is the circular convolution of f_Θ and f_ε, estimating f_Θ amounts to the deconvolution of f_Φ. It is mathematically convenient to deal with convolutions in the Fourier domain because, there, they can be treated in terms of the product between the characteristic functions of the involved variables.

In the circular setting, the characteristic function of a random variable Q, say $\varphi_Q(\ell)$, is defined only for integer ℓ, and it coincides with the ℓth trigonometric moment of Q, which can be also expressed in terms of ℓth sine and cosine moments $\varphi_Q(\ell) = \alpha_\ell + i\beta_\ell$, where $i^2 = -1$, $\alpha_\ell = \mathbb{E}[\cos(\ell Q)]$, and $\beta_\ell = \mathbb{E}[\sin(\ell Q)]$. It is also interesting to note that the trigonometric moments are the coefficients in the following Fourier series:

$$f_Q(q) = \frac{1}{2\pi} \sum_{\ell=-\infty}^{\infty} \varphi_Q(\ell) e^{-i\ell q} = \frac{1}{2\pi} \left\{ 1 + 2 \sum_{\ell=1}^{\infty} (\alpha_\ell \cos(\ell q) + \beta_\ell \sin(\ell q)) \right\}.$$

In the following paragraphs, we discuss the estimation of circular densities using two approaches proposed by Di Marzio et al. [5]. First, according to the fact that f_Φ is a convolution between f_Θ and f_ε, we use a kernel density estimator which employs a weight function known as circular deconvolution kernel. Then, we *correct* the naive density estimator by subtracting an estimate of the increase in bias due to the measurement error.

The Deconvolution Density Estimator

Let $\hat{\varphi}_\Phi(\ell) = n^{-1} \sum_{j=1}^n e^{i\ell\Phi_j}$ be the empirical characteristic function of Φ, and let $\varphi_\varepsilon(\ell)$ be the characteristic function of the error density, which satisfies $\varphi_\varepsilon(\ell) \neq 0$ for all ℓ. A deconvolution density estimator for f_Θ at $\theta \in [0, 2\pi)$ is

$$\tilde{f}_\Theta(\theta; \kappa) = \frac{1}{2\pi} \sum_{\ell=-\infty}^{\infty} \frac{\hat{\varphi}_\Phi(\ell)}{\varphi_\varepsilon(\ell)} \varphi_{K_\kappa}(\ell) e^{-i\ell\theta} = \frac{1}{n} \sum_{i=1}^{n} \widetilde{K}_\kappa(\Phi_i - \theta), \tag{2}$$

where $\widetilde{K}_\kappa$ is a weight function, known as deconvolution kernel, defined as

$$\widetilde{K}_\kappa(\phi) = \frac{1}{2\pi} \left\{ 1 + 2 \sum_{\ell=1}^{\infty} \frac{\gamma_\ell(\kappa)}{\lambda_\ell(\kappa_\varepsilon)} \cos(\ell\phi) \right\}$$

with $\gamma_\ell(\kappa)$ and $\lambda_\ell(\kappa_\epsilon) \neq 0$ (for each ℓ), respectively, being the ℓth Fourier coefficient of the circular kernel K_κ and f_ε (with concentration κ_ε).

We consider K_κ being a second sin-order kernel, i.e., a periodic density function which, for $\eta_s(K_\kappa) = \int_{-\pi}^{\pi} K_\kappa(\alpha) \sin^s(\alpha) d\alpha$, satisfies $\eta_0(K_\kappa) = 1$, $\eta_1(K_\kappa) = 0$, and $\eta_2(K_\kappa) = (1 - \gamma_2(\kappa))/2 \neq 0$.

The asymptotic properties of estimator (2) are collected in the following.

Result 1 Given a random sample $\Phi_1, \ldots, \Phi_n$ from f_Φ, assume (1). Then, for estimator $\tilde{f}_\Theta(\theta; \kappa)$, if

(i) f_Θ is twice continuously differentiable in a neighborhood of $\theta \in [0, 2\pi)$.
(ii) K_κ is a second sin-order kernel.
(iii) The concentration parameter κ increases with n in such a way that, for $\ell \in \mathbb{Z}^+$, $\lim_{n\to\infty} \gamma_\ell(\kappa) = 1$ and $\lim_{n\to\infty} n^{-1} \sum_{\ell=1}^{\infty} \gamma_\ell^2(\kappa) = 0$.

the following asymptotic bias and variance hold

$$\mathsf{E}[\tilde{f}_\Theta(\theta; \kappa)] - f_\Theta(\theta) = \frac{(1 - \gamma_2(\kappa))}{4} f''_\Theta(\theta) + o(1 - \gamma_2(\kappa)),$$

$$\mathsf{VAR}[\tilde{f}_\Theta(\theta;\kappa)] = \frac{\left(1+2\sum_{\ell=1}^{\infty}\gamma_\ell^2(\kappa)/\lambda_\ell^2(\kappa_\varepsilon)\right)}{2\pi n} f_\Theta(\theta) + o\left(\frac{\sum_{\ell=1}^{\infty}\gamma_\ell^2(\kappa)/\lambda_\ell^2(\kappa_\varepsilon)}{n}\right).$$

Notice that the bias of the estimator is the same as the bias of the error-free case estimator, while the accuracy of the estimator is also determined by the smoothness of f_ε which is said to be *supersmooth* if the rate of decay of the coefficients in its Fourier series representation is *exponential* (e.g., the wrapped Cauchy density), while it is *ordinary smooth* if this rate of decay is *polynomial* (e.g., the wrapped Laplace density).

The Bias-Corrected Density Estimator

Let $\hat{f}_\Phi(\theta;\kappa)$ be the *naive* kernel estimator of $f_\Theta(\theta)$, but based on $\Phi_1, \ldots, \Phi_n$

$$\hat{f}_\Phi(\theta;\kappa) = \frac{1}{n}\sum_{i=1}^{n} K_\kappa(\Phi_i - \theta). \tag{3}$$

By expanding $K_\kappa(\Phi_i - \theta)$ for Φ_i around Θ_i, one has

$$\hat{f}_\Phi(\theta;\kappa) \approx \hat{f}_\Theta(\theta;\kappa) + \frac{1}{n}\sum_{i=1}^{n}\sin(\Phi_i - \Theta_i)K'_\kappa(\Theta_i - \theta)$$
$$+ \frac{1}{2n}\sum_{i=1}^{n}\sin^2(\Phi_i - \Theta_i)K''_\kappa(\Theta_i - \theta).$$

Then, taking the expectation, and observing that $\mathsf{E}[\sin(\varepsilon_i)] = 0$, $\mathsf{E}[\sin^2(\varepsilon_i)] = (1-\lambda_2(\kappa_\varepsilon))/2$, a bias-corrected density estimator for f_Θ at $\theta \in [0, 2\pi)$ is defined by

$$\tilde{f}^*_\Theta(\theta;\kappa) = \hat{f}_\Phi(\theta;\kappa) - \frac{(1-\lambda_2(\kappa_\varepsilon))}{4n}\sum_{i=1}^{n} K''_\kappa(\Phi_i - \theta), \tag{4}$$

where the excess of bias conditional on the Θ_is is estimated on the basis of the sample $\Phi_1, \ldots, \Phi_n$. Differently from estimator (2), it requires only the knowledge of $\lambda_2(\kappa_\varepsilon)$ and is not limited to the cases where $\lambda_\ell(\kappa_\varepsilon) \neq 0$, $\ell \in \mathbb{Z}^+$.

The asymptotic properties of estimator (4) are collected in the following.

Result 2 Given a random sample $\Phi_1, \ldots, \Phi_n$ from f_Φ, assume (1). Then, for estimator $\tilde{f}^*_\Theta(\theta;\kappa)$, under assumptions $(ii) - (iii)$ of Result 1, if f_Θ and f_Φ have continuous derivatives up to order 2 and 4, respectively, and f_ε has finite second sin-moment, the following asymptotic bias and variance hold

$$\mathsf{E}[\tilde{f}^*_\Theta(\theta;\kappa)] - f_\Theta(\theta) = \frac{1-\gamma_2(\kappa)(2-\lambda_2(\kappa_\varepsilon))}{4} f''_\Phi(\theta) + o(1-\gamma_2(\kappa)),$$

$$\mathsf{VAR}[\tilde{f}^*_\Theta(\theta;\kappa)] = \frac{f_\Phi(\theta)}{2\pi n}\left\{1+2\sum_{\ell=1}^{\infty}\gamma_\ell^2(\kappa)\left[1+\frac{\ell^2(1-\lambda_2(\kappa_\varepsilon))}{4}\right]^2\right\}$$
$$+\,o\left(\frac{1}{n}\sum_{\ell=1}^{\infty}\gamma_\ell^2(\kappa)\right).$$

Notice that both bias and variance depend on the second Fourier coefficient of the error density. Moreover, as κ_ε increases, and then $\lambda_2(\kappa_\varepsilon)$ goes to 1, the leading terms in the asymptotic bias and variance are nearly the same as in the error-free case.

3 Errors-in-Variables Classification

Let (Θ, Y) be a pair of random variables, where Θ is a circular predictor and Y a binary response, and set $\delta(\theta) := P(Y = 1 \mid \Theta = \theta)$. Consider the densities $f_{\Theta,1}$ and $f_{\Theta,2}$ which describe, respectively, the covariate space for successes ($Y = 1$) and failures ($Y = 0$), and let π_1 be the proportion of successes in the population, and $\pi_2 = 1 - \pi_1$. Then, for $\theta \in [0, 2\pi)$, using the Bayes's theorem, we set

$$\delta(\theta) = \frac{\pi_1 f_{\Theta,1}(\theta)}{\pi_1 f_{\Theta,1}(\theta) + \pi_2 f_{\Theta,2}(\theta)}.$$

A standard approach to get a classification rule is based on replacing the unknown densities appearing in the above formula by their kernel density estimates. Then, for a point (θ, y), if the estimate of $\delta(\theta)$ satisfies $\hat{\delta}(\theta) \geq 0.5$, we predict a success ($\hat{y} = 1$); otherwise, we predict a failure ($\hat{y} = 0$).

Now, in the errors-in-variables context, Θ is not observable, whereas the sample available is $(\Phi_1, Y_1), \ldots (\Phi_n, Y_n)$, with Φ defined according to model (1). Assume that the i.i.d. sample has been reordered, so that the first n_1 pairs are successes and the last $n_2 = n - n_1$ are failures. Replacing π_j with n_j/n, $j \in \{1, 2\}$, an estimator of $\delta(\theta)$ which takes into account of the presence of the measurement error is

$$\ddot{\delta}(\theta;\kappa_1,\kappa_2) = \frac{n_1 \ddot{f}_{\Theta,1}(\theta;\kappa_1)}{n_1 \ddot{f}_{\Theta,1}(\theta;\kappa_1) + n_2 \ddot{f}_{\Theta,2}(\theta;\kappa_2)},$$

where $\ddot{f}_{\Theta,j}$s, $j \in \{1, 2\}$, are kernel density estimators (2) or (4) for $f_{\Theta,j}$s. Therefore, the classification rule which predicts $y = 1$ (and $y = 0$ otherwise) becomes $\ddot{\delta}(\theta;\kappa_1,\kappa_2) \geq 0.5$. In the particular case of $\pi_1 = \pi_2 = 0.5$, this rule becomes

$$\ddot{g}(\theta;\kappa_1,\kappa_2) := \ddot{f}_{\Theta,1}(\theta;\kappa_1) - \ddot{f}_{\Theta,2}(\theta;\kappa_2) \geq 0, \tag{5}$$

where $\ddot{g}(\theta;\kappa_1,\kappa_2)$ is the errors-in-variables estimator of $g(\theta) = f_{\Theta,1}(\theta) - f_{\Theta,2}(\theta)$. This topic has been studied in the Euclidean, error-free case by Di Marzio and Taylor [7].

When $f_{\Theta,j}$s are estimated using (2), the asymptotic properties of estimator $\ddot{g}$ appearing in Eq. (5) are collected in the following.

Result 3 Given the random sample $(\Phi_1, Y_1), \ldots, (\Phi_n, Y_n)$, assume (1) for Φ, and let Y having a binary nature, with the first n_1 pairs coming from population $f_{\Phi,1}$ and the last n_2 ones coming from $f_{\Phi,2}$. Then, if

(i) $f_{\Theta,1}$ and $f_{\Theta,2}$ are twice continuously differentiable in a neighborhood of θ.
(ii) K_{κ_j} for $j \in (1,2)$ is a second sin-order kernel.
(iii) κ_j for $j \in (1,2)$ increases with n in such a way that, for $\ell \in \mathbb{Z}^+$, $\lim_{n\to\infty} \gamma_\ell(\kappa_j) = 1$ and $\lim_{n\to\infty} n^{-1}\sum_{\ell=1}^{\infty}\gamma_\ell^2(\kappa_j) = 0$.

the following asymptotic bias and variance hold for estimator $\tilde{g}(\theta;\kappa_1,\kappa_2)$, $\theta \in [0, 2\pi)$:

$$\begin{aligned}
&\mathsf{E}[\tilde{g}(\theta;\kappa_1,\kappa_2)] - (f_{\Theta,1}(\theta) - f_{\Theta,2}(\theta)) \\
&= \frac{1}{4}\left\{(1-\gamma_2(\kappa_1))f''_{\Theta,1}(\theta) - (1-\gamma_2(\kappa_2))f''_{\Theta,2}(\theta)\right\} \\
&\quad + o(1-\gamma_2(\kappa_1) + 1 - \gamma_2(\kappa_2)),
\end{aligned}$$

$$\mathsf{VAR}[\tilde{g}(\theta;\kappa_1,\kappa_2)] = \sum_{j=1}^{2}\frac{f_{\Theta,j}(\theta)}{n_j}\nu(\widetilde{K}_{\kappa_j}) + o\left(\sum_{j=1}^{2}\frac{\nu(\widetilde{K}_{\kappa_j})}{n_j}\right),$$

where $\nu(\widetilde{K}_{\kappa_j}) = 1/(2\pi)\left(1 + 2\sum_{\ell=1}^{\infty}\gamma_\ell^2(\kappa_j)/\lambda_\ell^2(\kappa_\varepsilon)\right)$.

When $f_{\Theta,j}$s are estimated using (4), the asymptotic properties of estimator $\ddot{g}$ appearing in Eq. (5) are collected in the following.

Result 4 Given the random sample $(\Phi_1, Y_1), \ldots, (\Phi_n, Y_n)$, assume (1) for Φ, and let Y having a binary nature, with the first n_1 pairs drawn from $f_{\Phi,1}$ and the last n_2 ones drawn from $f_{\Phi,2}$. Then, under assumptions $(ii) - (iii)$ of Result 3, if $f_{\Theta,j}$s and $f_{\Phi,j}$, $j \in (1,2)$, have continuous derivatives up to order 2 and 4, respectively, and f_ε has finite second sin-moment, the following asymptotic bias and variance hold for estimator $\tilde{g}^*(\theta;\kappa_1,\kappa_2)$, $\theta \in [0, 2\pi)$:

$$\begin{aligned}
\mathsf{E}[\tilde{g}^*(\theta;\kappa_1,\kappa_2)] - (f_{\Theta,1}(\theta) - f_{\Theta,2}(\theta)) &= \frac{1}{4}(1-\gamma_2(\kappa_1)(2-\lambda_2(\kappa_\varepsilon)))f''_{\Phi,1}(\theta) \\
-\frac{1}{4}(1-\gamma_2(\kappa_2)(2-\lambda_2(\kappa_\varepsilon)))f''_{\Phi,2}(\theta) &+ o(1-\gamma_2(\kappa_1) + 1 - \gamma_2(\kappa_2)),
\end{aligned}$$

$$\mathsf{VAR}[\tilde{g}^*(\theta;\kappa_1,\kappa_2)] = \frac{1}{2\pi}\sum_{j=1}^{2}\frac{f_{\Phi,j}(\theta)}{n_j}\left\{1+2\sum_{\ell=1}^{\infty}\gamma_\ell^2(\kappa_j)\tau^2\right\}$$
$$+\,o\left(\sum_{j=1}^{2}\frac{1}{n_j}\sum_{\ell=1}^{\infty}\gamma_\ell^2(\kappa_j)\right),$$

where $\tau = 1 + (\ell^2(1-\lambda_2(\kappa_\varepsilon)))/4$.

Remark 1 We notice that when $\pi_1 \neq \pi_2$ the discriminant rule in Eq. (5) becomes $\ddot{g}(\theta;\kappa_1,\kappa_2) := \pi_1\ddot{f}_{\Theta,1}(\theta;\kappa_1) - \pi_2\ddot{f}_{\Theta,2}(\theta;\kappa_2) \geq 0$. Therefore, different a priori probabilities do not affect at all the properties of our methods but simply lead to distinct contributions of the populations estimators to the precision of the decision rule based on the difference.

4 Numerical Experiments

In this section, we aim to perform the supervised classification of groups coming from two populations using the classification rule in Eq. (5). We employ as density estimators the deconvolution ($\tilde{g}$) and bias-corrected ($\tilde{g}^*$) ones and compare them with the naive kernel density one ($\hat{g}$) in various simulative scenarios.

Our case studies consist of simple discrimination problems involving two groups of observations drawn from von Mises (vM) and wrapped Cauchy (wC) populations with equal or different means and concentrations. Specifically, the population models, also depicted in Fig. 1, are:

- M1:=vM($\pi-1,1$), vM($\pi+1,1$). M2:=vM($\pi,0.5$), vM($\pi,5$); M3:=vM($\pi,0.5$), vM($\pi+0.5,5$);
- C1:=wC($\pi-1,0.3$), wC($\pi+1,0.3$); C2:=wC($\pi,0.2$), wC($\pi,0.7$); C3:=wC($\pi,0.2$), wC($\pi+0.5,0.7$).

The kernel employed is the von Mises, and the two smoothing parameters, κ_1 and κ_2, are selected using the likelihood cross-validation. Finally, as for the error density, we assume wrapped Laplace or wrapped Cauchy models.

For each scenario, we draw 500 replicates of equally sized ($n_1 = n_2$) groups of observations. The accuracy of the estimators is evaluated according to the mean squared error (MSE) of the difference $\ddot{f}_{\Theta,1}(\theta_0;\kappa_1) - \ddot{f}_{\Theta,2}(\theta_0;\kappa_2)$, with θ_0 being the value where the true population densities $f_{\Theta,1}$ and $f_{\Theta,2}$ are equal, i.e., $f_{\Theta,1}(\theta_0) = f_{\Theta,2}(\theta_0)$. When more than one θ_0 occur, we consider the smallest value.

From Table 1, we notice that the proposed estimators improve on the naive approach in all scenarios but the simplest one, because in that case there is a specific bias deletion mechanism due to the sign of the second derivatives of the curves. Clearly, deconvolution approach is superior to the bias-corrected one, because it is

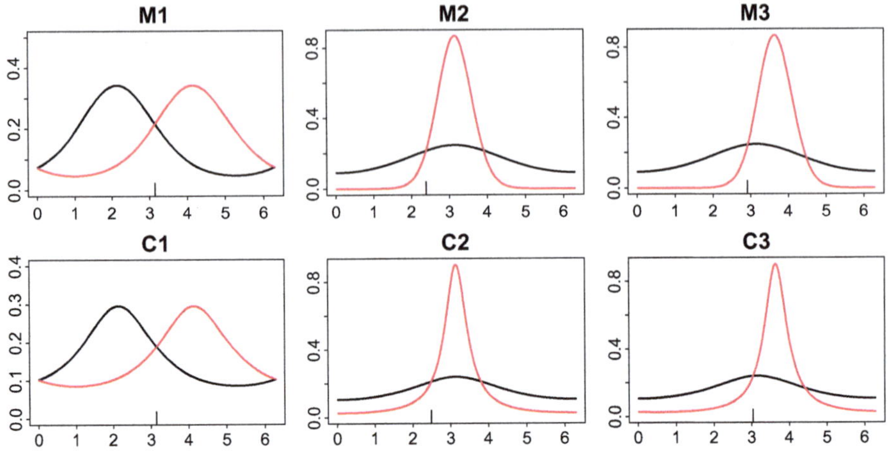

Fig. 1 Population models M1–M3 (top) and C1–C3 (bottom). Vertical bars refer to θ_0 values

Table 1 MSE × 1000 over 500 samples of size n_j, $j \in (1, 2)$, drawn from models M1–M3 (left panel) and C1–C3 (right panel) corrupted with a wrapped Laplace (wrapped Cauchy) error

n_j	Model	$\hat{g}$	$\tilde{g}$	$\tilde{g}^*$	Model	$\hat{g}$	$\tilde{g}$	$\tilde{g}^*$
100	M1	2.26 (1.98)	2.71 (2.62)	2.76 (3.57)	C1	1.79 (1.96)	2.06 (2.74)	2.33 (3.91)
	M2	7.93 (10.2)	7.28 (9.61)	7.91 (9.49)	C2	9.34 (9.19)	8.60 (9.09)	8.76 (8.81)
	M3	8.24 (9.72)	7.57 (9.21)	8.30 (9.33)	C3	13.2 (13.0)	12.1 (13.0)	12.3 (12.0)
500	M1	0.60 (0.63)	0.72 (0.91)	0.80 (1.64)	C1	0.49 (0.43)	0.57 (0.57)	0.68 (0.92)
	M2	3.34 (4.16)	2.58 (2.66)	2.88 (2.97)	C2	3.31 (3.48)	2.49 (2.18)	2.58 (3.36)
	M3	3.31 (4.16)	2.57 (2.72)	2.90 (3.51)	C3	5.12 (5.13)	3.59 (3.24)	3.88 (4.10)
1000	M1	0.36 (0.32)	0.44 (0.46)	0.48 (0.70)	C1	0.29 (0.27)	0.36 (0.36)	0.40 (0.63)
	M2	2.56 (2.80)	1.82 (1.54)	2.00 (1.82)	C2	2.16 (2.29)	1.42 (1.15)	1.58 (2.40)
	M3	2.47 (2.63)	1.84 (1.44)	2.00 (1.97)	C3	3.32 (3.66)	2.02 (1.88)	2.33 (2.42)

aimed to delete the whole bias due to measurement error, whereas this latter only subtract the leading term of the bias. As for the sample sizes of the groups, we see that as n_1 and n_2 increase, the performance of our estimators improves in accordance to their properties that are asymptotic in nature.

References

1. Delaigle, A.: An alternative view of the deconvolution problem. Statistica Sinica **18**, 1025–1045 (2008)
2. Delaigle, A.: Nonparametric kernel methods with errors-in-variables: constructing estimators, computing them, and avoiding common mistakes. Aust. N. Z. J. Stat. **56**, 105–124 (2014)
3. Di Marzio, M., Fensore, S., Panzera, A., Taylor, C.C.: Kernel density classification for spherical data. Stat. Probab. Lett. **144**, 23–29 (2019)

4. Di Marzio, M., Fensore, S., Panzera, A., Taylor, C.C.: Local binary regression with spherical predictors. Stat. Probab. Lett. **144**, 30–36 (2019)
5. Di Marzio, M., Fensore, S., Panzera, A., Taylor, C.C.: Density estimation for circular data observed with errors. Biometrics **78**, 248–260 (2022)
6. Di Marzio, M., Fensore, S., Panzera, A., Taylor, C.C.: Kernel regression for errors-in-variables problems in the circular domain. Stat. Meth. Appl. **32**, 1217–1237 (2023)
7. Di Marzio, M., Taylor, C.C.: Kernel density classification and boosting: an L_2 analysis. Stat. Comput. **15**, 113–123 (2005)
8. Fernandes, K., Cardoso, J.S.: Discriminative directional classifiers. Neurocomputing **207**, 141–149 (2016)
9. Ley, C., Verdebout, T.: Modern Directional Statistics, 1st edn. Chapman and Hall/CRC, Boca Raton (2017)
10. Ley, C., Verdebout, T.: Applied Directional Statistics: Modern Methods and Case Studies, 1st edn. Chapman and Hall/CRC, Boca Raton (2018)
11. Lopez-Cruz, P.L., Bielza, C., Larranaga P.: Directional naive Bayes classifiers. Pattern. Anal. Applic. **18**, 225–246 (2015)
12. SenGupta, A., Roy, S.: A simple classification rule for directional data. In: Advances in Ranking and Selection, Multiple Comparisons, and Reliability: Methodology and Applications. Springer, Berlin (2005)
13. SenGupta, A., Ugwuowo, F.I.: A classification method for directional data with application to the human skull. Commun. Stat.- Theory Methods **40**, 457–466 (2011)

Classification Trees Applied to Time-Lagged Data to Improve Quality in Official Statistics

Marco Di Zio, Romina Filippini, Gaia Rocchetti, and Simona Toti

Abstract Errors inevitably affect data. Statistical institutes deal with them with rules designing data relationships and imputation for missing information. In this paper, we report the use of classification trees to improve this process when using longitudinal time-lagged administrative data in the estimation process.

1 Introduction

National statistical institutes (NSIs) gather data for statistical purposes, and these data are frequently affected by errors that need to be resolved. In the emerging multisource data production system, survey data are often integrated with administrative data, making non-sampling errors particularly significant for the quality of the statistical figures produced by NSIs. A *statistical data editing and imputation* (E&I) step is typically performed to deal with errors and missing data. Logical and statistical rules connecting variables are used for resolving errors, and prediction of missing data (imputation) is carried out [1]. Although rules are proposed by subject matter experts, they may not be precise enough due to the complexity of the phenomenon under investigation. Machine learning can enhance the set of rules and decisions adopted for checking and imputing data by learning from the data itself. Some papers on the use of machine learning in the E&I phase are [2] and more recently [3, 4], and [5]. We focus on the use of classification trees [6] to check and improve the E&I process when, after some years, the actual data are available. This situation is common when using administrative data, since they are generally available with a time lag with respect to the reference time. For the estimation of a target variable at time t, only previous information may be used. However, once the time lag has passed, the actual data becomes available, allowing the procedure to be verified. This is the case of the estimation of the attained level

M. Di Zio · R. Filippini (✉) · G. Rocchetti · S. Toti
Italian National Institute of Statistics (ISTAT), Rome, Italy
e-mail: dizio@istat.it; filippini@istat.it; grocchetti@istat.it; toti@istat.it

G. Giordano et al. (eds.), *Statistical Models and Learning Methods for Complex Data*, Studies in Classification, Data Analysis, and Knowledge Organization,
https://doi.org/10.1007/978-3-031-84702-8_11

of education (ALE) of the Italian Permanent Census. To produce ALE estimates for all the Italian resident population, Istat adopted a mass imputation approach based on the integration of administrative, survey, and the 2011 Census data [7]. In this paper, we present a study on the application of classification trees to the updated administrative information to improve the data E&I process. The paper is structured as follows. Section 2 presents the data and the ALE estimation process used within the Italian Permanent Census. Section 3 illustrates how classification trees are used to enhance the official E&I procedure. Section 4 discusses the results of applying the revised procedure based on the suggestions from the classification trees. Some final remarks are reported in the last section.

2 Use of Administrative Data for the Attained Level of Education

Since October 2018, the so-called Permanent Census of Population has replaced the ten-yearly general census. The traditional strategy of complete field enumeration has been replaced with an approach that integrates administrative data and sample surveys. In this context, the Base Register of Individuals (BRI) forms the foundation of the Permanent Census. The BRI is primarily built through a massive integration of administrative data. However, the administrative data for the ALE are referred only to a subset of the population, specifically individuals enrolled in a school course from 2011 onward, and generally suffer from a time lag of 1 or 2 years. For this sub-population, information about the ALE of students and their attendance in courses (e.g., attending first year of primary education) is available.

Istat has developed a procedure for the prediction of ALE at the reference time t (henceforth Y_t); for details, see [7]. The information available for the prediction refers to demographic characteristics such as gender (G), age in 8 classes ($E8$), place of birth (P), country of citizenship at $31/12/t$ (C_t), and education characteristics as the ALE at $31/12/t-2$ (Y_{t-2}), the year of attendance of educational courses in the time period [t-2, t-1], e.g., 1st year, 2nd year, etc. (F_{t-1}), and the type of school (liceo, other) (L_{t-1}). The procedure first splits the observations into two subgroups:

- "Inactive" people: individuals not enrolled in any course in the last available academic year
- "Active" people: individuals attending a course in the last available academic year with longitudinal information on school enrolment

Information on ALE and year of attendance in educational courses in previous academic years is available for all individuals. Among "Active" people, two subgroups can be identified:

- "No-change": individuals with a zero probability of changing their educational level from $t-2$ to t

- "Change": individuals with a nonzero probability of obtaining a higher qualification than they held in the previous year

  The "No-change" subgroup is identified by one of the following conditions:

- Attending the 1st or 2nd year of the course (as they cannot conclude the course after 2 years)
- Already owning the educational level of the course they are attending
- Holding a PhD (the highest educational level)

The "Change" subgroup is composed of units that do not meet any of the above conditions.

For the "No-change" subgroup, the predicted ALE at time t is the same as the ALE in $t-2$. For each unit in the "Change" subgroup, the ALE is estimated by randomly selecting a value from the conditional probability distribution $h(Y_t|X)$ obtained from the models. The random draw aims to preserve the ALE probability distributions within domains identified by the covariates X.

The model $h(Y_t|X)$ is estimated using data from two years before the reference time and then applied to the reference year. The underlying hypothesis is that the conditional probabilities remain stable over this short period. Log-linear models are used for this procedure considering the variables $[C_t, E8_t, F_{t-1}, L_{t-1}, Y_t]$.

3 Classification Trees to Improve Quality of Prediction

When the data time lag is over, Istat gathers up-to-date administrative data, allowing the comparison between the ALE estimates and the actual administrative data considered as the gold standard. This comparison enables a more precise evaluation and potentially fine-tuning of the procedure by verifying the rules used for the deterministic corrections/imputation and the statistical model employed for imputation.

3.1 Fine-Tuning of the Deterministic Rules

The process of ALE estimation includes a deterministic E&I phase. For individuals in the "No-change" subgroup, the ALE observed in the previous year is simply confirmed. The rules adopted to assign observations to the "No-Change" class can be evaluated through a classification tree. To this end, the true ALE observed in the administrative source referred to year t (in this study $t = 2019$) is used as target variable. The same variables involved in the rules to identify the "No-change" subgroup are used as covariates. These include the ALE at $t-2$ (2017) and the school attendance in the academic year $(t-2/t-1)$, i.e., 2017/2018.

Table 1 Distribution of records with respect to the probability of the predicted value in leaves they belong to (absolute values in thousands and percentage values)

Probability range associated	$Y_{t,pred} = Y_{t-2,obs}$		$Y_{t,pred} > Y_{t-2,obs}$	
to the predicted value	a.v. (thousands)	%	a.v. (thousands)	%
(0.5, 0.90]	989	10.6	717	20.8
(0.90, 0.99]	4172	44.5	1632	47.2
(0.99, 1]	4214	45.0	1105	32.0
Total	9375	100.0	3455	100.0
$ALE_{t,pred} = ALE_{t,obs}$	8827	94.2	3271	94.7

The resulting classification tree shows probability distributions concentrated on a single value in most of the leaves (Table 1), indicating a high predictive power of the considered covariates. According to the tree estimates, in the subset of individuals who does not change their educational level from $t-2$ to t (i.e., Y_t predicted equal to Y_{t-2} (observed)), 45% belong to leaves with a predicted probability near 1 (>0.99), and about 90% belong to leaves with a predicted probability greater than 0.9.

The attributes describing leaves where ALE at time t is predicted to be equal to the observed ALE at time $t-2$ include all the conditions used in the deterministic rules to identify the "No-change" subgroup, thus confirming the adopted rules. Moreover, the predicted ALE matches the observed ALE in more than 94% of cases for both the "Change" and "No-change" subgroups.

Nevertheless, in some cases, the tree suggests the introduction of additional deterministic rules. These conditions can be evaluated as additional rules for identifying the "No-change" subgroup, specifically:

1. Individuals attending the third year of upper secondary education
2. Individuals not enrolled in any course with a master's degree
3. Individuals not enrolled in any course with a lower secondary education

The last two conditions suggest rules also for a subset of individuals in the "Inactive" group who were assumed to have a nonzero probability of attaining a new educational level. Individuals holding a master's degree, who wish to improve their educational level, have to enroll in a PhD course. Hence, it is reasonable to include individuals with a master's degree who are not enrolled in any course covered by administrative data, in the "No-change" subgroup. On the other hand, individuals with a lower secondary education seeking to enhance their educational level have various options: they may enroll in school courses covered or not covered by administrative data. Based on these results and after consulting with domain experts, we decided to modify the official procedure for estimating ALE by incorporating only the first two rules suggested by the classification tree. The results of the experimentation with the new deterministic rules are reported in Sect. 4.

3.2 Fine-Tuning of the Statistical Model for Imputation

Having reliable and current information from administrative sources is crucial for validating the accuracy of the prediction models. Prediction errors can be assessed by comparing the "true" values (actual observed values from administrative sources) with the predicted ones. The analysis of these discrepancies helps in understanding which profiles or characteristics have worse predictions, highlighting features that have a significant impact on model accuracy. This may allow a model refinement, for instance, by introducing additional explanatory variables to enhance the predictive power.

Classification trees are particularly effective for identifying patterns and making decisions. In our case, they are trained to classify profiles into different categories based on the observed differences in ALE. In particular, the aim is to identify profiles associated to a poor capacity of prediction to characterize errors in the official procedure. The difference between the observed and predicted ALE is used as response variable, with the same covariates used in the official prediction procedure introduced in Sect. 2. Trees are trained separately for each Italian region (NUTS 2). Given similar results across regions, we report those pertaining to the Emilia-Romagna region.

The model correctly predicts outcomes for the majority of individuals, achieving an accuracy rate of 97.3% (Table 2). This indicates that the model's predictions are coherent with the actual data for nearly all cases. Most of the errors are concentrated around a difference value of 1 (2.4%), suggesting that when mistakes occur, the model tends to assign an adjacent educational level compared to the actual one, deviating by at most one ALE level. If we focus on individuals with a difference greater than 1, we notice that they all share the same characteristics: the observed ALE from administrative sources often indicates a master's degree (level 7), whereas the predicted ALE frequently indicates upper secondary education (level 5). Additionally, all these individuals were enrolled in a master's degree program, attending the 5th year or beyond during the school year $(t-2/t-1)$. To investigate the reasons, we incorporate a new auxiliary information from the administrative source related to the type of master's degree attained.

Table 2 Number of individuals in leaves with respect to the estimated absolute difference between true and estimated ALE in Emilia-Romagna region (absolute values in thousands and percentage values)

Estimated absolute difference	Number of individuals in leaves	
in leaves	a.v. (thousands)	%
0	874,069	97.3
1	21,612	2.4
2	2478	0.3
3	3	0.0
Total	898,162	100.0

It results that over 40% of individuals in the group hold a degree from the Faculty of Medicine. The duration of a medical degree (MD) program, such as "Medicina e Chirurgia" (Medicine and Surgery) in Italy, typically spans 6 years. Hence, individuals in their 6th year of a MD program is not a 'non-course' student, unlike the many pursuing a master's degree. As a result, their probability of progressing to a higher educational level differs from that of students in other faculties. This observation suggests the introduction of a new covariate into the prediction model: a flag indicating enrollment in an MD program. Results of the experimentation with the new covariate are reported in Sect. 4.

4 Application of the Fine-Tuned Procedure

The procedure, updated with the changes suggested in Sect. 3, has been applied to the data. Specifically, the new deterministic rules to identify the "No-change" subgroup (NC) have been integrated into the procedure, and a new covariate indicating the attendance of a master's degree in Medicine is included in the model (MD). The ALE distributions estimated with the revised procedures (NC and MD) are very similar to those estimated with the official procedure (OFF) and observed from the administrative data (ADM) (Table 3).

Considering the absolute difference of the estimated percentages for each ALE category compared to those observed in the administrative data, both new estimates show values closer to those observed in the administrative sources (Table 4). Specifically, the mean of the difference between estimated and administrative data distribution (last row of Table 4) decreases from 0.10 of the official estimates to 0.05 of the procedure including the new deterministic rules and the new covariate on the medical degree enrollment. The main differences regard the lower secondary education and the master's degree. Concerning the frequency distributions, the

Table 3 ALE distributions obtained with administrative data (ADM), official estimates (OFF), modification of No-change rules (NC), introduction of the medical degree covariate (MD), and both modifications (TOT) (absolute values in thousands and percentage values)

	ADM	OFF	NC	MD	TOT	ADM	OFF	NC	MD	TOT
ALE	a.v. (thousands)					%				
Literate but no ed. att.	1003	1004	1004	1004	1004	11.6	11.6	11.6	11.6	11.6
Primary ed.	1632	1637	1637	1637	1637	18.8	18.9	18.9	18.9	18.9
Lower sec.	2690	2680	2686	2680	2686	31.0	30.9	31.0	30.9	31.0
Upper sec.	1796	1811	1806	1811	1806	20.7	20.9	20.8	20.9	20.8
Bachelor's d.	460	457	457	457	457	5.3	5.3	5.3	5.3	5.3
Master's d.	1076	1061	1076	1060	1075	12.4	12.2	12.4	12.2	12.4
PhD	16	24	9	24	9	0.2	0.3	0.1	0.3	0.1
Total	8674	8674	8674	8674	8674	100.0	100.0	100.0	100.0	100.0

Table 4 Absolute differences between the frequency percentage distribution of ALE computed with the official estimate (OFF), the No-change rules (NC), the new covariate (MD), and both modifications (TOT) *vs* the administrative source

ALE	OFF	NC	MD	TOT
Literate but no ed. att.	0.01	0.01	0.01	0.01
Primary ed.	0.06	0.06	0.06	0.06
Lower sec.	0.12	0.05	0.11	0.05
Upper sec.	0.17	0.11	0.17	0.12
Bachelor's d.	0.04	0.04	0.04	0.04
Master's d.	0.18	0.01	0.18	0.01
PhD	0.09	0.08	0.09	0.08
Mean	0.10	0.05	0.09	0.05

Table 5 ALE distribution for master's degree holders in administrative sources according to the official estimate (OFF), modified No-change rules (NC), new medical degree covariate (MD), and both modifications (TOT) (absolute values in thousands and percentage values

	OFF	NC	MD	TOT	OFF	NC	MD	TOT
ALE	a.v. (thousands)				%			
Upper sec.	33	33	32	32	3.1	3.1	3.0	3.0
Bachelor's d.	58	59	58	58	5.4	5.4	5.4	5.4
Master's d.	969	983	970	984	90.0	91.4	90.1	91.4
PhD	16	2	16	2	1.5	0.1	1.5	0.2
Total	1076	1076	1076	1076	100.0	100.0	100.0	100.0

highest impact is given by the new deterministic rules. Although the modifications introduced by these deterministic rules are relatively minor at a macrolevel, they are in line with the desired outcomes.

Considering prediction accuracy, the official estimate correctly assigns ALE to 91.1% of units, while the new estimates achieve slightly higher percentages, reaching 92.2% for NC and 91.9% for MD. The modifications introduce a small but valuable impact, especially on the imputation of master's degree. When considering individuals with a master's degree and examining the distribution of their estimated ALE (Table 5), we observe improved performance in the new estimates compared to the official ones. This is particularly evident in the reduction of the number of erroneously imputed PhD qualifications.

In conclusion, the new deterministic rules and the new covariate suggested by trees and used in the model improve the estimates, both in terms of distribution and prediction. Deterministic rules have an impact on the distributions, while the new covariate improves the prediction of master's degree qualifications.

5 Final Remarks

We have discussed a real and important case where classification trees can be effectively applied to enhance data quality by fine-tuning the data E&I procedure. Trees are used to learn rules and tune the imputation model using time-lagged data. This case, based on longitudinal information, is quite common when using administrative data, which by now have entered the production systems of official statistics. The paper describes the case of the attained level of education estimated in the Italian Permanent Census. Applying classification trees to actual data confirms the overall structure of the procedure and results in fine-tuning the models by suggesting new rules and covariates for the E&I process. These new rules are completely logical and should/could have been included since the beginning. However, the process is extremely complex and so these gaps can occur in practice. This is particularly true in scenarios where the proposed changes are appreciated at very high levels of disaggregation as they refer to particular subpopulations, as it is in this context. It is worthwhile noting that, especially in the case of registries, where microdata are reported with the aim of having the possibility of producing granular level estimates, improving predictions for particular subpopulations becomes a crucial target to pursue.

References

1. De Waal, T., Pannekoek, J., Scholtus, S.: Handbook of Statistical Data Editing and Imputation, vol. 563. John Wiley & Sons, London (2011)
2. Nordbotten, S.: Editing statistical records by neural networks. J. Official Stat. **11**, 14: 391–411 (1995)
3. Unece: Machine Learning for Official Statistics. United Nations Publication, Geneva (2021). Accessed 15 Dec 2023. https://unece.org/sites/default/files/2022-09/ECECESSTAT20216.pdf
4. Dagdoug, M., Goga, C., Haziza, D.: Imputation procedures in surveys using nonparametric and machine learning methods: an empirical comparison. J. Surv. Stat. Methodol. **11**(1), 141–188 (2023)
5. De Fausti, F., Di Zio, M., Filippini, R., Toti, S., Zardetto, D.: Multilayer perceptron models for the estimation of the attained level of education in the Italian Permanent Census. Stat. J. IAOS **38**(2), 637–646 (2022)
6. Breiman, L., Friedman, J.H., Olshen, R.A., Stone, C.J.: Classification and Regression Trees. Wadsworth & Brooks/Cole Advanced Books & Software, Monterey (1984)
7. Di Zio, M., Filippini, R., Rocchetti, G.: An imputation procedure for the Italian attained level of education in the register of individuals based on administrative and survey data. Rivista di Statistica Ufficiale 2–3 (2019)

Trimmed Factorial K-Means: A Clustering Application to a Cookie Dataset

Matteo Farnè and Furio Camillo

Abstract This paper explores trimmed factorial k-means (TFKM) in a clustering application to a cookie dataset. TFKM is a robust version of factorial k-means, where a robust covariance matrix input is used, and outliers in the identified reduced space are iteratively removed via a trimming procedure. The selected latent rank, number of clusters, and outlier proportion are those which maximize Hartigan's statistic. The TFKM partition is thoroughly compared to two alternatives, like a robust tandem procedure and trimmed k-means, via a simulation study. An Internet cookie example shows that TFKM provides on analyzed data the most parsimonious and informative partition by cluster homogeneity.

1 Introduction

Clustering multivariate data in high dimensions is a challenging task [1]. First, agglomerative hierarchical methods like Ward's method [26] may soon become computationally intractable as the dimension p increases. Second, hierarchical partitioning methods like k-means algorithm [12] may be very unstable in high dimensions, due to numerical instability and multicollinearity. Third, dimension reduction is imperative to avoid the curse of dimensionality and to obtain interpretable partitions [4]. Fourth, when the sample size n is also large, robust methods are often needed [7], due to the much likely presence of outliers able to distort the obtained partition. For these reasons, the need rises to develop clustering procedures able to combine robustness and dimension reduction.

As recently highlighted by [21], there are two prevalent approaches to clustering in high dimensions: least squares, based on seeking for the best compact partition in a (possibly reduced) data space, and mixture models, typically estimated by solving a maximum likelihood problem via the EM algorithm. While a very well-known

M. Farnè (✉) · F. Camillo
University of Bologna, Bologna, Italy
e-mail: matteo.farne@unibo.it; furio.camillo@unibo.it

G. Giordano et al. (eds.), *Statistical Models and Learning Methods for Complex Data*, Studies in Classification, Data Analysis, and Knowledge Organization,
https://doi.org/10.1007/978-3-031-84702-8_12

robust partitioning method of the first kind is trimmed k-means (TKM) [2], which is the robustified version of k-means, the second kind of methods presents several ongoing developments. Assuming the mixture of factor analyzers [13] as a reference modelling tool, a robustified version has been provided by imposing a skew-t distribution to the latent components, both with a different [15] and a common [14] loading matrix across clusters. Further studies include the use of trimmed maximum likelihood [23], the estimation of restricted mixtures of multivariate skew-normal [8] or skew-t [24], and unrestricted mixtures of multivariate skew-normal [25].

That said, before resorting to mixture modelling, relevant attempts had been made to combine traditional dimension reduction methods to existing clustering techniques of the first kind. The very first dimension reduction tool is principal component analysis (PCA) [10], which is a geometric procedure to represent a high-dimensional dataset by means of a much lower-dimensional one. At some point, when the data dimension started to grow, it became evident [16] that existing clustering methods could be applied to the first extracted principal components. However, it also became apparent that the directions identified by PCA may not be the optimal ones for clustering purposes, as they neglect the cluster structure present in the sample. This fact was pointed out in [3] as a motivation to propose reduced k-means (RKM), a method to cluster objects in a low-dimensional space. The strong consistency of RKM was proved under independent and identically distributed data in [19].

RKM, in turn, may still mask the most relevant directions for clustering purposes when the directions explaining most of the deviance are not close to the directions separating the objects the most. For this reason, [22] proposed factorial k-means (FKM), a method to identify the latent space most able to maximize the distinctiveness of projected objects. The strong consistency of FKM was proved in [20]. Proof is based on the original proof of the strong consistency of k-means clustering in [17].

Given these premises, in this paper, we present a robust version of factorial k-means, where outliers are iteratively removed via a trimming procedure in the low-dimensional space, thus simultaneously identifying radial outliers and designing better shaped clusters. This is obtained by minimizing the trimmed least squares criterion in the low-dimensional space. The resulting method is called trimmed factorial k-means (TFKM). TFKM algorithm is reported in [6], and a relevant application of a preliminary version of it is described in [5]. In this paper, we propose a new simulation study which shows the effectiveness of TFKM against competitors such as the trimmed k-means of [2] (using the R package *trimcluster* by Christian Hennig) and a robust tandem procedure which first estimates the robust principal components by the R package *robpca* as proposed in [11] and then applies the trimmed k-means of [2]. We then present a new illustrative example involving Internet cookie data.

The paper proceeds as follows. Section 2 briefly recalls trimmed factorial k-means algorithm. Section 3 reports the simulation study results. Section 4 describes a clustering application to an Internet cookie dataset, discussing the choices of number of clusters, latent rank, and outlier proportion, as well as the quality of obtained partitions. In the end, Sect. 5 concludes the paper.

2 Trimmed Factorial K-Means Procedure

Let us consider a $n \times p$ data matrix $\mathbf{X}$. The trimmed factorial k-means of [22] assumes that

$$\mathbf{XAA}' = \mathbf{U}\overline{\mathbf{Y}}\mathbf{A}' + \mathbf{E}, \tag{1}$$

where $\mathbf{A}$ is a $p \times r$ semi-orthogonal *coefficient matrix*, such that $\mathbf{A}'\mathbf{A} = \mathbf{I}_r$; $\mathbf{U}$ is a $n \times c$ *membership matrix* such that $\mathbf{U}_{ij} = 1, i = 1, \ldots, n, j = 1, \ldots, c,$ if observation i belongs to cluster j; $\overline{\mathbf{Y}}$ is a $c \times r$ *centroid matrix*; r is the *latent rank* and c is the *number of clusters*. Denoting the $n \times r$ factor score matrix by $\mathbf{F} = \mathbf{XA}$, we further assume that $(100\alpha)\%$ of the n true factor score vectors, with $\alpha \in [0, 0.5]$, are arbitrarily distant from the bulk of the rest of factor score vectors. Under this condition, model (1) can be recovered by minimizing $FKM(\mathbf{A}, \mathbf{U}, \overline{\mathbf{Y}}) = \|\mathbf{XAA}' - \mathbf{U}\overline{\mathbf{Y}}\mathbf{A}'\|^2 = \|\mathbf{XA} - \mathbf{U}\overline{\mathbf{Y}}\|^2$, which is the deviance within clusters in the reduced space, where by least squares we can obtain $\overline{\mathbf{Y}} = (\mathbf{U}'\mathbf{U})^{-1}\mathbf{U}'\mathbf{XA}$, imposing the constraint $\sum_{i=1}^{n}\sum_{j=1}^{c}\mathbf{U}_{ij} = [(1-\alpha)n]$, with $\sum_{j=1}^{c}\mathbf{U}_{ij} = \{0, 1\}$, for each $i = 1, \ldots, n$.

This problem can be numerically solved by adapting the original alternating least squares (ALS) algorithm of [22] to the framework of [18]. In particular, once initialized, $\mathbf{A}$, $\mathbf{U}$, and $\overline{\mathbf{Y}}$ are first recovered by the original ALS algorithm, which is the H-step, and a trimming procedure is subsequently applied by excluding the $[\alpha n]$ observations most apart from the respective cluster centroids in the reduced space, which is the C-step. The algorithm is reported in [6], to which we refer for technical details.

In practice, a grid of values for the latent rank r, the number of clusters c, and the outlier proportion α is set up, under the condition $r \leq c - 1$. For each triple (r, c, α), the algorithm is repeated $N = 1000$ times, and the final solution is chosen as the one with minimum objective $FKM(\mathbf{A}, \mathbf{U}, \overline{\mathbf{Y}})$ across the N trials. The optimal triple (r, c, α) is then identified by maximizing Hartigan's statistic [9] across chosen triples. We refer to [6] for full details.

3 Simulation Study

In this section, we aim to test which conditions provide the ideal ground for TFKM procedure to excel. We start by considering that a method based on robust principal components like the tandem procedure may be confounded by the simultaneous presence of two concurring orthogonal spaces behind the data. For this reason, we set for the $p \times 1$ data vector $\mathbf{x}_i$ the following mixture model:

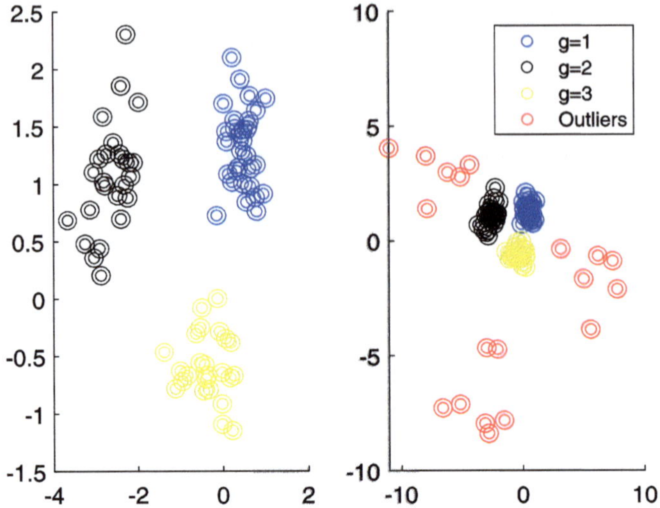

Fig. 1 Generated clustered $\mathbf{f}_1$ without outliers (left panel) and with outliers in red (right panel) for the scenario $\delta_\epsilon = 1.5$, $n = 100$, $p = 500$, $r = 2$, $c = 3$, $\alpha = 0.2$, $m = 10$, $w = 0.25$

$$\mathbf{x}_i = w\mathbf{A}_1\mathbf{f}_{1,i} + (1-w)\mathbf{A}_2\mathbf{f}_{2,i}, \tag{2}$$

where $w \in [0, 1]$ is a mixture weight, $\mathbf{A}_1$ and $\mathbf{A}_2$ are $p \times r$ semi-orthogonal matrices such that $\mathbf{A}_1'\mathbf{A}_1 = \mathbf{I}_r$, $\mathbf{A}_2'\mathbf{A}_2 = \mathbf{I}_r$, $\mathbf{A}_1'\mathbf{A}_2 = \mathbf{A}_2'\mathbf{A}_1 = \mathbf{0}_{r\times r}$, $\mathbf{f}_{2,i} = \delta_\epsilon \boldsymbol{\epsilon}_{2,i}$, with δ_ϵ nuisance parameter and $\boldsymbol{\epsilon}_{2,i} \sim \text{MVN}(\mathbf{0}, \mathbf{I}_r)$ random vector, and $\mathbf{f}_{1,i}$ is such that

$$\mathbf{f}_{1,i} = \begin{cases} \overline{\mathbf{Y}}_{l_i} + \delta_\epsilon \boldsymbol{\epsilon}_{1,i}, & \text{if } i \notin O, \\ (\overline{\mathbf{Y}}_{l_i} + \text{sgn}(\overline{\mathbf{Y}}_{l_i}) \cdot \mathbf{m}_i) + \delta_\epsilon \boldsymbol{\epsilon}_{1,i}, & \text{if } i \notin O, \end{cases}$$

where O is the outlier set with size $[\alpha n]$, $\overline{\mathbf{Y}}_{l_i}$ is the $r \times 1$ centroid vector associated to l_i, the cluster label of observation i, $\boldsymbol{\epsilon}_{1,i} \sim \text{MVN}(\mathbf{0}, \mathbf{I}_r)$ is a random vector, $\mathbf{m}$ is a $r \times 1$ perturbation vector, and $\cdot$ denotes the Hadamard product.

Defining another random vector $\boldsymbol{\epsilon}_{3,i} \sim \text{MVN}(\mathbf{0}, \mathbf{I}_r)$, the perturbation vector $\mathbf{m}$ is generated component-wise as $\mathbf{m}_i = m\,\text{sgn}(\boldsymbol{\epsilon}_{3,i})\text{Unif}(1.5, 2.5)$, where $m = 10, 15$ is the outlyingness parameter. The mixture weight is set to $w = 0.25$. The outlier proportion α is set to $0.1, 0.2$. The dispersion parameter δ_ϵ is set to $1.5, 2$. We consider three pairs (p, n), such as $(100, 500)$, $(300, 300)$, $(500, 100)$, and three pairs (r, c), such as $(2, 3)$, $(2, 4)$, $(3, 4)$. We thus obtain 72 possible scenarios (see Fig. 1 for an example).

$\mathbf{A}_1$ and $\mathbf{A}_2$ are obtained as follows. A random matrix is drawn by $\text{W}(\mathbf{I}_p/p, n-1)$; its spectral decomposition is derived; the first r eigenvectors become the columns of $\mathbf{A}_1$, the following r eigenvectors become the columns of $\mathbf{A}_2$. Note that, in theory, $\mathbf{A}_2$ could also contain any number of vectors from the orthogonal complement of $\mathbf{A}_1$, and model (2) could be adapted accordingly. Cluster labels and outlier identifiers are randomly generated. The $c \times r$ centroid matrix $\overline{\mathbf{Y}}$ has rows $(4, 4)$, $(8, -8)$, $(-2, -2)$ when $(r = 2, c = 3)$; the row $(-8, 8)$ is added when $(r = 2, c = 4)$; the column $(4, 8, -2, -8)$ is then added when $(r = 3, c = 4)$.

We launch 100 trials for each scenario. We run trimmed factorial k-means (TFKM), trimmed k-means (TKM), and the tandem approach combining ROBPCA and trimmed k-means (RTKM), setting the true parameters (r, c, α) (we leave to future research a criterion to select that triple in a data-driven way). We flag as unsuccessful any case where the confusion matrix, reporting in the columns true group labels and in the rows estimated group labels, presents at least two column-wise maxima in the same row. We then derive accordingly the nonsuccess rate across trials. In the following, we calculate the membership recovery rate and the outlier recovery rate for each successful trial, and we derive their average.

Table 1 reports simulation results for the scenario $(r = 2, c = 3, \alpha = 0.2)$. We can observe that RTKM presents nonsuccess rates systematically larger than 30%, further increasing with δ_ϵ and m. TKM presents very low nonsuccess rates when $\delta_\epsilon = 1.5$, around 50% when $\delta_\epsilon = 2$, even larger when p is large. On the contrary, TFKM consistently reports nonsuccess rates below 10%, slightly larger when p is large. Concerning membership recovery rate, we observe that TKM achieves excellent results when $\delta_\epsilon = 1.5$ and quite poor results (around 70%) when $\delta_\epsilon = 2$. TFKM instead always stands above 90%. RTKM behaves poorly under all conditions, with lows around 60%. Concerning outlier recovery rate, TFKM and TKM are doing almost perfectly under all conditions, while RTKM stands between 55% and 75%.

Concerning the comparison between TFKM and the two competitors, the metrics reported in Table 1 can be considered indicative also for the other five examined triples (r, c, α). In absolute terms, when $\alpha = 0.2$, the results get slightly worse for all methods as r and c increase. Also, the results for $\alpha = 0.1$ are in general slightly worse compared to $\alpha = 0.1$ under all (r, c) pairs.

These results tell us that TKM is very sensitive to dispersion within clusters, while RTKM is not working properly, with high outlier masking and cluster label misallocation, due to the confounding co-occurrence of two low-dimensional spaces behind the data.

4 An Internet Cookie Example

In this section, we conduct a clustering exercise on a cookie dataset (source: Nextplora). Our dataset contains 322 variables, composed of the number of accesses to 300 web pages by 1474 individuals during a three-month period (from 26.9.2019 to 09.12.2019) and 22 variables regarding the access time, weekday, device, and visibility status. The web pages are identified by the respective IAB category v2.0 (see https://42matters.com/docs/app-market-data/supported-iab-categories-v2). First, we operate a correlation screening to eliminate (almost)

Table 1 Simulation results for the scenario ($r = 2, c = 3, \alpha = 0.2$), under all the combinations of δ_ϵ, m, n and p: nonsuccess rate, membership recovery rate, outlier recovery rate

			Nonsuccess rate			Membership recovery rate			Outlier recovery rate		
			$n = 500$	$n = 300$	$n = 100$	$n = 500$	$n = 300$	$n = 100$	$n = 500$	$n = 300$	$n = 100$
Method	δ_ϵ	m	$p = 100$	$p = 300$	$p = 500$	$p = 100$	$p = 300$	$p = 500$	$p = 100$	$p = 300$	$p = 500$
TFKM	1.5	10	0.11	0.05	0.12	0.984	0.985	0.973	0.977	0.985	0.968
TKM	1.5	10	0.11	0.05	0.12	0.998	0.982	0.938	0.990	0.999	0.965
RTKM	1.5	10	0.11	0.05	0.12	0.657	0.688	0.726	0.672	0.726	0.737
TFKM	1.5	15	0.09	0.04	0.19	0.984	0.991	0.978	1.000	1.000	1.000
TKM	1.5	15	0.00	0.01	0.11	0.998	0.983	0.950	1.000	1.000	1.000
RTKM	1.5	15	0.36	0.42	0.51	0.678	0.707	0.761	0.731	0.810	0.831
TFKM	2	10	0.11	0.09	0.15	0.956	0.952	0.943	0.971	0.973	0.951
TKM	2	10	0.50	0.51	0.50	0.701	0.707	0.753	0.941	0.943	0.932
RTKM	2	10	0.53	0.54	0.60	0.612	0.622	0.652	0.575	0.601	0.651
TFKM	2	15	0.11	0.10	0.11	0.964	0.962	0.960	1.000	1.000	1.000
TKM	2	15	0.48	0.53	0.58	0.700	0.704	0.729	0.998	1.000	1.000
RTKM	2	15	0.55	0.54	0.57	0.613	0.626	0.609	0.633	0.671	0.652

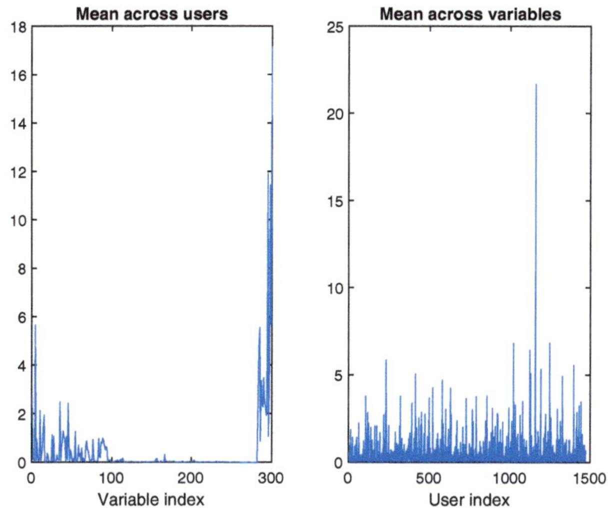

Fig. 2 Cookie dataset: mean value for each variable across users (left) and for each user across variables (right)

perfect identity relations, through which 300 variables survive in the end, of which 283 IAB categories and 17 general usage intensity variables.

Looking at the mean values for each variable and each user (see Fig. 2), we can understand the data content: the small indices represent the most popular IAB categories, referring, for instance, to IAB197 (variable 5, Parenting Children Aged 4–11), IAB366 (variable 36, Classic Rock), IAB101 (variable 46, Logistics and Transportation Industry), while large indices represent general usage intensity variables. Concerning users, it is clear that we can find several outlying web surfers very likely to affect a non-robust clustering algorithm, the most prominent of which is user 1160.

We apply trimmed factorial k-means, with the twofold aim to obtain a meaningful and informative partition of web users, and a latent space over which web users are mostly separated. We consider the possible values $r = 1, 2, 3, 4$ for the latent rank, $c = 2, 3, 4, 5$ for the number of clusters, and $\alpha = 0, 0.05, 0.1, 0.15, 0.2, 0.25$ for the outlier proportion.

The optimal values identified by maximizing Hartigan's statistic for trimmed factorial k-means (TFKM) are $r = 1$, $c = 3$, $\alpha = 0.1$. The recovered factor is able to explain 59.8% of the overall deviance in the data. It clearly represents a level factor, where the first IAB categories and the web usage variables appear with positive weight, while the remaining IAB categories present nearly zero weight. The identified partition presents one large and very compact cluster of low Internet users (826 people, silhouette index = 0.728), a second "transition" cluster of moderate users which is the least homogenous (337 people, silhouette index = 0.533), a third compact cluster of intense users (164 people, silhouette index = 0.595), and 147 outliers. The average silhouette index across the three clusters is 0.662.

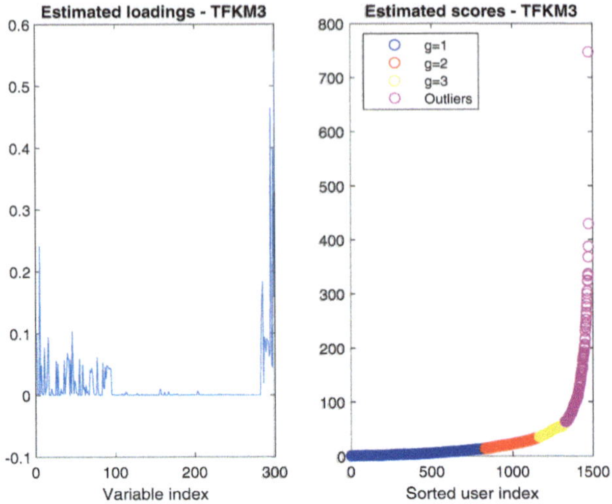

Fig. 3 Loadings (left) and factor scores (right) recovered by trimmed factorial k-means partition with $r = 1, c = 3, \alpha = 0.1$

Estimated loadings and factor scores are reported in Fig. 3, where we can catch that the average factor score into each cluster is 5.49 for the low users, 23.15 for the moderate users, 48.01 for the intense users, and 144.77 for outliers. The cutoff values are around 14.23 and 35.46, respectively, while the outlier threshold is around 63.49. The most prominent outlier is user 1160, consistently with Fig. 2.

5 Conclusions

In this paper, we have presented a robust clustering method suitable to high dimensions, called trimmed factorial k-means. We have developed a simulation study where a normal mixture model is tested. We have empirically shown that this procedure compares favorably to trimmed k-means when the objects are more dispersed within clusters and is strongly superior to a robust tandem approach which combines ROBPCA and trimmed k-means, both for cluster label and outlier recovery. We have presented an Internet cookie dataset, where trimmed factorial k-means allows to obtain a meaningful partition in three clusters according to the intensity of Internet usage, a set of outliers excluded from any group, and a single latent factor which explains around 60% of the deviance and is able to separate Internet users the most. A qualitative analysis shows that recovered outliers are consistent with reasonable expectations.

References

1. Bouveyron, C., Brunet-Saumard, C.: Model-based clustering of high-dimensional data: a review. Comput. Stat. Data Anal. **71**, 52–78 (2014)
2. Cuesta-Albertos, J., Gordaliza, A., Matrán, C.: Trimmed k-means: an attempt to robustify quantizers. Ann. Stat. **25**, 553–576 (1997)
3. De Soete, G., Carroll, J.: K-means clustering in a low-dimensional Euclidean space. In: Diday, E., Lechevallier, Y., Schader, M., Bertrand, P., Burtschy, B. (eds.) New Approaches In Classification And Data Analysis, pp. 212–219. Springer, Berlin (1994)
4. Ding, C.: Dimension reduction techniques for clustering. In: Liu, L., Özsu, M.T. (eds.) Encyclopedia of Database Systems. Springer, Boston (2009)
5. Farnè, M., Vouldis, A.: Banks' business models in the euro area: a cluster analysis in high dimensions. Ann. Oper. Res. **305**, 23–57 (2021)
6. Farnè, M.: Trimmed factorial k-means. In: Book Of Abstracts And Short Papers, pp. 148–151 (2023)
7. García-Escudero, L.A., Gordaliza, A., Matrán, C., Mayo-Iscar, A.: A review of robust clustering methods. Adv. Data Anal. Classif. **4**, 89–109 (2010)
8. Lin, T., McLachlan, G., Lee, S.: Extending mixtures of factor models using the restricted multivariate skew-normal distribution. J. Multivar. Anal. **143**, 398–413 (2016)
9. Hartigan, J.: Clustering Algorithms. Wiley, London (1975)
10. Hotelling, H.: Analysis of a complex of statistical variables into principal components. J. Educ. Psychol. **24**(6), 417 (1933)
11. Hubert, M., Rousseeuw, P., Vanden Branden, K.: ROBPCA: a new approach to robust principal component analysis. Technometrics **47**, 64–79 (2005)
12. MacQueen, J.: Classification and analysis of multivariate observations. In: 5th Berkeley Symposium on Mathematical Statistics and Probability, pp. 281–297 (1967)
13. McLachlan, G.J., Peel, D., Bean, R.W.: Modelling high-dimensional data by mixtures of factor analyzers. Comput. Stat. Data Anal. **41**(3–4), 379–388 (2003)
14. Murray, P., McNicholas, P., Browne, R.: A mixture of common skew-t factor analysers. Statistics **3**, 68–82 (2014)
15. Murray, P., Browne, R., McNicholas, P.: Mixtures of skew-t factor analyzers. Comput. Stat. Data Anal. **77**, 326–335 (2014)
16. Hubert, L., Arabie, P.: The analysis of proximity matrices through sums of matrices having anti-Robinson forms. Br. J. Math. Stat. Psychol. **47**, 1–40 (1994)
17. Pollard, D.: Strong consistency of k-means clustering. Ann. Stat. **9**(1), 135–140 (1981)
18. Rousseeuw, P., Van Driessen, K.: An algorithm for positive-breakdown regression based on concentration steps. In: Data Analysis: Scientific Modeling and Practical Application, pp. 335–346. Springer, Berlin, Heidelberg (2000)
19. Terada, Y.: Strong consistency of reduced k-means clustering. Scand. Stat. Theory Appl. **41**, 913–931 (2014)
20. Terada, Y.: Strong consistency of factorial k-means clustering. Ann. Inst. Stat. Math. **67**, 335–357 (2015)
21. Saporta, G.: Clusterwise Methods: a synthesis and new developments. In: Homenagem A Fernando Da Costa Nicolau, pp. 23–26 (2018)
22. Vichi, M., Kiers, H.: Factorial k-means analysis for two-way data. Comput. Stat. Data. Anal. **37**, 49–64 (2001)
23. Yang, L., Xiang, S., Yao, W.: Robust fitting of mixtures of factor analyzers using the trimmed likelihood estimator. Commun. Stat. Simul. Comput. **46**, 1280-1291 (2017)
24. Wang, W., Castro, L., Chang, Y., Lin, T.: Mixtures of restricted skew-t factor analyzers with common factor loadings. Adv. Data Anal. Classif. **13**, 445–480 (2019)
25. Wang, W.L., Lin, T.I.: Model-based clustering via mixtures of unrestricted skew normal factor analyzers with complete and incomplete data. Stat. Methods Appl. **32**(3), 787-817 (2023)
26. Ward, J.: Hierarchical grouping to optimize an objective function. J. Am. Stat. Assoc. **58**, 236–244 (1963)

Visualization of Proximity and Role-Based Embeddings in a Regional Labor Flow Network

Sara Geremia, Fabio Morea, and Domenico De Stefano

Abstract This study uses graph representation learning techniques to study the organizational inter-connections established through labor mobility flows in Friuli Venezia Giulia, a medium-sized region in northern Italy ranked as one of the leading regions for innovation. The regional labor flow network is constructed from regional data on the commencement or termination of employment contract. The methods employed, Node2Vec and Role2Vec, allow for a visualization of the community structures and the centrality of universities and research institutions in the network. The study demonstrates the potential of such techniques for analyzing complex networks and uncovering hidden structures by identifying the role of specific firms/organizations, which can be of interest to public decision-makers.

1 Introduction

The mobility of workers creates a network of connections that reflects the interconnectivity between employers, the so-called labor flow network (LFN). Such network data can reveal insights into the structure of the relationships between employers, which can be used to identify communities of employers that share geographic location, industry, and workforce characteristics [1]. The mobility of individual workers, from one job to another, plays a role in knowledge spillover within economies [2]. The analysis of LFNs' topological features and their community structures shows that LFNs can be useful in identifying high-performance firms (i.e., employers) [3]. Therefore, examining regular patterns in such networks is a key step in understanding the role and position of some targeted employers in the labor

S. Geremia (✉) · D. De Stefano
University of Trieste, Trieste, Italy
e-mail: sara.geremia@phd.units.it; ddestefano@units.it

F. Morea
Area Science Park, Trieste, Italy
e-mail: fabio.morea@areasciencepark.it

G. Giordano et al. (eds.), *Statistical Models and Learning Methods for Complex Data*, Studies in Classification, Data Analysis, and Knowledge Organization,
https://doi.org/10.1007/978-3-031-84702-8_13

market. The role of large public sector organizations, such as universities, in the economic context under study can be determined by their centrality and relationship with industries employing a high number of experienced professionals [4].

This work aims to investigate the structure of an LFN in Friuli Venezia Giulia (FVG), a medium-sized region in northern Italy ranked as one of the leading regions for innovation. FVG stands out as one of the few regions in Italy classified as a strong innovator, with a demonstrated increase in innovation performance over time [5]. We construct the LFN based on a dataset provided by the Regional Observatory on Policies and the labor market of FVG, which encodes the commencement or termination of employment contracts as single "events" Morea and De Stefano [6]. In particular, the data are filtered to include only a subset of professional groups, encompassing highly skilled workers whose expertise is potentially related to the most innovative sectors in the region.

Previous studies on FVG have shown that knowledge and innovation diffusion are guided by the role of academia and the leading Public Research Organizations (PROs) operating at the local level [7]. For this reason, we are especially interested in exploring the structure of such a peculiar regional LFN, as well as the position and role of the regional universities and prominent PROs.

Understanding the structure of such LFN involves dealing with graph data containing rich relational information.

Traditional machine learning algorithms require hand-engineered feature representation which is labor-intensive and relies on domain-specific knowledge. Representation learning (RL) provides an alternative learning approach to represent graph data using low-dimensional vectors automatically [8]. The learned embeddings can be used with data visualization and clustering techniques to generate representations of graphs useful for discovering communities, hub nodes, and other hidden structures.

The graph RL task can be performed to assess the potential of universities and research institutions as drivers of economic development and innovation in FVG. The interest is in investigating whether exploring the network by looking at both relational proximity and regular patterns yields valuable insights. The two graph RL we adopted are well suited for such tasks. The identification of groups of equivalent nodes could also be performed using other approaches, such as generalized blockmodeling [9]. However, this approach allows for a global positional analysis and can be problematic for large and dense networks.

1.1 Data

The present study utilizes employment contract data in FVG for constructing an LFN, wherein nodes represent employers and edges between two nodes i and j signify the transition of at least an employee p from employer i to employer j. The employment contract data are organized as "events," where each event represents the initiation or termination of an employment contract. The dataset is anonymized and includes 1,155,342 employment events from 74,317 local units of companies, universities, and PROs in the FVG region, between 2014 and 2021. For the purpose of this paper, the data are filtered out to include only a subset of professional groups, namely, ISCO-21 (science and engineering occupations) and ISCO-25 (information and communication technology occupations), as defined by the International Standard Classification of Occupations 2008 [10]. The dataset comprises approximately 60,164 events involving 1,890 employers and 16,474 employees. To ensure data reliability, the raw dataset is cleaned and aggregated (e.g., adding implicit contract terminations, identifying actual workplaces for employment agencies, and grouping professional profiles). We then constructed a network of "transitions," defined as a cessation event followed by a start event involving the same employee and two different employers. Transitions involving the same employer or those not fully traceable within the region, such as self-employment, leaving the labor force, or finding a new job in a different region, were excluded. The final network contains 1,084 nodes (unique employers involved in regional transitions) and 1,641 edges (identifiable transitions), weighted by the number of transitions and the relevance of each transition. Specifically, edges' weight is assumed to be proportional to the experience that employee p transfers between employer i and employer j to account for the knowledge spillover effect. The weight of this transfer is given by $w_{ij}^{p} = \min(D_i^p, D_j^p, w_{\max})$, where D_i^p is the duration of the latest contract of p with i and D_j^p is the duration of the latest contracts of p with j (both expressed in years or fraction of years). The parameter $w_{\max}$ models a nonlinear pattern of knowledge transfer between i and j, suggesting that knowledge spillovers are more effective when they are derived from recent and relevant experience. Hence, we set $w_{\max}$ equal to 5 years. The edge weight w_{ij} is given by $w_{ij} = \sum_{p=1}^{P} w_{ij}^{p}$, where P is the number of employees transferred between i and j.

2 Methods

In this work, two methods for graph RL are employed: Node2Vec [11] and Role2Vec [12]. The methods differ in their approaches to preserving the graph's structure, as they are based on two distinct definitions of node structural similarity, known as structural equivalence and regular equivalence [13], respectively. Two nodes are structurally equivalent if they are relationally close: structural equivalence captures similarities in the way nodes are connected to the same neighboring nodes. Two nodes are regularly equivalent if they occupy similar roles or positions in the network: regular equivalence captures similarities in the way nodes contribute to the overall structure of the network. Understanding these differences is crucial because the choice between structural and regular equivalence reflects the emphasis placed on different aspects of the graph topology.

Node2vec might be more suited when the goal is to capture similarities in how nodes interact with each other. It is designed to map structurally similar nodes to

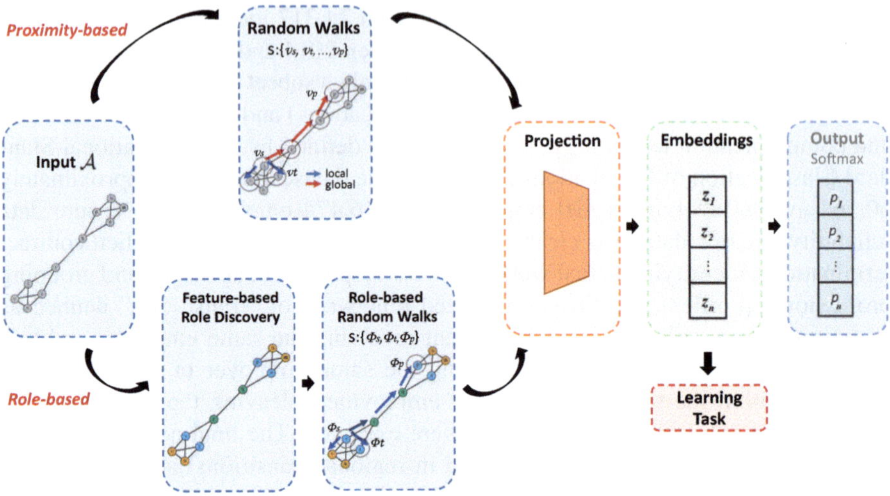

Fig. 1 Skip-gram Node2vec and Role2vec architecture

similar proximity-based embeddings, considering both local and global structural information. This is achieved through the use of a random walk strategy, generating sequences of L nodes, $(S : v_1, v_2, ..., v_L)$, that capture neighborhood relationships. Node2Vec shares the skip-gram model architecture (Fig. 1) with the widely known Word2Vec [14] and optimizes embeddings so that, given a node, it can predict its neighbors. The objective is to maximize the log-probability of observing a network neighborhood for each node [11]: $\max \sum_{i \in V} \sum_{j \in c_{v_i}} \log P[v_j | v_i]$, where c_{v_i} represents v_i's context nodes, i.e., nodes that are topologically related to v_i. The conditional probability $P[v_j | v_i]$ is often modelled using a softmax function: $P[v_j | v_i] = \frac{\exp(\boldsymbol{z_i} \cdot \boldsymbol{\beta_j})}{\sum_{k \in V} \exp(\boldsymbol{z_i} \cdot \boldsymbol{z_k})}$, where $\boldsymbol{z_i}$ denotes v_i embedding and $\boldsymbol{\beta_j}$ the embedding of context node v_j. The node2vec embedding dimension is set to 50.

Role2Vec is used to incorporate global regular equivalence information, which means that nodes that share regular patterns in the graph should be mapped to similar role-based embeddings. Notably, nodes may be in different regions of the graph while possessing similar embeddings. The latent representations are learned using a feature-based random walk approach, where walks find similar nodes identified by structural properties and higher-order graph features (e.g., triangles, 4-cycles, etc.). A feature-based random walk is a random sequence of adjacent nodes sharing the same role, $S : \phi_{v_1}, \phi_{v_2}, ..., \phi_{v_L}$, with ϕ function that maps each node to a role. S is induced by a sequence of indices $(v_1, v_2, ..., v_L)$ generated by a random walk. Role2vec uses the generated sequences to train a skip-gram model and learn 50-dimensional role-based embeddings.

Both Node2vec and Role2Vec have been shown to achieve state-of-the-art performance on various graph RL tasks, but the choice of method depends on the nature of the dataset and the task at hand. The performance evaluation in this work is conducted without true labels for a supervision task; thus, it is based on visualization and cluster analysis. We applied the popular K-means clustering approach to the 50-dimensional embeddings and used the within-cluster dispersion (S_k) to assign clustering scores and select the optimal number of clusters k. The results of the models are evaluated by exploring the network latent representations re-embedded with Uniform Manifold Approximation and Projection (UMAP) [15]. UMAP is a dimensionality reduction technique that takes local structure into account, to increase the data representation quality in terms of clusterability.

3 Results and Final Remarks

The nodes in the LFN network exhibit a strength, i.e., the sum of edge weights of the adjacent edges for each node that varies greatly. In Fig. 2, the y-axis displays the logarithmic values of the node strength. Institutes of research and higher education, i.e., Universities of Trieste (UniTS) and Udine (UniUD), SISSA, OGS, and Elettra, are highlighted with colors as they represent the largest research institutions in terms of both employee count and their impact on the regional economic system [16]. The scatterplot shows that such institutions are among the nodes with the highest coreness and strength.

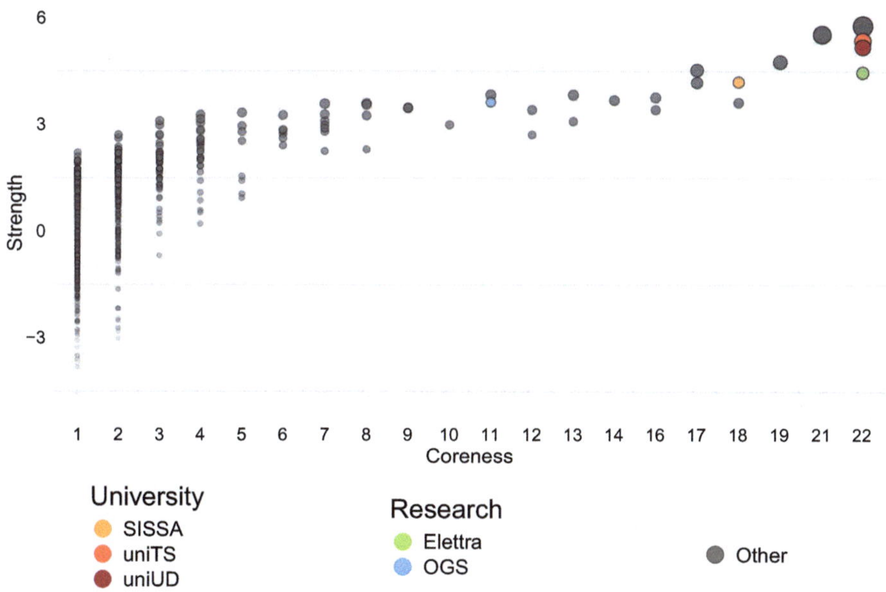

Fig. 2 Visualization of node coreness and strength. Node size represents the strength, while node color represents university and public research organizations

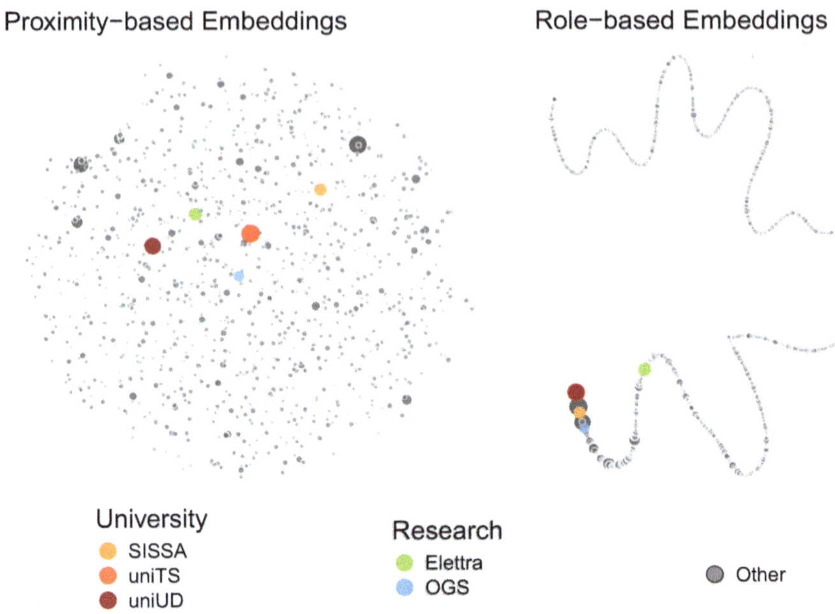

Fig. 3 Two-dimensional UMAP visualization of node embeddings generated from the LFN using Node2vec (left panel) and Role2Vec (right panel). Node size represents the strength, while node color represents university and public research organizations

Figure 3 compares the UMAP visualizations of node embeddings learned with the methods. The distances between nodes in the embedding space reflect structural (left panel) and regular (right panel) equivalence in the original graph. The size of the nodes indicates the strength; thus, employers employing a high number of experienced professionals are shown with bigger nodes. The color of the nodes highlights universities and PROs in the network. The left plot seems to suggest the absence of a community structure: colored and large-sized nodes are scattered in distinct regions of the graph, with no discernible patterns. The right plot shows the universities very close together in space, along with the PROs, illustrating the similarity of their roles in the network. It reveals the centrality of universities and PROs, indicating that although they share qualified employers with different organizations they play the same role as central hubs of their respective neighborhoods.

Figure 4 helps in summarizing the results of the k-means clustering. The S_k vs k plot in panel a) shows that there is not a clear "elbow" (a point of inflexion on the curve); hence, we choose the number of clusters k equal to four since it is the point after which S_k begins to decrease more smoothly. Panel b) shows the UMAP visualization for data partition in four groups. The results of the cluster analysis demonstrate that colored nodes and large-sized nodes cluster together in the role-

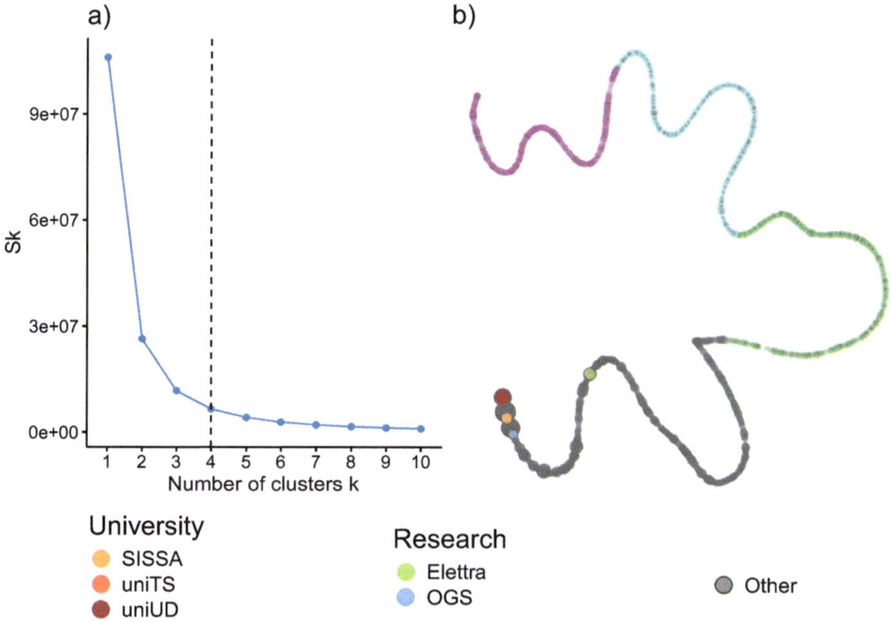

Fig. 4 (**a**) Elbow method for choosing the optimal number of clusters k. (**b**) Two-dimensional UMAP for the visualization of Role2vec embeddings partition in four clusters. Node size represents the strength, node color represents university and public research organizations, while node border color represents the cluster assignment

based embedding space. These findings provide additional evidence of the cohesive roles and influential positions held by universities and PROs.

For the analyses, we used the role2vec and node2vec Python packages to generate the embeddings, the UMAP package to reduce the embedding dimensions for visualization purposes, and the ggplot2 R package for creating the visualizations.

Overall, the study highlights the potential of graph RL techniques to analyze complex networks and uncover hidden structures and patterns, which provide new outlooks on economic development and innovation. It also suggests that examining different embedding algorithms tailored to specific tasks would be valuable in addressing different research inquiries. Results show that universities and leading PROs operating in the FVG region share graduates and/or experienced labor force with different, nonoverlapping, organizations since they are quite different in terms of structural equivalence. On the other hand, when regular equivalence comes into play, such differences are reduced. This suggests that their relational role in the overall LFN is similar. Therefore, policymakers may find it interesting that directing efforts to these organizations allows them to reach different other economic actors while following the same structural patterns. Moreover, given that employment contract data is available in the same format across all Italian regions, this analysis can be reproduced in different contexts. Analyzing the LFN in FVG allows us to understand labor mobility and knowledge spillovers in a medium-sized region with notable innovation activities. It would be interesting to compare FVG results with those from other regions, such as the larger innovative regions of Lombardia and Emilia Romagna or regions hosting large universities and PROs like Lazio.

In future applications, exploring the impact of incorporating node attributes in the RL task could yield valuable insights.

Acknowledgments We acknowledge financial support under the National Recovery and Resilience Plan (NRRP), Mission 4, Component 2, Investment 1.1, Call for tender No. 104 published on 2.2.2022 by the Italian Ministry of University and Research (MUR), funded by the European Union - NextGenerationEU- Project Title Models and methods for the analysis of collaboration networks, CUP J53D23011540006, Principal Investigator (PI) prof. Domenico De Stefano. The pubblication has been produced with co-funding from the European Union-NextGenerationEU.

References

1. Park, J., Wood, I.B., Jing, E., Nematzadeh, A., Ghosh, S., Conover, M.D., Ahn, Y.Y.: Global labor flow network reveals the hierarchical organization and dynamics of geo-industrial clusters. Nat. Commun. **10**(1) (2019). https://doi.org/10.1038/s41467-019-11380-w. ISSN: 20411723
2. Eriksson, R.H.: Localized spillovers and knowledge flows: how does proximity influence the performance of plants? Econ. Geogr. **87**(2), 127–152 (2011). https://doi.org/10.1111/j.1944-8287.2011.01112.x. ISSN: 00130095
3. Guerrero, O.A., Axtell, R.L.: Employment growth through labor flow networks. PLoS ONE **8**(5), (2013). https://doi.org/10.1371/journal.pone.0060808. ISSN: 19326203
4. Smallbone, D., Kitching, J., Blackburn, R., UKCES: Anchor institutions and small firms in the UK: a review of the literature on anchor institutions and their role in developing management and leadership skills in small firms (2015)
5. Directorate-General for Research European Commission and H.-Es-Sadki N. Innovation, Hollanders: Regional Innovation Scoreboard 2023. Publications Office of the European Union (2023). https://ec.europa.eu/assets/rtd/ris/2023/ec%7B%5C_%7Drtd%7B%5C_%7Dris-regional-profiles-italy.pdf
6. Morea, F., De Stefano, D: Innovation Patterns within a Regional Economy through Consensus Community Detection on Labour Market Network. In Proceedings of the Statistics and Data Science Conference, pp. 6–11, Pavia University Press (2023)
7. De Stefano, D., Zaccarin, S.: Modelling Multiple interactions in science and technology networks. Ind. Innov. **20**(3), 221–240 (2013). https://doi.org/10.1080/13662716.2013.791130. ISSN: 13662716
8. Cui, P., Wu, L., Pei, J., Zhao, L., Wang, X.: Graph representation learning. Graph Neural Netw. Found. Front. Appl. **14**(3), 17–26 (2022). https://doi.org/10.1007/978-981-16-6054-2_2
9. Doreian, P., Batagelj, V., Ferligoj, A.: Generalized Block-Modeling, pp. xv + 384. Cambridge University Press, Cambridge (2005). ISSN: 0176-4268. ISBN: 0521840856. J. Classif. **24**(2), 308–311 (2007). https://doi.org/10.1007/s00357-007-0016-8
10. European Union: Commission recommendation of 29 October 2009 on the use of the International Standard Classification of Occupations (ISCO-08), 11 (2009)
11. Grover, A., Leskovec, J.: Node2vec: scalable feature learning for networks. In: Proceedings of the ACM SIGKDD International Conference on Knowledge Discovery and Data Mining, pp 855–864 (2016). https://doi.org/10.1145/2939672.2939754. ISSN: 2154-817X

12. Ahmed, N.K., Rossi, R., Lee, J.B., Willke, T.L., Zhou, R., Kong, X., Eldardiry, H.: Role-based Graph Embeddings. IEEE Transactions on Knowledge and Data Engineering. **34**(5), 2401–2415 (2022). https://doi.org/10.1109/TKDE.2020.3006475
13. Rossi, R.A., Jin, D., Kim, S., Ahmed, N.K., Koutra, D., Lee, J.B., : On proximity and structural role-based embeddings in networks: misconceptions, techniques, and applications. ACM Trans. Knowl. Discov. Data **14**(5) (2020). https://doi.org/10.1145/3397191. ISSN: 1556472X
14. Mikolov, T., Chen, K., Corrado, G., Dean, J.: Efficient estimation of word representations in vector space (2013)
15. McInnes, L., Healy, J., Melville, J.: UMAP: Uniform Manifold Approximation and Projection for Dimension Reduction (2018)
16. SiS FVG: The Scientific and Innovation System of Friuli Venezia Giulia (2024)

Bridging the Gap: Investigating Correlation Clustering and Manifold Learning Connections

Daniyal Kazempour and Peer Kröger

Abstract Correlation clustering and manifold learning, traditionally seen as distinct tasks, possess underlying similarities often overlooked. This paper elucidates on some of these shared traits, as well as on differences, focusing on local principal component analysis (LPCA) as a correlation clustering method and locally linear embedding (LLE) as a manifold learning approach. We elaborate on how both methods capture intrinsic data structures and handle complex distributions, suggesting convergence in objectives. While the findings are not generalizable over all manifold and correlation clustering methods, the gained insights serve as a basis for future investigations.

1 Introduction

Correlation clustering and manifold learning have long been perceived as distinct tasks within the realm of machine learning and data analysis. While correlation clustering primarily focuses on grouping similar data points based on potential correlations in subsets of features, manifold learning aims to uncover the underlying structure or manifold embedded in high-dimensional data. Despite their apparent differences, these two tasks share certain common traits. In this paper, we delve into the parallels between correlation clustering and manifold learning, shedding light on their shared characteristics. Drawing upon exemplary methods from each of these tasks, we highlight the similarities through the lens of local principal component analysis (LPCA) [7] as a representative correlation clustering technique and locally linear embedding (LLE) [12] as a prominent manifold learning approach. By examining their underlying principles and computational strategies, we demonstrate how both LPCA and LLE operate on the premise of capturing intrinsic data structures while accounting for local similarities. Through comparative analyses of

---

D. Kazempour (✉) · P. Kröger
Christian-Albrechts-University Kiel, Kiel, Germany
e-mail: dka@informatik.uni-kiel.de; pkr@informatik.uni-kiel.de

G. Giordano et al. (eds.), *Statistical Models and Learning Methods for Complex Data*, Studies in Classification, Data Analysis, and Knowledge Organization,
https://doi.org/10.1007/978-3-031-84702-8_14

the objective functions of both methods, we showcase how LPCA and LLE exhibit analogous behaviors in terms of preserving local structures. Our findings contribute to bridging the perceived gap between correlation clustering and manifold learning.

2 Correlation Clustering

Correlation clustering is an unsupervised machine learning method tailored at the discovery of clusters that reside in lower-dimensional subspaces. In more formal terms:

Definition 1 (Correlation Clustering) Given a d-dimensional dataset $\mathcal{D} \in \mathbb{R}^d$. Correlation clustering is a partitioning C of $\mathcal{D}$ with every cluster $C_i \in C$ resembling objects $x \in C_i$ which exhibit in their individual lower $d - l$-dimensional *cluster subspace* $S_{i_{clus}} \in \mathbb{R}^{d-l} \in \mathcal{S}$ a high similarity (or in other terms: low variance), while exhibiting in its orthogonal l-dimensional *correlation subspace* $S_{i_{corr}} \in \mathbb{R}^l \in \mathcal{S}$ a high variance and hence a potential correlation. Each cluster C_i of a correlation clustering C can be represented as a tuple $(C_i, S_i = (S_{i_{clus}}, S_{i_{corr}}))$.

The particular necessities of correlation clustering methods arise in high-dimensional data settings with the following purposes: (1) discovery of relevant dimensions (2) mitigating the effect of irrelevant dimensions that may compromise the discovered patterns (clusters). There is an extensive amount of literature proposing different correlation clustering algorithms such as in [1–3, 5, 6]. The clusters in the majority of algorithms reside on *linear* subspaces spanned by an orthogonal basis. However, there exist methods that discover clusters residing on nonlinear subspaces [8] or where the orthogonality property of the subspaces is lifted such as in [11].

2.1 Nonlinear Dimensionality Reduction via LPCA

Besides the so far elaborated correlation clustering algorithms, there exist methods which are developed independently of this research strain but share a striking amount of similar aspects with correlation clustering algorithms. We chose the upcoming algorithm on purpose since it facilitates the investigation of the relations between correlation clustering and manifold learning. One of these methods is local principal component analysis (LPCA) [7]. The prime motivation behind LPCA is that a classic *global* PCA (1) is limited to a linear approximation of the data and (2) as such neglects potentially locally linear correlations. The overall approach of LPCA is that a given dataset is partitioned via vector quantization (i.e., k-means [10]) and through a distance function the closest object is assigned to the l-dimensional subspaces, where l denotes the eigenvectors with the corresponding l-largest eigenvalues. The distance function that is used in LPCA is according to [7]:

$$d(x, r^{(i)}) \equiv \left\| x - r^{(i)} - \sum_{j=1}^{m} z_j e_j^{(i)} \right\|^2 = (x - r^{(i)})^T P^{(i)T} P^{(i)} (x - r^{(i)})$$
$$\equiv (x - r^{(i)})^T \Pi^{(i)} (x - r^{(i)})$$

Here, $r^{(i)}$ denotes a reference vector (centroid), z_j denotes the encoded object, $e_j^{(i)}$ denotes the eigenvector of a partition (i), and $P^{(i)}$ denotes the $m \times n$ matrix with the rows representing the trailing eigenvectors, obtained from a decomposition of the covariance matrix $\Sigma^{(i)}$. Lastly, the matrix $\Pi^{(i)}$ denotes the projection orthogonal to the local l-dimensional subspace.

Contrary to correlation clustering algorithms such as ORCLUS [2], LPCA finds for each cluster C_i a l-dimensional subspace exhibiting the highest variance. While this may seem at first glance non-similar to correlation clustering, it becomes evident when remembering that the obtained subspaces are spanned by orthogonal vectors, meaning that if we find the l-dimensional *correlation subspace* $S_{i_{corr}}$ of high-variance, we automatically obtain the $d - l$-dimensional subspaces of low variance, hence a clusters respective *cluster subspace* $S_{i_{clus}}$.

Besides the distance function, it is also interesting that the authors of [7] elaborate on an encoding and decoding (reconstruction) of the data as well as an objective function. The encoding of the vectors in a latent space is expressed by

$$z = (e_1^{(i)} \cdot (x - r^{(i)}), \ldots, e_l^{(i)} \cdot (x - r^{(i)}))$$

As the decoding, the computation rule is

$$\hat{x} = r^{(i)} + \sum_{j=1}^{l} z_j e_j^{(i)}$$

As a measure for the reconstruction error, the authors use the reconstruction error normalized by the variance of the data:

$$\mathcal{E}_{norm} = \frac{E[||x - \hat{x}||^2]}{E[||x - E[x]||^2]}$$

In the work of LPCA [7], the authors propose a method that comes with the concepts of encoding, decoding, and minimizing reconstruction errors for local PCA, which was at that time existent for autoencoders (autoassociators) and PCA. Besides these observations, we ask: what are the commonalities between LPCA and the correlation clustering task? First, the data is partitioned in both (correlation clustering and LPCA) in such a way that the emerging clusters are located on or around lower-dimensional arbitrarily oriented subspaces. Since correlation clustering methods, so far elaborated on, do that, LPCA does that as well by incorporating vector quantization in its algorithm. For a second point, LPCA learns nonlinear manifolds that are backed up by the statement of the authors in [7]: "We have applied locally linear models to the task of dimension reduction, finding superior results to both PCA and the global nonlinear model built by five-layer autoassociators." This implies that LPCA is capable of learning the dimensionality of nonlinear manifolds as well since autoassociators (autoencoders) are capable of detecting inherently nonlinear lower-dimensional embeddings. The point that LPCA approximates a manifold in a dataset through piecewise linear tangent planes (subspaces) where each tangent plane represents a local Euclidean space is also stated in [7]: "Our proposal is to construct local models, each pertaining to a different disjoint region of the data space. [...] we construct linear models by PCA. If the local regions are small enough, the data manifold will not curve much over the extent of the region, and the linear model will be a good fit (low bias)." This statement from the paper again supports the claim that LPCA can (1) learn nonlinear manifolds and (2) approximate them through piecewise linear Euclidean spaces.

As a conclusive statement, it is interesting to see the high similarity of local PCA to correlation clustering. The prime difference between both is that in LPCA we are interested in the *correlation subspace* while in correlation clustering algorithms we are interested in the *cluster subspace*. While both subspaces are orthogonal and serve different purposes (approximate nonlinear manifold vs. finding compact partitions [clusters]), we obtain in all tasks both subspaces, e.g., via an eigenvalue decomposition on objects of a cluster C_i.

3 Manifold Learning

In the previous section, the term *manifold* emerged; however, it has not been elaborated on so far. Among the literature [4, 13] of the machine learning domain, the term *manifold* is mostly used without specific formal introduction. For this purpose, we first elaborate on what we understand in this work by the term manifold, and we present below a definition which is taken from [9]:

Definition 2 (Topological Manifold) Suppose $\mathbf{M}$ is a topological space. We say that $\mathbf{M}$ is a topological manifold of dimension n or a topological n-manifold if it has the following properties:

1. $\mathbf{M}$ is a *Hausdorff space*: for every pair of distinct objects , $p, q \in \mathbf{M}$, there are disjoint open subsets $U, V \subseteq \mathbf{M}$ such that $p \in U$ and $q \in V$.
2. $\mathbf{M}$ is *second-countable*: there exists a countable basis for the topology of $\mathbf{M}$.
3. $\mathbf{M}$ is *locally Euclidean of dimension n*: each object of $\mathbf{M}$ has a neighborhood that is homeomorphic to an open subset of $\mathbb{R}^n$.

For manifold learning, a central assumption has to be made, namely:

Assumption 1 The data of a given dataset $\mathcal{D} \in \mathbb{R}^d$ is a sample from one (or multiple) arbitrarily shaped manifold(s) **M** characterized by potentially nonlinear, (arbitrarily shaped) function(s) f.

With the definition of manifolds and Assumption 1, we can now investigate in how far correlation clustering can be regarded as a manifold learning task. The first point (1) of Definition 2 states that *distinct objects* are located on disjoint open subsets or in other terms: two distinct objects are located on different subspaces U and V, since objects of different subspaces and thus clusters are actually disjoint subsets, unless overlapping of clusters is permitted. The objects in their disjoint subsets are represented as a linear combination through the eigenvectors representing the subspaces of dimensionality n.

For the second point (2), it is stated that there exists a countable basis for the topology of **M**. Since each object can be characterized by a local basis of arbitrary cardinality of its local neighborhood, so can the topology $\mathcal{T}$ be described by a set of local basis. Indeed with all of the subspaces from correlation clustering, manifolds can be characterized by sets of arbitrarily oriented subspaces.

The probably most relevant point is (3) which states that the manifold **M** is *locally Euclidean*. This is indeed the case for correlation clusters since each correlation cluster with its individual respective arbitrarily oriented subspace represents a linear Euclidean coordinate system.

Further, we derive the following second assumption:

Assumption 2 Correlation clustering is a piecewise linear approximation (PLA) of arbitrarily shaped manifolds of arbitrary precision. As such, the task of correlation clustering can be regarded as a manifold learning task.

We shall see in the upcoming sections on LLE, which is a manifold learning algorithm that correlation clustering algorithms like LPCA share similarities with manifold learning methods such as LLE.

3.1 *Locally Linear Embedding*

In this section, we deal with the locally linear embedding (LLE) method [12]. The idea behind LLE is to capture the local affine structure of some given data. For each object, x_i an approximation through a linear combination of neighbor objects is constructed. Afterward, a lower-dimensional representation y_i is computed which yields a low deviation from the original objects x_i. Roughly speaking, the algorithm undergoes the following three steps:

1. Compute the local k-nearest neighbors $\mathcal{N}(i)$.
2. Compute weights w_{ij} that best reconstruct each object x_i through its neighbors by minimizing:

$$\sum_i ||x_i - \sum_j w_{ij} x_j||^2$$

3. Compute the lower-dimensional vectors y_i that are best reconstructed via the weights w_{ij} by minimizing:

$$\sum_i ||y_i - \sum_j w_{ij} y_j||^2$$

The third step refers to the embedding costs of a full-dimensional vector x_i to a lower-dimensional embedding y_i. The equation in the third step can be solved via a sparse eigenvector problem. In fact, the dimensionality reduction step optimizes for a minimization of the trace:

$$tr[(\mathbf{Y} - \mathbf{WY})^T(\mathbf{Y} - \mathbf{WY})] = tr[\mathbf{Y}^T(\mathbf{I} - \mathbf{W})^T(\mathbf{I} - \mathbf{W})\mathbf{Y}]$$

Here, $\mathbf{Y}$ denotes the eigenvector matrix $U \in \mathbb{R}^{n \times l}$.

By minimizing the trace, each full-dimensional object in ambient space is mapped to a low-dimensional embedding which represents the global internal coordinates of the underlying manifold. In the optimization step by minimizing the trace, the weights $\mathbf{W}$ are set fix, while the eigenvector matrix $\mathbf{Y}$ is optimized. It is noteworthy at this point that while the weights w_{ij} for each object x_i are computed from the *local* neighborhood, that the embedded lower-dimensional coordinates y_i are computed by applying *globally* an eigensolver that links objects in the dataset via connected components derived from the weight matrix $\mathbf{W}$. This means that while LLE captures the locality of a dataset, it yields a single unified lower-dimensional space on which the objects are embedded. To look for parallels with correlation clustering, we observe certain similarities and dissimilarities. As a commonality, we observe that it follows the goal of approximating nonlinear manifolds through (locally) linear embeddings. As such, it means that LLE approximates manifolds through piecewise linear tangent planes. Also, the projection of these locally linear combinations is performed to individual (local) subspaces, yielding for each object and its neighborhood an individual l-dimensional arbitrarily oriented subspace. As such, LLE, like local PCA, approximates manifolds through piecewise linear patches where each patch represents a local Euclidean (sub)space. Through the optimization of $\sum_{i=1}^{N} ||y_i - \sum_{k=1}^{N} w_{ik} y_k||^2$, one can see that LLE is tailored at minimizing the projection loss for each of its local patches. In LPCA, the projection loss for each of the individual subspaces is also achieved by performing an eigendecomposition on the covariance matrix Σ of every cluster. On the contrary, in a final step, LLE yields a global coordinate system for all of the objects in the dataset. Contrary to that correlation clustering methods, like LPCA do no provide a single unified subspace onto which all objects are projected. Instead for each cluster, an individual arbitrarily oriented correlation subspace is computed.

First Experiment While from a formal point of view correlation clustering and manifold learning seem to exhibit similarities, we want to investigate how far these similarities are observable in practice. Exemplarily, we chose a more recent method compared to LPCA, ORCLUS [2] as a correlation clustering algorithm (since it relies on the same optimization objective as LPCA and has PCA and k-means as its basis) and LLE ($kNN = 10$ and $l = 1$). As a dataset, we have a two-dimensional dataset with three linear clusters of different sizes and variance with labels obtained from clustering by ORCLUS with $k = 3$ and $l = 1$. We decided to pursue our investigations based on the aspect of preserving distances. We hypothesize that if both methods are similar, then we would observe, per cluster, the same lower-dimensional perseverance of distances. In Fig. 1, we can observe in each column the different clusters and in the first two rows normalized pairwise distances between one-dimensional embeddings of ORCLUS and LLE. From a purely visual approach, both embeddings (ORCLUS [top row] and LLE [middle row]) exhibit high similarity, except for cluster 2. To highlight similar and non-similar regions, within the heatmaps, we subtracted per cluster, the distance matrices of ORCLUS and LLE from each other and took the absolute value. The result matrices in Fig. 1 (bottom row) show that in the case of the left (cluster 1) and right (cluster 3) clusters, we barely notice any differences, meaning that the distances in one dimension via ORCLUS and LLE have been embedded similarly. In the case of the middle cluster (cluster 2), we observe deviations between both distance matrices.

4 Conclusion

As a summary, we distil the following insights by comparing LLE as a representative of manifold learning methods and LPCA as a representative of correlation clustering methods:

- Both approximate nonlinear manifolds via piecewise linear approximation yielding locally Euclidean subspaces (manifolds).
- Both capture local patterns (clusters), LLE via its local neighborhood and LPCA via the vector quantization; hence, both can be used as well for clustering.
- LLE provides a single subspace (manifold) on which the data can be projected, while LPCA provides for each cluster its individual subspace.

While this work provides first endeavors to investigate commonalities and differences between manifold learning and correlation clustering, we hope the

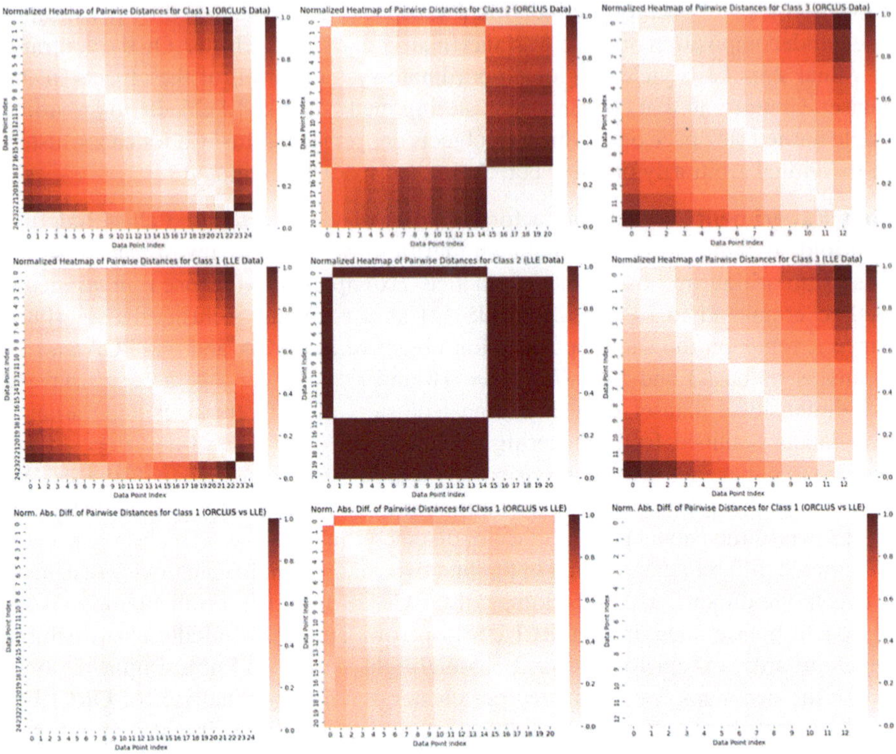

Fig. 1 Normalized pairwise distances per cluster (from left to right). Top row: ORCLUS embeddings; middle row: LLE embeddings; bottom row: difference of pairwise distances between ORCLUS and LLE

insights stimulate further research in this direction, revealing commonalities and differences between both tasks, viewing both tasks as not isolated but related.

References

1. Achtert, E., Böhm, C., Kriegel, H.-P., Kröger, P., Zimek, A.: Robust, complete, and efficient correlation clustering. In: Proceedings of the 2007 SIAM International Conference on Data Mining, pp. 413–418. SIAM (2007)
2. Aggarwal, C.C., Yu, P.S.: Finding generalized projected clusters in high dimensional spaces. In: Proceedings of the 2000 ACM SIGMOD International Conference on Management of Data, pp. 70–81 (2000)
3. Böhm, C., Kailing, K., Kröger, P., Zimek, A.: Computing clusters of correlation connected objects. In: Proceedings of the 2004 ACM SIGMOD International Conference on Management of Data, pp. 455–466 (2004)
4. Goodfellow, I., Bengio, Y., Courville, A.: Deep Learning. MIT Press, Cambridge (2016)

5. Günnemann, S., Färber, I., Virochsiri, K., Seidl, T.: Subspace correlation clustering: finding locally correlated dimensions in subspace projections of the data. In: Proceedings of the 18th ACM SIGKDD International Conference on Knowledge Discovery and Data Mining, pp. 352–360 (2012)
6. Haralick, R., Harpaz, R.: Linear manifold clustering in high dimensional spaces by stochastic search. Pattern Recognit. **40**(10), 2672–2684 (2007)
7. Kambhatla, N., Leen, T.K.: Dimension reduction by local principal component analysis. Neural Comput. **9**(7), 1493–1516 (1997)
8. Kazempour, D., Mauder, M., Kröger, P., Seidl, T.: Detecting global hyperparaboloid correlated clusters: a hough-transform based multicore algorithm. Distrib. Parallel Databases **37**(1), 39–72 (2019)
9. Lee, J.M.: Smooth manifolds. In: Introduction to Smooth Manifolds, pp. 1–31. Springer, Berlin (2013)
10. Lloyd, S.: Least squares quantization in PCM. IEEE Trans. Inform. Theory **28**(2), 129–137 (1982)
11. Plant, C.: Sonar: signal de-mixing for robust correlation clustering. In: Proceedings of the 2011 SIAM International Conference on Data Mining, pp. 319–330. SIAM, (2011)
12. Roweis, S.T., Saul, L.K.: Nonlinear dimensionality reduction by locally linear embedding. Science **290**(5500), 2323–2326 (2000)
13. Souvenir, R., Pless, R.: Manifold clustering. In: Tenth IEEE International Conference on Computer Vision (ICCV'05) Volume 1, vol. 1, pp. 648–653. IEEE, Piscataway (2005)

Improving Performance in Neural Networks by Dendrite-Activated Connection

Carlo Metta, Marco Fantozzi, Andrea Papini, Gianluca Amato, Matteo Bergamaschi, Andrea Fois, Silvia Giulia Galfré, Alessandro Marchetti, Michelangelo Vegliò, Maurizio Parton, and Francesco Morandin

Abstract We introduce a novel computational unit for neural networks featuring multiple biases, challenging the conventional perceptron structure. Designed to emphasize preserving uncorrupted information as it transfers from one unit to the next, this unit applies activation functions later in the process, incorporating specialized biases for each unit. We posit this unit as an improved design for neural networks and support this with (1) empirical evidence across diverse datasets; (2) a class of functions where this unit utilizes parameters more efficiently; and (3)

All authors can be contacted at CuriosAI.

C. Metta
ISTI-CNR Pisa, Pisa, Italy
e-mail: carlo.metta@isti.cnr.it

M. Fantozzi · A. Fois · F. Morandin
University of Parma, Parma, Italy
e-mail: andrea.fois@unipr.it; francesco.morandin@unipr.it

A. Papini
Chalmers University of Technology, Gothenburg, Sweden
e-mail: andreapa@chalmers.se

G. Amato · A. Marchetti · M. Vegliò · M. Parton (✉)
University of Chieti-Pescara, Pescara, Italy
e-mail: gianluca.amato@unich.it; alessandro.marchetti@unicampus.it; michelangelo.veglio@unich.it; maurizio.parton@unich.it

M. Bergamaschi
University of Padova, Padova, Italy
e-mail: matteo.bergamaschi@math.unipd.it

S. G. Galfré
University of Pisa, Pisa, Italy
e-mail: silvia.galfre@di.unipi.it

G. Giordano et al. (eds.), *Statistical Models and Learning Methods for Complex Data*, Studies in Classification, Data Analysis, and Knowledge Organization,
https://doi.org/10.1007/978-3-031-84702-8_15

biological analogies suggesting closer mimicry to natural neural processing. Source code is available at https://github.com/CuriosAI/dac-dev.

1 Introduction

Historically, the perceptron's structure, the artificial neural network's fundamental computational unit, has rarely been questioned, due to its straightforward biological inspiration. In time, the early sigmoid activation function was replaced by ReLU and variants, and the biological analogy became less stringent, shifting focus on the desirable mathematical properties of the class of functions computed by the networks, like representation power and nonvanishing gradients. This has brought us to the current situation in which most units output their signal through ReLU, a nonlinear activation function that effectively destroys some information.

We propose a fundamental redesign of the standard computational unit, where outputs carry complete information to the next layers before activation occurs, mimicking activation at the dendrites instead of the axon base. We call this new computational unit "DAC" or "dendrite-activated connection," reflecting its biologically inspired mechanism.

DAC units have been announced during ClaDAG 2023 conference and extensively described in the arXiv paper. Because of an eight-page limit, we focus here on selected results from the arXiv paper: image recognition via convolutional experiments (Sect. 3), a representation theorem for piecewise-linear functions (Sect. 4), and a biological rationale for pre-activation and multiple biases (Sect. 5). Fully connected experiments, geometric interpretation, and comprehensive theoretical analysis from the arXiv paper will appear in a forthcoming paper. See also [10] and [13] for a different interpretation of DAC units. Source code is available at https://github.com/CuriosAI/dac-dev.

2 Model

In a standard neural network, units are post-activated: they compute a linear combination of their inputs and then apply a nonlinear filter to the result:

$$\begin{cases} z_i = \sum_{j \in \mathcal{I}_i} w_{i,j}\, y_j & \text{linear aggregation} \\ y_i = \varphi(b_i + z_i) & \text{nonlinear filter} \end{cases} \tag{1}$$

where, for unit i, $\mathcal{I}_i$ are the input nodes, b_i the bias, and φ the activation function. We propose to consider pre-activated units with unshared biases:

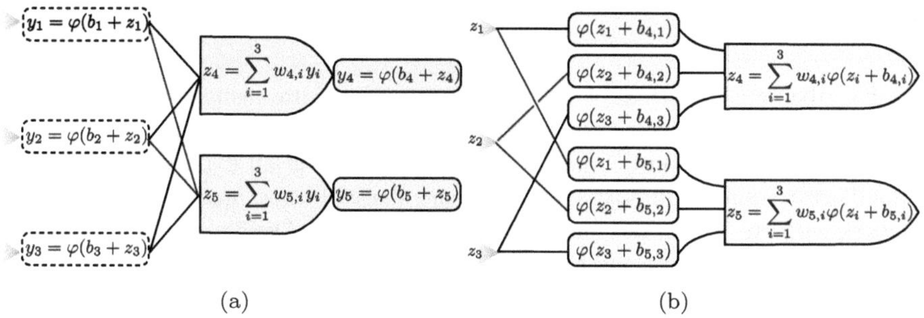

Fig. 1 Two computational units 4, 5 with three input nodes 1, 2, 3. Bullets and rectangles represent linear aggregation and nonlinear filters, respectively. On the left, standard units 4 and 5 share the same biases b_1, b_2, b_3 in the activation of their inputs. On the right, biases of DAC units 4 and 5 depend on input and output nodes: $b_{4,1}, b_{4,2}, b_{4,3}$ can now be different from $b_{5,1}, b_{5,2}, b_{5,3}$, and so, from the point of view of the output units, they are *unshared.* (**a**) Standard unit. (**b**) DAC unit

$$\begin{cases} y_{i,j} = \varphi(b_{i,j} + z_j) & \text{nonlinear filter} \\ z_i = \sum_{j \in \mathcal{I}_i} w_{i,j}\, y_{i,j} & \text{linear aggregation} \end{cases} \tag{2}$$

In (1), $w_{i,j}$ is a connection-specific weight, while b_i is specific to each unit. In (2), each connection still has its own weight $w_{i,j}$, but it also has its own nonlinear filter with a specific bias $b_{i,j}$. (Compare Fig. 1a, b.)

Connections between units correspond to dendrites in biological neurons, and it is known that activation at the level of dendrites can actually occur in biological neural networks; see Sect. 5. With this motivation, we refer to a connection as in (2) as a dendrite-activated connection (DAC).

Remark 1 (Complexity) A DAC unit can have almost twice as many parameters as a standard unit and a correspondingly greater representation capacity. In our experiments, DAC neural networks are compared to standard ones with a similar number of parameters or computational complexity, not with the same number of units or channels. In this paper, we favor floating point operations (FLOP) over the number of parameters, as advocated by other sources, including [18]. We provide plots with the number of parameters in the arXiv paper. Note that, in convolutions larger than 1×1, biases can be efficiently *partially shared*, by making them depend solely on the input channel and output kernel. In the case of a 3×3 kernel, this approach leads to a modest increase of approximately $+11\%$ in the number of parameters and an additional 5.5% more FLOPs compared to the baseline.

3 Experiments

We describe here the effect of using pre-activations with unshared biases on classification tasks with diverse convolutional architectures: without skip connections (VGG-like) and with skip connections (ResNet-like). In all experiments, we train a standard structure with post-activations and shared biases, *baseline*, and a similar network with pre-activations and unshared biases, *DAC*. Details on implementation, training hyperparameters, and more experiments on fully connected architectures as well can be found on the arXiv paper.

Plain Convolutional Networks on CIFAR We designed two VGG-like architectures for CIFAR-10 and CIFAR-100, in line with modern revisitations like [5]. The first structure (referred as 20/16) has 20 layers including the output, and the 19 convolutional layers start with 16 kernels that become 32 and 64, after 7 and 13 layers. The second structure (14/32) has 14 layers, starts with 32 kernels, and doubles after 5 and 9. Each experiment was replicated five times with a fivefold cross-validation scheme. For the two architectures 20/16 and 14/32, the test error on the baselines was 10.66% and 8.13% on CIFAR-10 and 37.23% and 32.36% on CIFAR-100. DAC networks, using pre-activations and unshared biases, obtained a systematic and statistically significant improvement in the test error: $-0.65\% \pm 0.17\%$ and $-0.43\% \pm 0.09\%$ on CIFAR-10 and $-1.16\% \pm 0.22\%$ and $-1.81\% \pm 0.19\%$ on CIFAR-100.

ResNet Networks on CIFAR We used as baseline the architecture proposed for CIFAR in the original ResNet papers [7, 8], with 20, 32, 44, and 56 layers ($n = 3, 5, 7, 9$) both with the v1 post-activated [7] architecture and the v2 pre-activated [8] architecture. The structure is similar to the VGG of the previous section, but with skip connections every two layers.

Figure 2 summarizes the results in terms of best test accuracy. Among the 16 comparisons, 15 are in favor of DAC models, with 9 of them statistically significant (p-value below 5%). This confirms that using unshared biases improves the performances of residual convolutional networks on CIFAR, with only a marginal increase in the model complexity. In fact, if the improvement of the test error between baseline and DAC was due only to the increased size in the latter, we would expect the decreasing blue and orange lines, which represent models of increasing depth, to overlap in the figure. However, while there is some overlap for CIFAR-10, v2 and CIFAR-100, v1, the lines are distinctly separate in the other cases.

Table 1 presents a simpler metric: the lowest test error rate (the minimum over the epochs of the average of the five replicates). The performance of the baselines aligns with typical values for networks of similar complexity found in the literature. Both the figure and table suggest that using unshared biases in ResNet architectures leads to measurable performance improvements across most versions and benchmark datasets.

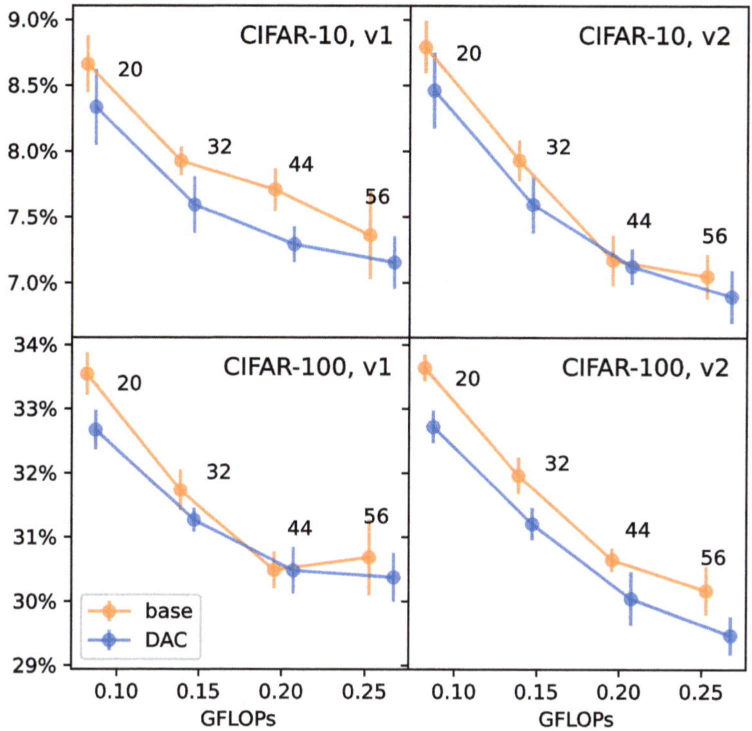

Fig. 2 ResNet, average test error. Compared performances of ResNet networks with shared biases (baseline, orange) and unshared biases (DAC, blue), on CIFAR-10 (top) and CIFAR-100 (bottom), with architectures v1 (left) and v2 (right). Floating numbers are the layers count. Test error (vertical axis) is averaged over five replicates. Error bars are 95% confidence intervals for the true mean value. Complexity (horizontal axis) in GFLOPs per forward pass

ResNet Networks on Imagenette and Imagewoof Imagenette and Imagewoof are subsets of ImageNet that are frequently utilized for model benchmarking. They offer a simpler and faster alternative to ImageNet while preserving many of its challenges. We used the 20 and 32 layers ResNet structure of the CIFAR-10 experiments, with the same training hyperparameters.

Table 2 summarizes the resulting lowest test error rates (the minimum over the epochs of the average of the five replicates). The performance of the baselines aligns with typical values for networks of similar complexity found in the literature. Using unshared biases (DAC) shows marked performance improvements, again at the cost of a marginal increase in size and complexity. In particular, it can be observed that the baseline ResNet v1 (post-activated) is much worse than baseline v2 and DACs (both pre-activated). In some cases, using unshared biases improves the baselines despite having fewer layers.

Table 1 Comparison of minimum test errors in percent points on CIFAR-10 and CIFAR-100 (± one standard error)

	CIFAR-10, v1		CIFAR-10, v2	
Layers	Base	DAC	Base	DAC
20	8.64±0.12	**8.28**±0.19	8.73±0.10	**8.33**±0.07
32	7.90±0.04	**7.57**±0.11	7.85±0.09	**7.56**±0.15
44	7.69±0.09	**7.27**±0.06	7.14±0.09	**7.10**±0.05
56	7.34±0.17	**7.13**±0.11	7.02±0.09	**6.89**±0.05
	CIFAR-100, v1		CIFAR-100, v2	
Layers	Base	DAC	Base	DAC
20	33.50±0.16	**32.54**±0.04	33.50±0.11	**32.62**±0.10
32	31.66±0.12	**31.09**±0.09	31.79±0.12	**31.05**±0.12
44	**30.40**±0.11	30.42±0.19	30.50±0.07	**29.93**±0.20
56	30.53±0.23	**30.32**±0.19	29.89±0.22	**29.32**±0.17

Table 2 Comparison of minimum test errors in percent points on Imagenette and Imagewoof (± one standard error)

	Imagenette, v1		Imagenette, v2	
Layers	Base	DAC	Base	DAC
20	13.42±0.26	**11.69**±0.11	11.98±0.17	**11.79**±0.13
32	13.14±0.45	**11.75**±0.08	11.40±0.09	**11.28**±0.11
	Imagewoof, v1		Imagewoof, v2	
Layers	Base	DAC	Base	DAC
20	23.20±0.26	**22.61**±0.25	22.61±0.17	**21.70**±0.22
32	23.06±0.24	**21.32**±0.29	21.75±0.17	**20.65**±0.14
GFLOPs	0.509	0.542	0.509	0.542
	0.865	0.918	0.865	0.918

Remark 2 The DAC framework is versatile and can be applied to any model due to its general nature, not just the convolutional architectures discussed in this section. We plan to test DAC together with GloNet [1, 4]; on diverse time-series settings [6, 14]; in a reinforcement learning setting [3, 11, 12, 15, 16]. These experiments will appear in a series of papers on DAC units.

4 Approximation of Piecewise-Linear Functions

To support the statement that DAC makes better use of parameters, we prove that piecewise-linear functions $\mathbb{R} \to \mathbb{R}$ with k components can be represented with $2k$ parameters by DAC and not less than $3k + 1$ parameters by a standard neural network. In this example, DAC obtains the best theoretical parameter efficiency. A complete theoretical analysis, including a universal representation theorem, backpropagation equations, and a geometric description, can be found in the arXiv paper.

Theorem 4.1 *Let k be a positive integer, and let PL_k be the set of continuous piecewise-linear functions $\mathbb{R} \rightarrow \mathbb{R}$ consisting of exactly k linear components. Then:*

1. *There exists a one-layer DAC neural network f_k representing PL_k. The input is replicated k times, and there are 2k parameters.*
2. *A standard neural network with one layer cannot represent PL_k, with or without input replication.*
3. *For a standard neural network, the minimal structure needed to represent PL_k is two layers and $3k + 1$ parameters.*

Proof

1. Let $f_{k,\mathbf{w},\mathbf{b}}$ be a one-layer DAC neural network with one unit, weight and bias vector $\mathbf{w} \in \mathbb{R}^k$ and $\mathbf{b} \in \mathbb{R}^k$, respectively, and input $x \in \mathbb{R}$ replicated k times. By Formula (2), DAC output is

$$f_{k,\mathbf{w},\mathbf{b}}(x) = \sum_{j=1}^{k} w_j \varphi(b_j + x),$$

and $\text{PL}_k = \{f_{k,\mathbf{w},\mathbf{b}} | \mathbf{w}, \mathbf{b} \in \mathbb{R}^k\}$.
2. Without input replication, a one-layer standard neural network $\mathbb{R} \rightarrow \mathbb{R}$ is just $f(x) = \varphi(wx + b)$. With input replication, the representational power does not change, because

$$f_{k,\mathbf{w},b}(x) = \varphi(b + \sum_{j=1}^{k} w_j x) = \varphi\Big(b + (\sum_{j=1}^{k} w_j)x\Big).$$

3. By point (2), we need at least two layers with a standard neural network. Since we need to represent k different slopes, we need at least k units in the first layer, for a total of $2k$ parameters (k weights and k biases). The second layer is then a single unit, adding k additional weights and one bias. Thus, a standard neural network needs at least $2k + k + 1 = 3k + 1$ parameters to represent the whole PL_k.

□

Remark 3 Note that PL_k depends on exactly $2k$ parameters (specifically, k slopes and k intercepts), allowing DAC to realize PL_k with the theoretically minimal number of parameters.

5 Biological Interpretation

In this section, we draw parallels between DAC units and biological neurons, suggesting that DAC units more closely mimic natural neural processes. In fact, DAC reflects to some extent a recent shift in the understanding of the biological neuron. The early soma-centric representation of the neuron today has been discarded in favor of a more realistic and complex model that incorporates *active* dendrites [9, 19]. A typical biological neuron consists of many input branches called dendrites, a main body called soma, and the axon, which branches at its end in many terminals, where synapses connect to the dendrites of other neurons. The input signals originate in the dendrites and flow through the soma. They are integrated at the region where the axon connects to the soma. If a specific threshold is reached, the neuron fires, sending the signal down the axon to the synapses. Until recently, the biological neuron model was primarily soma-centric, depicted as a point neuron where dendrites merely passed signals and all processing occurred in the soma. This elementary representation was the inspiration of the traditional perceptron in artificial neural networks. Current biological models are more complicated, and the central role of dendrites in signal modulation is better understood [17]. Dendrites in fact present voltage-gated ion channels [9] able to produce local electrical events termed dendritic spikes. Dendrites actually present at least four groups of ion channels [19]: the synaptic receptors, activated by neurotransmitters, the passive leak channels, the active subthreshold ion channels, able to produce transmembrane currents also when the threshold for the action potential is not reached, and supra-threshold ion channel, active when the threshold is reached.

As described, a dendrite or a group of dendrites can perform significant local, nonlinear signal integration before the signal reaches the axon. This process is mimicked by a DAC unit through its use of pre-activation and unshared biases, representing a more nuanced approach to signal integration akin to that of biological dendrites.

Acknowledgments Carlo Metta is supported by EU Horizon 2020: G.A. 871042 SoBig-Data++, NextGenEU - PNRR-PEAI (M4C2, investment 1.3) FAIR and "SoBigData.it". Gianluca Amato, Fraancesco Morandin, and Maurizio Parton are partially supported by INdAM groups GNAMPA, GNCS, and GNSAGA. Alessandro Marchetti is a PhD student, National PhD in AI, XXXVII cycle, health and life sciences, UCBM. Computational resources provided by CLAI lab, Chieti-Pescara.

References

1. Angileri, F., Lombardi, G., Fois, A., Faraone, R., Metta, C., Salvi, M., Bianchi, L.A., Fantozzi, M., Galfrè, S.G., Pavesi, D., Parton, M., Morandin, F.: In: A Systematization of the Wagner Framework: Graph Theory Conjectures and Reinforcement Learning. Lecture Notes in Computer Science, vol. 15243, pp. 325–338. LNAI (2025). https://doi.org/10.1007/978-3-031-78977-9_21

2. CurioSAI: Increasing biases can be more efficient than increasing weights (2023). https://github.com/CuriosAI/dac-dev
3. Di Cecco, A., Papini, A., Metta, C., Fantozzi, M., Galfré, S.G., Morandin, F., Parton, M.: In: SwitchPath: Enhancing Exploration in Neural Networks Learning Dynamics, 2025. Lecture Notes in Computer Science, vol. 15243, pp. 275–291. LNAI (2025). https://doi.org/10.1007/978-3-031-78977-9_18
4. Di Cecco, A., Metta, C., Fantozzi, M., Morandin, F., Parton, M.: Glonets: Globally connected neural networks. In: IDA 2024, Lecture Notes in Computer Science, vol. 14641, pp. 53–64 (2024). https://doi.org/10.1007/978-3-031-58547-0_5
5. Ding, X., Zhang, X., Ma, N., Han, J., Ding, G., Sun, J.: RepVGG: making VGG-style ConvNets great again. pp. 13733–13742 (2021)
6. Gregnanin, M., De Smedt, J., Gnecco, G., Parton, M.: Signature-based community detection for time series. In: Complex Networks 2023, Studies in Computational Intelligence, vol. 1142, pp. 146–158 (2023). https://doi.org/10.1007/s41109-024-00651-1
7. He, K., Zhang, X., Ren, S., Sun, J.: Deep residual learning for image recognition. In: IEEE CVPR, pp. 770–778 (2016)
8. He, K., Zhang, X., Ren, S., Sun, J.: Identity mappings in deep residual networks. In: ECCV 2016, Lecture Notes in Computer Science, vol. 9908, pp. 630–645 (2016)
9. Larkum, M.E.: Are dendrites conceptually useful? Neuroscience **489**, 4–14 (2022). https://doi.org/10.1016/j.neuroscience.2022.03.008
10. Metta, C., Fantozzi, M., Papini, A., Amato, G., Bergamaschi, M., Galfrè, S.G., Marchetti, A., Vegliò, M., Parton, M., Morandin, F.: Increasing biases can be more efficient than increasing weights. In: IEEE/CVF, Winter Conference on Applications of Computer Vision WACV, pp. 2798–2807 (2024). https://doi.org/10.1109/WACV57701.2024.00279
11. Morandin, F., Amato, G., Fantozzi, M., Gini, R., Metta, C., Parton, M.: SAI: a sensible artificial intelligence that plays with handicap and targets high scores in 9×9 go. In: ECAI 2020, vol. 325, pp. 403–410 (2020). https://doi.org/10.3233/FAIA200119
12. Morandin, F., Amato, G., Gini, R., Metta, C., Parton, M., Pascutto, G.: SAI: a sensible Artificial Intelligence that plays Go. In: IJCNN, pp. 1–8 (2019). https://doi.org/10.1109/IJCNN.2019.8852266
13. Metta, C., Fantozzi, M., Papini, A., Amato, G., Bergamaschi, M., Fois, A., Galfrè, S.G., Marchetti, A., Vegliò, M., Parton, M., Morandin, F.: Increasing biases can be more efficient than increasing weights. Advances in Data Analysis and Classification. Springer Nature (2025). https://doi.org/10.1007/s11634-025-00649-2
14. Parton, M., Fois, A., Vegliò, M., Metta, C., Gregnanin, M.: Predicting the failure of component x in the Scania dataset with graph neural networks. In: IDA 2024, Lecture Notes in Computer Science, vol. 14642, pp. 251–259 (2024). https://doi.org/10.1007/978-3-031-58553-1_20
15. Pasqualini, L., Parton, M.: Pseudo random number generation through reinforcement learning and recurrent neural networks. Algorithms **13**(11), 307 (2020). https://doi.org/10.3390/a13110307
16. Pasqualini, L., Parton, M., Morandin, F., Amato, G., Gini, R., Metta, C., Fantozzi, M., Marchetti, A.: Score vs. winrate in score-based games: which reward for reinforcement learning? In: ICMLA, pp. 573–578 (2022). https://doi.org/10.1109/ICMLA55696.2022.00099
17. Poirazi, P., Papoutsi, A.: Illuminating dendritic function with computational models. Nat. Rev. Neurosci. **21**(6), 303–321 (2020). https://doi.org/10.1038/s41583-020-0301-7
18. Schwartz, R., Dodge, J., Smith, N.A., Etzioni, O.: Green AI. Commun. ACM **63**(12), 54–63 (2020). https://doi.org/10.1145/3381831
19. Sinha, M., Narayanan, R.: Active dendrites and local field potentials: biophysical mechanisms and computational explorations. Neuroscience **489**, 111–142 (2022)

Regression Models with Compositional Regressors in Case of Structural Zeros

Francesco Porro

Abstract In many real-life situations, it may happen to consider a regression model with compositional explanatory variables. Compositional data describe parts of some whole, having the feature to sum to a fixed value, so they are commonly presented as vectors of proportions, percentages, or frequencies. In the compositional framework, the presence of structural zeros is problematic, since a composition is not allowed to have a part equal to zero. In the recent years, a few techniques have been introduced in the literature to address the issue. In this paper, three methods (the replacement, the conditioning, and projection approach) are described and illustrated by an example of application.

1 Introduction and Preliminaries

The main feature of compositional data is that the relevant information they convey is in the ratios between the parts and not in their absolute values or in their sum. Quite recently, the methodologies developed for compositional data have been applied in many different fields: physics, chemistry, finance, social sciences, and economics, just to mention some of them (cf. [7]). The pillar of the compositional data framework is the definition of a composition.

Definition 1 A D-part composition is a class of equivalence containing all the compositionally equivalent vectors (with strictly positive components) in $\mathbb{R}^D$, where two compositions $\mathbf{x} = (x_1, x_2, \ldots, x_D)$ and $\mathbf{y} = (y_1, y_2, \ldots, y_D)$ are said to be compositionally equivalent if there exists a positive constant $\lambda \in \mathbb{R}^+$ such that $\mathbf{x} = \lambda \cdot \mathbf{y}$.

The most suitable space to represent the D-part compositions is the simplex $\mathbb{S}^D$:

F. Porro (✉)
Dipartimento di Matematica, Università degli Studi di Genova, Genova, Italy
e-mail: francesco.porro@unige.it

G. Giordano et al. (eds.), *Statistical Models and Learning Methods for Complex Data*, Studies in Classification, Data Analysis, and Knowledge Organization,
https://doi.org/10.1007/978-3-031-84702-8_16

$$\mathbb{S}^D = \{(x_1, x_2, \ldots, x_D) \in \mathbb{R}^D \ : \ x_i > 0 \ \forall i; \ \sum_{i=1}^{D} x_i = \kappa\}, \tag{1}$$

where the arbitrary constant κ is often set to 1, in order to identify the vectors of proportions (which sum to 1) as representatives of an equivalence class. For further details, see [7, 8], and references therein. The next definition of *closure* is important.

Definition 2 Let $\mathbf{x} = (x_1, x_2, \ldots, x_D)$ be a vector of $\mathbb{R}^D$. The closure (to κ) of $\mathbf{x}$ is a D-part composition defined as

$$\mathscr{C}(\mathbf{x}) = \left(\frac{\kappa \cdot x_1}{\sum_{i=1}^{D} x_i}, \frac{\kappa \cdot x_2}{\sum_{i=1}^{D} x_i}, \ldots, \frac{\kappa \cdot x_D}{\sum_{i=1}^{D} x_i} \right).$$

The closure is a sort of normalization that aims to rescale the values of the components of a vector, setting their sum equal to the constant κ.

A classical compositional dataset is a matrix $\mathbf{X}$ containing a sample of observations, each one being a D-part composition. Therefore, $\mathbf{X}$ can be represented by a matrix with n rows (being n the number of observation) and D columns (being D the dimension of the compositions):

$$\mathbf{X} = \begin{pmatrix} \mathbf{x}_1 \\ \mathbf{x}_2 \\ \vdots \\ \mathbf{x}_n \end{pmatrix} = \begin{pmatrix} x_{11} & x_{12} & \ldots & x_{1D} \\ x_{21} & x_{22} & \ldots & x_{2D} \\ \vdots & \vdots & \ddots & \vdots \\ x_{n1} & x_{n2} & \ldots & x_{nD} \end{pmatrix}.$$

It is well known that the standard statistical descriptive measures, based on the real Euclidean structure, should be used with attention in a compositional dataset, since they can lead to erroneous conclusions (see [7]). To overcome this issue, the compositional approach introduced alternative statistical tools and methods, based on a different kind of geometry. This geometry labelled for the first time in [6] the *Aitchison geometry on the simplex* is very suited for visualizing, analyzing, and interpreting the compositional data, and it can be equipped with a consistent distance, norm, and inner product.

The handling of D-part compositions can sometimes not be very easy. For this reason, a usual practice in compositional data analysis is the application of transformations, which map compositions into real vectors (belonging to suitable spaces) for exploiting the usual Euclidean structure. In the literature, many transformations have been proposed: the most common are based on the logarithm of specific ratios (called logratios). Two of them need to be introduced for the remaining of the paper: the centered (*clr*) and the isometric (*ilr*) logratio transformations. More details of their features and the description of other transformations used in compositional data analysis can be found in [7] and [8].

Definition 3 The centered logratio transformation (clr) of the D-part composition $\mathbf{x} = (x_1, x_2, \ldots, x_D)$ is the mapping from $\mathbb{S}^D$ to $\mathbb{R}^D$ given by

$$clr(\mathbf{x}) = \ln\left(\frac{x_1}{g(\mathbf{x})}, \frac{x_2}{g(\mathbf{x})}, \ldots, \frac{x_D}{g(\mathbf{x})}\right),$$

where $g(\mathbf{x})$ denotes the geometric mean of the D parts, that is: $g(\mathbf{x}) = \sqrt[D]{\prod_{i=1}^{D} x_i}$.

Definition 4 For the D-part composition $\mathbf{x} = (x_1, x_2, \ldots, x_D)$, the isometric logratio (ilr) transformation associated to the Aitchison-orthonormal basis in $\mathbb{S}^D$, denoted by $\{\mathbf{e}_i\}$ (with $i = 1, \ldots, D-1$), is the mapping from $\mathbb{S}^D$ to $\mathbb{R}^{D-1}$ given by

$$ilr(\mathbf{x}) = \big(\langle \mathbf{x}, \mathbf{e}_1\rangle_a, \langle \mathbf{x}, \mathbf{e}_2\rangle_a, \ldots, \langle \mathbf{x}, \mathbf{e}_{D-1}\rangle_a\big),$$

where $\langle\cdot,\cdot\rangle_a$ is the Aitchison inner product in $\mathbb{S}^D$: $\langle \mathbf{x}, \mathbf{y}\rangle_a = \frac{1}{2D}\sum_{i=1}^{D}\sum_{j=1}^{D}\left(\ln\frac{x_i}{x_j}\ln\frac{y_i}{y_j}\right)$.

All the *ilr*-transformations are characterized by two important features: i) they reduce the problem dimension, since they map a D-part composition into a vector in $\mathbb{R}^{D-1}$; ii) they preserve both the distances and the angles: in the simplex, the Aitchison distance of two compositions is equal to the distance of the corresponding *ilr*-transformed vectors in $\mathbb{R}^{D-1}$ (see [7] for further details). As stated in the definition, an ilr-transformation is based on a given Aitchison-orthonormal basis in $\mathbb{S}^D$. The transformation that was originally defined in [3] maps a D-part composition $\mathbf{x}$ in a vector $\mathbf{z} = (z_1, z_2, \ldots, z_{D-1})$ in $\mathbb{R}^{D-1}$ with components given by

$$z_i = ilr_i(\mathbf{x}) = \sqrt{\frac{D-i}{D-i+1}}\ln\frac{x_i}{\sqrt[D-i]{\prod_{j=i+1}^{D} x_j}} \qquad i = 1, 2, \ldots, D-1. \tag{2}$$

There is a link between the *ilr* and the *clr* transformation, since it holds that

$$\mathbf{z} = ilr(\mathbf{x}) = V^T\ln(\mathbf{x}) = V^T clr(\mathbf{x}), \tag{3}$$

where V is a $D \times (D-1)$ matrix such that $V^TV = I_{D-1}$, being I_D the identity matrix of dimension D (see [4] for details).

2 Regression Models with Compositional Explanatory Variables and Structural Zeros

In this paper, we will focus on regression models with D explanatory variables $X_1, X_2, \ldots, X_D$ that sum to a constant (generally equal to one). The main issue of these models is that the regressors cannot be directly used since compositional data are by definition singular: the constraint about their sum provides the linear dependency of the regressors and a singular covariance matrix. The standard approach in these cases is to apply a *ilr*-transformation to the original explanatory variables and to consider the corresponding *ilr*-transformed variables as new regressors (see [5]). In this way, the linear dependence of the compositional regressors can be discarded, and usual regression methods can be used. Moreover, the suggestion is to fit D models, each one with a different first regressor, since only the first coefficient β_1 can be interpreted (refer to [5] for details). Even if the procedure "to permute the indices and construct D regression models, each time with a different component first, can be undesirable"—as stated in [9]—in this paper, we basically follow this procedure described in [5], adapting it to the situation with structural zeros. Some modifications are needed, because in presence of zeros, the procedure in [5] cannot be directly applied, since no logratio transformation can be carried out. It is well known that in compositional data analysis an important role is played by the different kinds of zeros in the data. There can be *rounded zeros*, *count zeros*, and *structural zeros* (see [7] for further details about this classification). While some procedures exist to deal with the first two kinds of zeros, the structural zeros are quite complicated to be managed, and there are only a few procedure to take them into account. To our best knowledge, there are three different approaches to address this issue: the first one is due to [1], while the other two are described in [9]. Before the explanation of the methods, a description of the setting is needed.

Given a compositional dataset—with positive or structural zero parts only—for any composition $\mathbf{x}$, we will denote by $M \subset \{1, 2, \dots, D\}$ the set of the indices of structural zeros of $\mathbf{x}$ and by $M^c = \{i_1, i_2, \dots, i_m\} = \{1, 2, \dots, D\} \setminus M$ its complementary set, consisting of the indices related to the nonzero components. The subcomposition with all positive parts is denoted by $\mathbf{x}_{M^c} = \mathscr{C}(x_{i_1}, x_{i_2}, \dots, x_{i_m})$. The set of all possible structural 0-patterns M is denoted by $\mathscr{M}$, and for each $M \in \mathscr{M}$, let

$$d_M(\mathbf{x}) = \begin{cases} 1 & \text{if the set of indices of structural 0s of } \mathbf{x} \text{ is } M \\ 0 & \text{otherwise} \end{cases}$$

denotes the dummy variable that assumes value 1 if the 0-pattern of $\mathbf{x}$ is M. Clearly, if all the parts of the composition $\mathbf{x}$ are positive, $d_M(\mathbf{x}) = 1$ if and only if $M = \emptyset$.

(i) ***The replacement approach.*** This strategy is the first method proposed in the literature, and it is the most intuitive one. The idea is to take all the zero values and replace them with a small fixed value. This approach can be easily implemented, but it has the relevant drawbacks due to the arbitrary nature of the replaced values and the inconsistency in case of structural zeros, as they are *true* zeros.

(ii) ***The conditioning approach.*** It consists in treating the observations with different structural 0-patterns as different subgroups within the data, so that the regression coefficients are modeled conditionally on the 0-patterns. This method requires to compute for each compositional observation $\mathbf{x}$ with at least one null part, the *ilr*-transformation of the corresponding subcomposition with nonzero parts $\mathbf{x}_{M^c}$ (obtained by removing the zero parts) and to use it to fit the model for the corresponding 0-pattern. In other words, we split the observations in groups depending on the 0-pattern M. For each group, we compute the subcompositions with nonzero parts $\mathbf{x}_{M^c}$, and we fit a regression model, following the procedure in [5]. The regression coefficients obtained by this method are different for each structural 0-pattern and hence only estimated by using observations with that particular 0-pattern. Denoting with $\boldsymbol{\beta}|_{M^c}$ the parameter vector associated to the 0-pattern M, and with $\mathbf{b}|_{M^c} = ilr^{-1}(\boldsymbol{\beta}|_{M^c})$, all the regression models obtained by the conditioning approach (each one depending on a different 0-pattern M) can be summarized as

$$\mathbb{E}(Y|\mathbf{X}) = \sum_{M\in\mathcal{M}} d_M(\mathbf{x})\langle \mathbf{b}|_{M^c}, \mathbf{x}_{M^c}\rangle_a = \sum_{M\in\mathcal{M}} d_M(\mathbf{x})\langle ilr(\mathbf{b}|_{M^c}), ilr(\mathbf{x}_{M^c})\rangle.$$

(iii) ***The projection approach.*** It is more parsimonious than the conditioning one, since the regression parameters are shared across all the different 0-patterns. Following this method, a *generalized isometric logratio transformation* from the simplex $\mathbb{S}^D$ to $\mathbb{R}^{D-1}$ is proposed as an extension of the usual one. This new transformation can be applied also to a compositions with zero parts, since the logratios are calculated using the projections onto the orthogonal complement of the structural 0 parts. In other words, given a D-part composition $\mathbf{x}$, we consider the subcomposition $\mathbf{x}_{M^c}$ and its *clr*-tranformation $clr(\mathbf{x}_{M^c})$. This vector can be projected (by the projection operator P_M) for defining a generalization of the *clr*-transform, labelled $\widetilde{clr}$, having the components defined as

$$\widetilde{clr}(\mathbf{x})_i = P_M clr(\mathbf{x})_i = \begin{cases} clr(\mathbf{x}_{M^c})_i & \text{if } i \in M^c \\ 0 & \text{otherwise} \end{cases} \quad \text{with } i = 1, 2, ..., D.$$

By using the link between *clr* and *ilr* transformations reported in formula (3), we obtain a generalized *ilr*-transformation $\widetilde{ilr}(\mathbf{x}) = V^T\widetilde{clr}(\mathbf{x})$. At this point, the procedure of [5] can be applied, by directly using the components of $\widetilde{ilr}(\mathbf{x})$ as regressors. The model obtained by the projection approach can be written as

$$\mathbb{E}(Y|\mathbf{X}) = \langle \boldsymbol{\beta}, \widetilde{ilr}(\mathbf{x})\rangle = \sum_{i=1}^{D-1} \beta_i \widetilde{ilr}(\mathbf{x})_i = \langle clr(\mathbf{b}), \widetilde{clr}(\mathbf{x})\rangle = \sum_{M\in\mathcal{M}} d_M(\mathbf{x})\langle \mathbf{b}_{M^c}, \mathbf{x}_{M^c}\rangle_a,$$

where $\boldsymbol{\beta}$ is the coefficient vector of the model and $\mathbf{b} = ilr^{-1}(\boldsymbol{\beta})$.

3 Examples of Applications Through an Artificial Dataset

The examples described in this section are based on a simulated dataset, from a Dirichlet distribution with $K = 4$ and parameter vector $\boldsymbol{\alpha} = (0.4, 0.3, 0.2, 0.1)$. Each observation is a D-part composition, with $D = 4$. To simulate structural zeros, the $n = 60$ observations (stored in the matrix $\boldsymbol{X}$) have been split into three groups, each one containing 20 of them: in the first group, only the first part has been set equal to zero, while in the second one, the first two parts have been set to zero. No replacements have been made for the observations in the third group. After the application of the closure operator, the result is that the compositions in the first group share the same 0-pattern $M_1 = \{1\}$, those in the second group are characterized by the 0-pattern $M_2 = \{1, 2\}$, while the remaining ones have all positive parts ($M_3 = \emptyset$).

The values of the response variable Y have been computed by $Y = \boldsymbol{X\beta} + \epsilon$, where the erratic terms $\epsilon_i \sim N(0, 1)$, (for $i = 1, 2, ..., 60$) and the parameter vector $\boldsymbol{\beta} = (40, 30, 20, 10)$. Figure 1 shows the scatterplots of the response variable and each one of the four parts: X_1, X_2, X_3, and X_4. The structural zeros are denoted in red. The three approaches presented in the previous sections have been applied to the dataset, and the corresponding models have been fitted.

The first model (Model1) has been obtained by replacing the zeros by the fixed value $\delta = 0.001$. The new obtained matrix (containing no more zero values) has been considered as input matrix for the procedure in [5]. Table 1 reports the

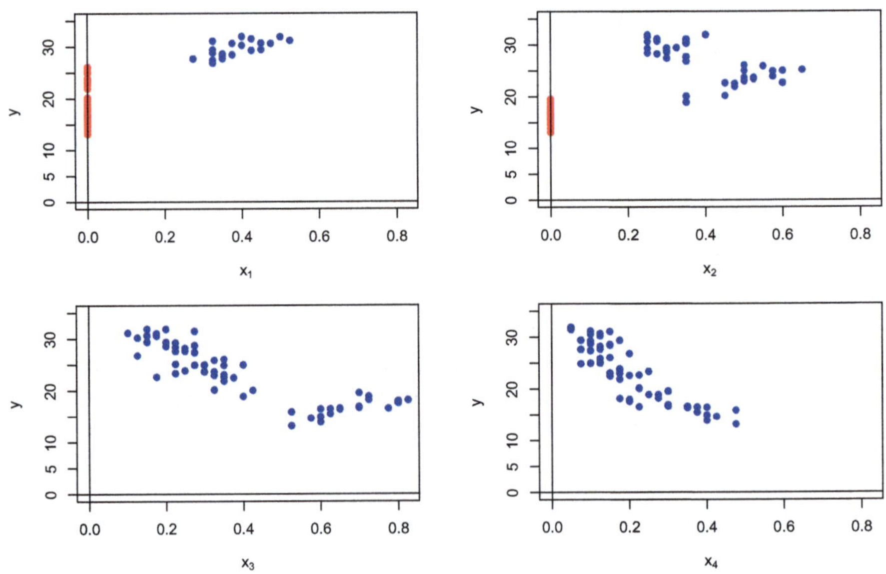

Fig. 1 Scatterplots of the response variable Y versus the four parts (X_1, X_2, X_3, and X_4): the points (in red) with abscissa equal to zero are in only two plots (in the upper panel)

Table 1 Output for the Model1 obtained by replacement of the 0s

	Estimate	Std. error	t-value	$Pr(>\|t\|)$
(Intercept)	27.14518	0.35382	76.721	<2e-16***
ilr(replaced_X_1)	1.15484	0.07595	15.206	<2e-16***
ilr(replaced_X_2)	1.02831	0.07455	13.794	<2e-16***
ilr(replaced_X_3)	0.43698	0.44145	0.990	0.326
ilr(replaced_X_4)	−2.62012	0.44757	−5.854	2.62e-07***

Res. Std. error = 1.34 on 55 dof; $R^2 = 0.94$; $R^2_{adj} = 0.94$; F-stat = 327 on 3.56 DF; p-value <2.2e-16

Table 2 Output for the Model2 obtained by the conditioning approach for the different 0-patterns

0-pattern = M_1	Estimate	Std. error	t-value	$Pr(>\|t\|)$
(Intercept)	18.9745	0.6852	27.690	1.39e-15***
$ilr(X_1)$	5.4132	0.9057	5.977	1.50e-05***
$ilr(X_2)$	−1.3307	0.9426	−1.412	0.176
$ilr(X_3)$	−4.0825	0.7622	−5.356	5.24e-05***
0-pattern = M_2	Estimate	Std. error	t-value	$Pr(>\|t\|)$
(Intercept)	14.5619	0.5179	28.12	2.52e-16***
$ilr(X_1)$	3.9094	0.8766	4.46	0.000303***
$ilr(X_2)$	−3.9094	0.8766	−4.46	0.000303***
0-pattern = M_3	Estimate	Std. error	t-value	$Pr(>\|t\|)$
(Intercept)	26.4729	0.8360	31.664	7.33e-16***
$ilr(X_1)$	3.6714	1.1462	3.203	0.00554**
$ilr(X_2)$	−0.2186	1.4662	−0.149	0.88333
$ilr(X_3)$	−1.6087	0.8779	−1.833	0.08554
$ilr(X_4)$	−1.8440	0.7470	−2.469	0.02521*

parameter estimates, their standard errors, and the values of the t-statistic with the corresponding p-value. As explained in [5], the output summarizes four different regression models. For this approach, the only not significant variable is X_3. For all the approaches, the value of the estimate indicates how much the response variable Y changes in average by a unit change of the logratio in formula (2) related to the compositional part of interest. Then, according to Model1, a unit increase of the logratio related to X_4 produces a decrease for Y, while an unit increase of the logratios related to the other regressors increases (in different ways) the value of Y.

Table 2 shows the summary of the Model2 obtained by the conditioning approach. As aforementioned explained, this model consists of three different "submodels," one for each 0-pattern (M_1, M_2, M_3) in the dataset. It is worth noting that the three submodels have different numbers of explanatory variables, depending on how many non-null parts the submodel has. For the 0-patterns M_1 and M_3, the variable X_2 is not significant. Moreover, for the 0-pattern M_3, the variable X_3 is significant only at level 0.1. For all the 0-patterns, X_1 has a positive effect on Y and X_2 a negative one.

Table 3 Output for the Model3 obtained by the projection approach

	Estimate	Std. error	t-value	$Pr(>\|t\|)$
(Intercept)	17.4809	0.6195	28.216	<2e-16***
$\widetilde{ilr}(X_1)$	9.0013	0.9324	9.654	1.61e-13***
$\widetilde{ilr}(X_2)$	4.7245	0.9902	4.771	1.35e-05***
$\widetilde{ilr}(X_3)$	−6.5122	0.8411	−7.742	2.06e-10***
$\widetilde{ilr}(X_4)$	−7.2136	0.8402	−8.585	8.5e-12***

Res. Std. error $= 2.25$ on 56 dof; $R^2 = 0.84$; $R^2_{adj} = 0.84$; F-stat $= 104.2$ on 3.56 DF; p-value <2.2e-16

Finally, the projection approach has been applied, and the output is stored in Table 3: here, the effects of X_1 and X_4 have the same signs as those of the two previous models, while this does not hold for X_2 and X_3.

4 Conclusions

In this paper, three approaches dealing with structural zeros for compositional regressors have been described and applied to an artificial dataset: the first one consists of the replacement of the null values, while the other two require more sophisticated procedures. The results of these three approaches show relevant differences (in terms of coefficient values and significance of the variables). An interesting future research project should be a deeper investigation into the characteristics of these approaches. Moreover, a more complete comparative study on compositional regressors with structural zeros deserves to be performed by including also different methods, such as the α-transformation approach (see [2] for further details).

Acknowledgments This work was supported by group GNAMPA of INdAM, by the founding of the project “Modelli di regressione e modelli grafici probabilistici per dati composizionali”: “INdAM-GNAMPA Project,” grant code [CUP E53C23001670001].

References

1. Aitchison, J.: The Statistical Analysis of Compositional Data. Chapman and Hall, London, New York (1986)
2. Clarotto, L., Allard, D.,Menafoglio, A.: A new class of α-transformations for the spatial analysis of compositional data. Spat. Stat. **47**, 100570 (2022)
3. Egozcue,J.J., Pawlowsky-Glahn, V., Mateu-Figueras, G., Barcelo-Vidal, C.: Isometric logratio transformations for compositional data analysis. Math. Geol. **35**, 279–300 (2003)
4. Egozcue,J.J., Barcelo-Vidal, C., Martin-Fernandez, J.A., Jarauta-Bragulat, E., Diaz-Barrero, J.L., Mateu-Figueras, G.: Elements of simplicial linear algebra and geometry. In: Pawlowsky-Glahn, V., Buccianti, A. (eds.) Compositional Data Analysis: Theory and Applications. Wiley, Chichester (2011)

5. Hron, K., Filzmoser, P., Thompson, K.: Linear regression with compositional explanatory variables. J. Appl. Stat. **39**(5), 1115–1128 (2012)
6. Pawlowsky-Glahn, V., Egozcue, J.J.: Geometric approach to statistical analysis on the simplex. Stoch. Environ. Res. Risk Assess. **15**, 384–398 (2001)
7. Pawlowsky-Glahn, V., Egozcue,J.J., Tolosana-Del Gado, R.: Modeling and Analysis of Compositional Data. John Wiley & Sons, UK (2015)
8. Filzmoser, P., Hron, K., Templ, M.: Applied Compositional Data Analysis. Springer, New York (2018)
9. Verbelen, R., Antonio, K., Claesken, G.: Unravelling the predictive power of telematics data in car insurance pricing. J. R. Stat. Soc. C: Appl. Stat. **67**(5) 1275–1304 (2018)

Multidimensional Robinson Dissimilarities

Pascal Préa

Abstract A dissimilarity d on a set X is said to be *Robinson* (Robinson, Am Antiquity 16:293–301, 1951) if there exists a total order, said *compatible*, on X such that $x < y < z \implies d(x,z) \geq \max\{d(x,y), d(y,z)\}$. Roughly speaking, d is Robinson if the points of X can be represented on a line.

We define k-dimensional Robinson dissimilarities, which generalize the possibility, for a metric set (X, d), to be embedded into a k-dimensional Euclidean space. This generalization is more flexible than the classical embedding, and we show that every dissimilarity on an n-set X is $(\log n)$-dimensional Robinson. We give an $O(n^3)$ algorithm which builds such an embedding. This algorithm is based on an incremental algorithm to recognize Robinson dissimilarities which, after an initial sorting step in $O(n^2 \log n)$, runs in $O(n^2)$.

1 Introduction

Given a finite set X, a *dissimilarity* on X is a symmetrical function $X \times X \longmapsto \mathbb{R}^+$ such that $\forall x \in X, d(x,x) = 0$ (we say that (X, d) is a dissimilarity space). Dissimilarities generalize distances (a distance is dissimilarity with the triangular inequality).

Given a dissimilarity on a set X, a fundamental problem is to derive "geometrical" properties of X from d or to characterize dissimilarities from which such properties can be obtained. For instance, Robinson dissimilarities [14] correspond to points on a line: A dissimilarity d on a set X is *Robinson* if there exists a total order on X, which is said *compatible*, such that

P. Préa (✉)
Aix Marseille Univ, CNRS, Université de Toulon, LIS, Marseille, France

École Centrale Méditerranée, Marseille, France
e-mail: pascal.prea@lis-lab.fr

G. Giordano et al. (eds.), *Statistical Models and Learning Methods for Complex Data*, Studies in Classification, Data Analysis, and Knowledge Organization,
https://doi.org/10.1007/978-3-031-84702-8_17

$$\forall x, y, z \in X, x < y < z \implies d(x, z) \geq \max\{d(x, y), d(y, z)\}$$

In this case, (X, d) is a called a *Robinson space.* These dissimilarities were invented to solve seriation problems in archaeology, and they are now a classical tool for seriation problems in any field (see [13] for examples of applications). Robinson dissimilarities are a generalization of ultrametrics, and they are in one-to-one correspondence with pyramids [9, 10], the standard model with overlapping classes, which is a generalization of hierarchies. Moreover, they play an important role to recognize tractable cases for the travelling salesperson problem [7].

In this paper, we generalize Robinson dissimilarities to *k(-dimensional)-Robinson* dissimilarities, which represent the fact for X to be embedded into a k-dimensional space. This embedding is less strict than the usual Euclidean embedding. We show that, if d is a dissimilarity on a set X with $|X| = n$, then d is $(O(\log n))$-Robinson, and we give a $O(n^3)$ algorithm which builds such an embedding. For an ordinary dissimilarity, this embedding is a tractable alternative to Robinson approximation, which is an NP-hard problem [2, 8]. This algorithm is based on an $O(n^2 \log n)$ incremental algorithm to recognize Robinson dissimilarities which is presented in Sect. 3. There already exist many algorithms to recognize Robinson dissimilarities (see [6] for references), but this algorithm is the first incremental one for that task.

2 Multidimensional Robinson Dissimilarities

Let (X, d) a dissimilarity space and $k \in \mathbb{N}$. We say that (X, d) is *k-Robinson* if there exist k orders $<_1, <_2, \ldots, <_k$ such that

$$\forall x, y, z, t \in X, (\forall 1 \leq i \leq k,\ y, z \in [x, t]_{<_i}) \implies d(y, z) \leq d(x, t)$$

where, for $x, y \in X$ and a order $<$ on X, $[x, y]_<$ is the set $\{t \in X : x \leq t \leq y\}$.

We say that (X, d) is *k-quasi-Robinson* if there exist k orders $<_1, <_2, \ldots, <_k$ such that

$$\forall x, y, z \in X, (\forall 1 \leq i \leq k,\ y \in [x, z]_{<_i}) \implies d(x, z) \geq \max\{d(x, y), d(y, z)\}$$

If $k = 1$, it is equivalent for a dissimilarity space to be Robinson or 1-quasi-Robinson. For $k \geq 2$, if (X, d) is k-Robinson, then (X, d) is k-quasi-Robinson, but the converse is false (see Fig. 1). Notice in addition that, if (X, d) is k-Robinson (resp. k-quasi-Robinson), then (X, d) is $k + 1$-Robinson (resp. $k + 1$-quasi-Robinson). The smallest k such that (X, d) is k-Robinson is the *Robinson dimension* of (X, d).

If a metric space (X, d) can be embedded into a $\mathbb{R}^k$, then (X, d) is k-Robinson. But the Robinson dimension of (X, d) is generally smaller. For instance, if $|X| = n$ and d is the constant dissimilarity, then its Robinson dimension is 1 (the space

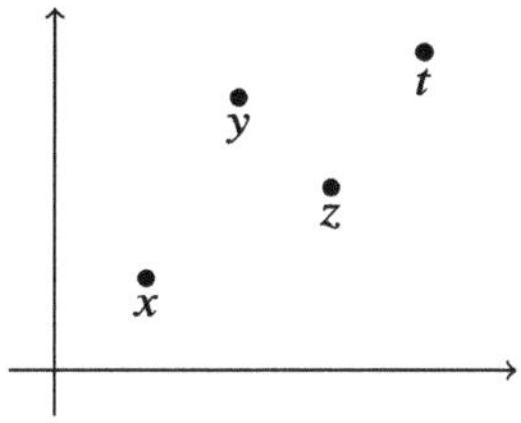

Fig. 1 A set X with four points x, y, z, t. If (X, d) is 2-quasi-Robinson with the two orders represented by the two axis, then no condition is imposed on $d(y, z)$, and we can set $d(y, z) > d(x, t)$. If (X, d) is 2-Robinson (with the same orders), then $d(y, z) \leq d(x, t)$

(X, d) is Robinson) although it needs an $(n-1)$-dimensional Euclidean space to be embedded. Moreover, we have:

Proposition 1 *The Robinson dimension of a dissimilarity space* (X, d) *with* $|X| = n$ *is* $\leq \lceil \log_2 \lceil \frac{n}{3} \rceil \rceil + 1$.

Proof The proof relies on the following elementary observation:

Claim Let $Y \cup Z$ be a bipartition of a set X, σ be permutation of Y, π be a permutation of Z, and $x, y, z \in X$. If $y \in [x, z]_{\sigma\pi} \cap [x, z]_{\sigma^{-1}\pi^{-1}}$, then $x, y, z \in Y$ or $x, y, z \in Z$.

Suppose that X can be partioned into $X_1 \cup X_2 \cup \ldots \cup X_p$ where each (X_i, d) is Robinson and admits σ_i as compatible order. We suppose, with no loss of generality, that $p = 2^k$ (an empty set is Robinson). We consider the following permutations of X:

$\xi_0 = \sigma_1\sigma_2 \ldots \sigma_{p-1}\sigma_p$
$\xi_1 = (\sigma_1\sigma_2 \ldots \sigma_{2^{k-1}})^{-1}(\sigma_{2^{k-1}+1} \cdots \sigma_p)^{-1}$
$\xi_2 = (\sigma_1 \ldots \sigma_{2^{k-2}})^{-1}(\sigma_{2^{k-2}+1} \cdots \sigma_{2^{k-1}})^{-1}(\sigma_{2^{k-1}+1} \cdots \sigma_{3.2^{k-2}})^{-1}$
$(\sigma_{3.2^{k-2}+1} \cdots \sigma_p)^{-1}$
$\ldots$
$\xi_i = (\sigma_1 \ldots \sigma_{2^{k-i}})^{-1}(\sigma_{2^{k-i}+1} \cdots \sigma_{2.2^{k-i}})^{-1} \ldots (\sigma_{2^i-2.2^{k-i}+1} \cdots \sigma_{2^i-1.2^{k-i}})^{-1}$
$(\sigma_{2^i-1.2^{k-i}+1} \cdots \sigma_p)^{-1}$
$\ldots$
$\xi_k = \sigma_1^{-1}\sigma_2^{-1} \ldots \sigma_{p-1}^{-1}\sigma_p^{-1}$

Let $x, y, z \in X$. By the claim, if y is between x and z for ξ_0 and ξ_1, then either $x, y, z \in X_1 \cup X_2 \cup \ldots \cup X_{p/2}$ or $x, y, z \in X_{p/2+1} \cup \ldots \cup X_{p-1} \cup X_p$. When $x, y, z \in X_1 \cup X_2 \cup \ldots \cup X_{p/2}$, if y is between x and z for ξ_0 and ξ_2, then either $x, y, z \in X_1 \cup X_2 \cup \ldots \cup X_{p/4}$ or $x, y, z \in X_{p/4+1} \cup \ldots \cup X_{p/2}, \ldots$

So, if y is between x and z for $\xi_0, \xi_1, \ldots, \xi_k$, then x, y, and z are all in the same X_i. As each X_i is Robinson, X is $k+1$-Robinson. Since every 3-set is Robinson, the result follows.

From Proposition 1, it is possible to represent a dissimilarity space into a low-dimensional space. In Fig. 2, we have represented the dissimilarity of Table 1 into a three-dimensional space.

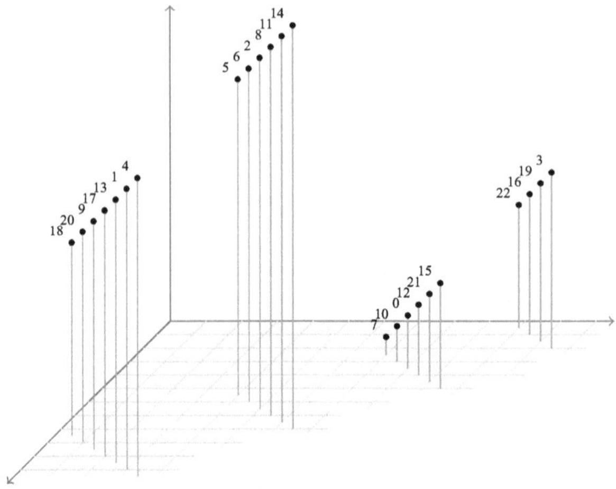

Fig. 2 A three-dimensional representation of Giraudoux's writings, compatible with the distance of Table 1. The different novels and dramas are denoted by their line number in Table 1. This representation is built from a decomposition of this dataset into the four Robinson subsets $\{18, 20, 9, 17, 13, 1, 4\}$, $\{5, 6, 2, 8, 11, 14\}$, $\{7, 10, 0, 12, 21, 15\}$, and $\{22, 16, 19, 3\}$ (each subset is given in compatible order)

The following algorithm 1 is based on the same idea that the proof of Proposition 1 and returns an approximate value for the Robinson dimension of a dissimilarity space.

Algorithm 1: Approximate-Robinson-dimension

Input: (X, d), a dissimilarity space
Output: An upper bound on the Robinson dimension of (X, d)
begin
  $X' \leftarrow X$; $k \leftarrow 0$;
  SORT-LINES(X, d) ;
  while $X' \neq \emptyset$ **do**
    $S \leftarrow$ MAXIMAL-ROBINSON-SUBSPACE(X', d) ;
    $X' \leftarrow X' \setminus S$;
    $k \leftarrow k + 1$;
  return $\lceil \log_2 k \rceil + 1$;

The function SORT-LINES(X, d), for every $x \in X$, sorts the points of X by increasing values of their distance from x. This function runs in $O(n^2 \log n)$ where $n = |X|$. The function MAXIMAL-ROBINSON-SUBSPACE returns a subset S of X', maximal for inclusion and such that (S, d) is Robinson. This can be easily implemented by a greedy algorithm. We will see in Sect. 3 that, after SORT-LINES, such a greedy version of MAXIMAL-ROBINSON-SUBSPACE runs in $O(|X'|^2)$.

So, as there is at most $n/3$ iterations of the **while** loop, Algorithm 1 runs in $O(n^3)$.

Table 1 A distance between the main writings (novels and dramas) of Jean Giraudoux [4]. This distance is based on the difference of vocabulary between these different writings [5]

0	14	19	24	25	25	26	29	31	35	27	33	28	24	34	35	22	25	36	23	37	29	38	0:	Provinciales
	0	7	16	16	19	18	17	22	29	20	28	21	17	34	34	21	20	34	23	34	28	34	1:	L'École des indifferents
		0	26	32	29	26	33	27	34	31	31	31	20	35	40	20	23	38	22	37	40	40	2:	Simon le pathétique
			0	38	34	40	46	34	38	38	36	36	23	41	36	20	26	39	19	37	41	46	3:	Suzanne et le Pacifique
				0	21	21	35	26	42	37	37	29	20	38	48	22	24	44	21	43	36	46	4:	Siegfried et le Limousin
					0	19	25	33	40	24	40	30	18	44	55	24	23	47	26	45	41	50	5:	Juliette au pays des hommes
						0	22	27	42	25	36	29	18	42	48	23	21	46	24	41	34	42	6:	Bella
							0	30	39	23	38	32	16	40	49	22	21	45	24	40	42	43	7:	Églantine
								0	28	30	26	21	23	28	31	16	22	29	19	32	30	31	8:	Siegfried
									0	35	19	25	28	19	23	17	27	24	18	19	37	25	9:	Amphitryon 38
										0	35	32	18	40	47	23	19	46	22	39	42	43	10:	Les aventures de Jérôme Bardini
											0	28	25	21	19	14	27	23	15	19	30	16	11:	Judith
												0	20	28	34	16	21	29	17	33	26	30	12:	Intermezzo
													0	26	30	19	12	33	20	32	25	28	13:	Combat avec l'ange
														0	21	16	29	22	18	22	32	25	14:	La guerre de Troie n'aura pas lieu
															0	14	28	21	20	20	33	19	15:	Électre
																0	16	14	13	15	15	13	16:	Le Cantique des cantiques
																	0	28	21	27	25	24	17:	Choix des élues
																		0	16	22	29	21	18:	Ondine
																			0	15	16	14	19:	L'Apollon de Bellac
																				0	37	16	20:	Sodome et Gomorrhe
																					0	28	21:	La Folle de Chaillot
																						0	22:	Pour Lucrèce

3 An Incremental Algorithm to Recognize Robinson Dissimilarities

A *PQ-tree* [3] on a set X is a tree with leaf set X which represents a set of permutations on X. The internal nodes are of two types: the *P-nodes* and the *Q-nodes*. The children of a P-node can be ordered in any way, but the children of a Q-node can only be reversed. For instance, the PQ-tree of Fig. 3 represents all the permutations $(\pi, 4, 5, 6, 7)$ and $(7, 6, 5, 4, \pi)$, where π is any permutation on $\{1, 2, 3\}$.

PQ-trees play an important role for Robinson dissimilarities:

Proposition 2 ([11, 13]) *Let (X, d) be a Robinson space, the set of all compatible permutations of (X, d) can be represented by a PQ tree that we denote by $\mathcal{T_P}(X, d)$.*

In order to implement MAXIMAL-ROBINSON-SUBSPACE, we need a function ADD-AND-TEST which takes as entry a dissimilarity space (X, d), a set $S \subset X$ such that (S, d) is Robinson, the PQ-tree $\mathcal{T_P}(S, d)$, and a point $x \in X \setminus S$. ADD-AND-TEST returns the PQ-tree $\mathcal{T_P}(S \cup \{x\}, d)$ (If $(S \cup \{x\}, d)$ is not Robinson, then $\mathcal{T_P}(S \cup \{x\}, d) = \textbf{None}$). The algorithm ADD-AND-TEST can be sketched as follows:

Step 1 Compute the sets $B^S_\delta := B_\delta(x) \cap S$, where $B_\delta(x) := \{t \in X : d(x, t) \leq \delta\}$ is the *ball* of *center* x and *radius* δ .

Step 2 Insert the sets B^S_δ into $\mathcal{T_P}(S, d)$. We get a PQ-tree $\mathcal{T}^x_\mathcal{P}(S, d)$ which represents all permutations σ of $\mathcal{T_P}(S, d)$ such that all the sets B^S_δ are intervals of σ.

Step 3 Add the point x to $\mathcal{T}^x_\mathcal{P}(S, d)$. We get the PQ-tree $\mathcal{T_P}(S \cup \{x\}, d)$. This will be done in two steps:

(a) Consider only the points of S the closest from x.
(b) Consider the other points of S.

If we have already computed the balls $B_\delta(x)$ (this computation, for all points of X, is SORT-LINES(X, d) and takes $O(n^2 \log n)$), then Step 1 takes $O(n)$. Since the sets B^S_δ form an inclusion chain ($B^S_\delta \subseteq B^S_{\delta'}$ when $\delta < \delta'$) and there exist at most n different sets B^S_δ, Step 2 runs in $O(n)$ [12]. The rest of this section will be mainly devoted to Step 3.

Let (X, d) be a Robinson space, a subset B of X is a *block* if B is an interval for all compatible orders of (X, d). Notice that every ball is a block. Given a PQ-tree $\mathcal{T}$ on X and a node α of $\mathcal{T}$, we denote by $X(\alpha)$ the set of the leaves under α.

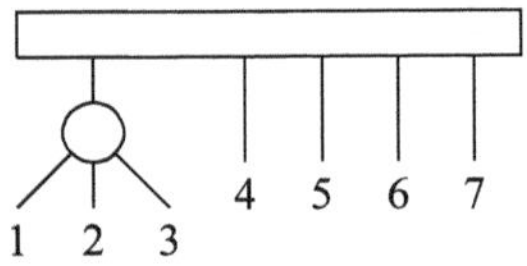

Fig. 3 A PQ-tree on $X = \{1, \ldots, 7\}$. We use the standard notation to represent the P-nodes by circles and the Q-nodes by rectangles

By Proposition 2, we have:

Proposition 3 *If $(S \cup \{x\}, d)$ is Robinson, then each set B^S_δ is a block of (S, d).*

We denote by $B_1(x)$ the smallest ball (centered at x) with more than one element and we set $B^S_1 := B^S_1(x) \setminus \{x\}$ and $\delta := d(x, B^S_1)$. For Step 3-(a), we have to insert x near or inside B^S_1, which is a block.

We now detail this insertion:

(A) We first treat the case where $B^S_1 = S(\alpha)$ for some node α of $\mathcal{T}^x_\mathcal{P}(S, d)$. We denote by γ the parent of α (if α is not the root of $\mathcal{T}^x_\mathcal{P}(S, d)$). This case is divided into the following subcases:

1. diam$(\alpha) < \delta$ or diam$(\alpha) = \delta$ and α is a Q-node:
   We add to $\mathcal{T}^x_\mathcal{P}(S, d)$ a new node α' with two children α and the leaf x. If α was the root of $\mathcal{T}^x_\mathcal{P}(S, d)$, α' is the root of the new tree; otherwise, α' replaces α among the children of γ.
2. α is a P-node and diam$(\alpha) = \delta$:
   We add the leaf x among the children of α.
3. α is a P-node and diam$(\alpha) > \delta$:
   If α has two children β_1 and β_2, then α is transformed into a Q-node with children, in this order, β_1, x, β_2.
4. α is a Q-node (with children, in this order, $\beta_1, \ldots, \beta_p$) and diam$(\alpha) > \delta$:
   a. If there exists exactly one node β_i such that $d(\beta_i, \beta_{i+1}) \geq \delta$, we add the leaf x between β_i and β_{i+1}.
   b. If there exists exactly one node β_i such that $\forall j \neq i, d(\beta_i, \beta_j) = \delta$ (it is impossible that many such nodes β_i exist):
      i. If β_i is a P-node of diameter δ, then the leaf x is added among the children of β_i.
      ii. If diam$(\beta_i) < \delta$ or diam$(\beta_i) = \delta$ and β_i is a Q-node, then we create a new node β with two children β_i and x, and, among the children of α, we replace β_i by β.
   c. If there exist many nodes β_i such that $d(\beta_i, \beta_{i+1}) = \delta$, the leaf x is inserted after one of these β_i.

(B) If B^S_1 is not equal to $S(\alpha)$ for some node α of $\mathcal{T}^x_\mathcal{P}(S, d)$, then there exist a Q-node α with children, in this order, $\beta_1, \beta_2, \ldots, \beta_p$ and a, b with $1 \leq a < b \leq p$ and $(a, b) \neq (1, p)$ such that $B^S_1 = S(\beta_a) \cup S(\beta_{a+1}) \cup \ldots \cup S(\beta_b)$. This case is divided into the following subcases:

1. $d(\beta_a, \beta_b) \leq \delta$:
   The leaf x is inserted between β_{a-1} and β_a or between β_b and β_{b+1}.
2. $d(\beta_a, \beta_b) \geq \delta$ (this case similar to case A-4):

a. If there exists exactly one node β_i ($a \leq i < b$) such that $d(\beta_i, \beta_{i+1}) > \delta$, we add the leaf x between β_i and β_{i+1}. If there exist no such node β_i:
b. If there exists nodes β_i ($a \leq i < b$) such that $d(\beta_i, \beta_{i+1}) = \delta$, then for one of the nodes β_i, we add the leaf x between β_i and β_{i+1}.
c. If there exists nodes β_i ($a \leq i \leq b$) such that $\forall a \leq j \leq b$ such that $j \neq i$, we have $d(\beta_i, \beta_j) = \delta$, then, for one of the nodes β_i:
    i. If β_i is a P-node of diameter δ, then the leaf x is added among the children of β_i.
    ii. If $\mathrm{diam}(\beta_i) < \delta$ or $\mathrm{diam}(\beta_i) = \delta$ and β_i is a Q-node, then we create a new node β with two children β_i and x, and, among the children of α, we replace β_i by β.

Otherwise, $(S \cup \{x\}, d)$ is not Robinson.

Notice that Cases A-4-(c), B-1, B-2-(b), and B-2-(c) are ambiguous (the place where to insert x is not precisely determined) and that some cases are not exclusive: Cases B-2-(b) and B-2-(c) can occur simultaneously, as Cases B-1 and B-2. To fix these ambiguities, we first notice that they only occur when B_1^S is the set of leaf of a Q-node or of some children of a Q-node. So, for every child β_i of α, there exists $j \neq i, i+1$ such that $d(\beta_j, \beta_i) \neq d(\beta_j, \beta_{j+1})$, and we have $d(\beta_j, x) \neq d(\beta_j, \beta_i)$ or $d(\beta_j, x) \neq d(\beta_j, \beta_{i+1})$. Thus, we can determine where to place x and we get a PQ-tree $\mathcal{T}'_\mathcal{P}$.

Notice that we have consider only a few values $d(x, t)$ for $t \in X$, and so, if (X, d) is Robinson, $\mathcal{T}'_\mathcal{P}$ represents a superset of the compatible permutations of (X, d). In order to consider the other values of d, we insert the balls $B_\delta(x)$ into $\mathcal{T}'_\mathcal{P}$. If (X, d) is Robinson, we get the PQ-tree $\mathcal{T}_\mathcal{P}(X, d)$; otherwise, we get a contradiction from which we can conclude that $(S \cup \{x\}, d)$ is not Robinson. There are at most n different balls $B_\delta(x)$, and they form a chain for inclusion, so, as for Step 2, the insertion of all these balls can be done in $O(n)$ [12], and we have:

Proposition 4 *If, for all $x \in X$, the values $d(x, t)$ are given in increasing order, it is possible to incrementally recognize Robinson dissimilarities in $O(n^2)$.*

Proposition 5 *It is possible to incrementally recognize Robinson dissimilarities in $O(n^2 \log n)$.*

4 Conclusion

In this paper, we define the notions of k-Robinson dissimilarities and of Robinson dimension, which generalize both the Robinson dissimilarities and the embedment into an Euclidean space, and we give a $O(n^3)$ algorithm which computes an upper bound for the Robinson dimension of a dissimilarity space (X, d) with $|X| = n$. This algorithm is based on a $O(n^2 \log n)$ incremental algorithm to recognize Robinson dissimilarities. This algorithm is first incremental one for that task.

Nevertheless, many questions remain open:

- Are these algorithms optimal? Or is it possible to improve them? What is the complexity of determining the Robinson dimension?
- Is the upper bound $\lceil \log_2 \lceil \frac{n}{3} \rceil \rceil + 1$ for the Robinson dimension tight?
- Is it possible to get a structural characterization of the k-Robinson dissimilarities, at least for $k = 2$ or $k = 3$? What kind of classes yield the balls, the two-balls, and the clusters of these dissimilarities[1]?
- Is it possible to get a polynomial-time approximation scheme for TSP on k-Robinson dissimilarities for small values of k, as it is the case for the Euclidean TSP [1]?

References

1. Arora, S.: Polynomial-time approximation schemes for euclidean TSP and other geometric problems. J. ACM **45**, 753–782 (1998)
2. Barthélemy, J.-P., Brucker, F.: NP-hard approximation problems in overlapping clustering. J. Classif. **18**, 159–183 (2001)
3. Booth, K.S., Lueker, G.S.: Testing for the consecutive ones property, interval graphs and graph planarity using PQ-tree algorithm. J. Comput. Syst. Sci. **13**, 335–379 (1976)
4. Brucker, F.: Classification en classes empiétantes, modèles et algorithmes. Habilitation Thesis. Université de Metz (2008)
5. Brunet, É: Une mesure de la distance intertextuelle: la connexion lexicale. Revue Informatique et Statistique dans les Sciences Humaines **24**, 82–116 (1988)
6. Carmona, M., Chepoi, V., Naves, G., Préa, P.: Modules in Robinson spaces. SIAM J. Discr. Math. **38**, 190–224 (2024)
7. Çela, E., Deineko, V., Woeninger, H.J.: Recognising permuted Demidenko matrices. Oper. Res. Lett. **51**, 494–500 (2023)
8. Chepoi, V., Fichet, B., Seston, M.: Seriation in the presence of errors: NP-hardness of $l\infty$-fitting Robinson structures to dissimilarity matrices. J. Classif. **26**, 279–296 (2009)
9. Diday, E.: Orders and overlapping clusters by pyramids. In: de Leeuw, J., Heiser, W., Meulman, J., Critchley, F. (eds.) Multidimensionnal Data Analysis, pp. 201–234. DSWO (1986)
10. Durand, C., Fichet, B.: One-to-one correspondences in pyramidal representation: an unified approach. In: Bock, H.H. (ed.) Classification and Related Methods of Data Analysis, pp. 85–90. North-Holland, Amsterdam (1988)
11. Mirkin, B., Rodin, S.: Graphs and Genes. Springer, Berlin (1984)
12. Naves, G.: Efficient insertion of a chain into a PQ-tree. Personal Communication (2023)
13. Préa, P., Fortin, D.: An optimal algorithm to recognize Robinsonian dissimilarities. J. Classif. **31**, 351–385 (2014)
14. Robinson, W.S.: A method for chronologically ordering archaeological deposits. Am. Antiquity **16**, 293–301 (1951)

[1] Given a dissimilarity space (X, d), $x, y \in X$ and $\delta \in \mathbb{R}^+$, the *two-ball* $B(x, y)$ is the set $B_{d(x,y)}(x) \cap B_{d(x,y)}(y)$ and a *cluster* of *diameter* δ is a maximal set $S \subset X$ such that, $\forall t, t' \in S$, $d(t, t') \leq \delta$. If (x, d) is Robinson, then the set of all balls, the set of all two-balls and the set of all clusters are interval hypergraphs.

Composite Selection Criteria for the Number of Components of a Finite Mixture Model

Monia Ranalli and Roberto Rocci

Abstract Finite mixture models are frequently used to capture the underlying clustering structure of the data. They provide a model-based approach to unsupervised classification, where the mixture components are usually interpreted as clusters. However, the number of clusters is frequently unknown and should be learned from the data. When the finite mixture model is rather complex and then estimated by a composite likelihood approach, classical likelihood-based selection criteria cannot be used and should be adapted. In this paper, we compare different methods to choose such number of components when the composite likelihood is adopted. The comparison is made with particular reference to the case of ordinal data.

1 Introduction

The finite mixture of Gaussians (FMG) is being increasingly used as a model-based clustering [7] for continuous data $\mathbf{y}$, whose the pdf is given by

$$f(\mathbf{y}; \boldsymbol{\theta}) = \sum_{g=1}^{G} p_g \phi\left(\mathbf{y}; \boldsymbol{\mu}_g, \boldsymbol{\Sigma}_g\right) \tag{1}$$

where $\phi\left(\mathbf{y}; \boldsymbol{\mu}_g, \boldsymbol{\Sigma}_g\right)$ is the P-variate Gaussian density with mean $\boldsymbol{\mu}_g$ and covariance matrix $\boldsymbol{\Sigma}_g$ and $p_1, p_2, \ldots, p_G$ is the set of positive weights that sum to 1. From a clustering point of view, each component, i.e., a P-variate Gaussian density with component-specific parameters, is usually seen as a cluster, while the corresponding weight represents the unconditional group membership probability.

M. Ranalli (✉) · R. Rocci
Department of Statistics, Sapienza University, Rome, Italy
e-mail: monia.ranalli@uniroma1.it; roberto.rocci@uniroma1.it

G. Giordano et al. (eds.), *Statistical Models and Learning Methods for Complex Data*, Studies in Classification, Data Analysis, and Knowledge Organization,
https://doi.org/10.1007/978-3-031-84702-8_18

FMG works on continuous variables, but frequently data are ordinal. Adopting the underlying response variable approach (URV, [9]), the observed ordinal variables are considered partial manifestation of some continuous latent variables distributed as a FMG (see, e.g., [4, 8, 10–12]). However, the URV approach implies the presence of multidimensional integrals into the likelihood, making its computation time-consuming, and in some cases infeasible, as the number of variables increases. To overcome such a computational issue, a composite likelihood approach is used. It consists in considering some marginal and/or conditional distributions, instead of the full joint, to build the objective function to be maximized.

In the estimation procedure, we assume that the number of mixture components is fixed; in practice, it is usually unknown, and thus, it should be selected through the data. In the sequel, several criteria are considered and compared theoretically and empirically through a simulation study. The reference model considered is the one introduced by [10] and summarized in Sect. 2 that is a finite mixture model for the unsupervised classification of ordinal data. In Sect. 3, we review the main methods to choose the number of components compared by a simulation study in Sect. 4. In the last section, we draw some conclusions.

2 Clustering Ordinal Data

Let $x_1, \ldots, x_P$ be ordinal variables and $c_i = 1, \ldots, C_i$ the associated categories for $i = 1, \ldots, P$. There are $R = \prod_{i=1}^{P} C_i$ possible response patterns $\mathbf{x}_r = (x_1 = c_1, \ldots, x_P = c_P)$, with $r = 1, \ldots, R$. The ordinal variables are generated by thresholding $\mathbf{y}$ that is a multivariate continuous random variable distributed as a FMG (1). The link between $\mathbf{x}$ and $\mathbf{y}$ is expressed by a threshold model defined as

$$x_i = c_i \Leftrightarrow \gamma_{c_i-1}^{(i)} \leq y_i < \gamma_{c_i}^{(i)}. \tag{2}$$

Let $\boldsymbol{\psi} = \left\{p_g, \ldots, p_{G-1}, \boldsymbol{\mu}_1, \ldots, \boldsymbol{\mu}_G, \boldsymbol{\Sigma}_1, \ldots, \boldsymbol{\Sigma}_G, \boldsymbol{\Gamma}\right\}$ be the set of model parameters, where $\boldsymbol{\Gamma}$ is the set of thresholds $\gamma_{c_i}^{(i)}$, the probability of response pattern $\mathbf{x}_r$ is

$$\Pr(\mathbf{x}_r; \boldsymbol{\psi}) = \sum_{g=1}^{G} p_g \int_{\gamma_{c_1-1}^{(1)}}^{\gamma_{c_1}^{(1)}} \cdots \int_{\gamma_{c_P-1}^{(P)}}^{\gamma_{c_P}^{(P)}} \phi(\mathbf{y}; \boldsymbol{\mu}_g, \boldsymbol{\Sigma}_g) d\mathbf{y} = \sum_{g=1}^{G} p_g \pi_r(\boldsymbol{\mu}_g, \boldsymbol{\Sigma}_g, \boldsymbol{\Gamma}) \tag{3}$$

where $\pi_r(\boldsymbol{\mu}_g, \boldsymbol{\Sigma}_g, \boldsymbol{\Gamma})$ is the probability of response pattern $\mathbf{x}_r$ in cluster g. Thus, for a random i.i.d. sample of size N, the log-likelihood is

$$\ell(\boldsymbol{\psi}; \mathbf{X}) = \sum_{r=1}^{R} n_r \log \left[\sum_{g=1}^{G} p_g \pi_r \left(\boldsymbol{\mu}_g, \boldsymbol{\Sigma}_g, \boldsymbol{\Gamma}\right) \right], \tag{4}$$

where n_r is the observed sample frequency of response pattern $\mathbf{x}_r$ and $\sum_{r=1}^{R} n_r = N$.

The maximization of (4) is quite time-consuming and could become infeasible, when P increases due to the presence of multidimensional integrals. For this reason, [10] suggest to estimate model parameters through an EM framework maximizing the following composite log-likelihood given by the sum of all possible log-likelihoods based on the bivariate marginals:

$$
\begin{aligned}
c\ell(\boldsymbol{\psi}; \mathbf{X}) &= \sum_{i=1}^{P-1} \sum_{j=i+1}^{P} \ell(\boldsymbol{\psi}; (x_i, x_j)) \\
&= \sum_{i=1}^{P-1} \sum_{j=i+1}^{P} \sum_{c_i=1}^{C_i} \sum_{c_j=1}^{C_j} n_{c_i c_j}^{(ij)} \log \left[\sum_{g=1}^{G} p_g \pi_{c_i c_j}^{(ij)}(\boldsymbol{\mu}_g, \boldsymbol{\Sigma}_g, \boldsymbol{\gamma}) \right], \qquad (5)
\end{aligned}
$$

where $n_{c_i c_j}^{(ij)}$ is the observed joint frequency of categories c_i and c_j for variables x_i and x_j, respectively, while $\pi_{c_i c_j}^{(ij)}(\boldsymbol{\mu}_g, \boldsymbol{\Sigma}_g, \boldsymbol{\gamma})$ is the corresponding probability obtained by integrating the bivariate marginal (y_i, y_j) of the normal distribution with parameters $(\boldsymbol{\mu}_g, \boldsymbol{\Sigma}_g)$ between their threshold parameters. The estimators obtained have been proven to be consistent, asymptotically unbiased, and normally distributed [13]. For identification reasons, a component is fixed as a reference group; thus, its mean vector is set to $\mathbf{0}$, and the main diagonal of the Cholesky decomposition of the covariance matrix is fixed to 1.

3 The Choice of *G*

In practice, the number of components of the model previously presented is often unknown, and it should be learned from the data. In practice, the number of components of the model previously presented is often unknown, and it should be learned from the data. Here, we compare the two information criteria *C-AIC* [14] and *C-BIC* [5], based on the composite log-likelihood function penalized for model complexity, with the cross-validation (CV) technique [1] and the elbow method.

3.1 Information Criteria

Using automatic selection criteria, the number of components G is chosen by minimizing the C-AIC or the C-BIC. They combine the goodness of fit for a given model s (minus twice the composite log-likelihood) and the penalty term for the model complexity together:

$$\text{C-AIC}(s) = -2c\ell_s(\hat{\boldsymbol{\psi}}_s; \boldsymbol{X}) + 2\text{tr}\left(\hat{\boldsymbol{H}}_s^{-1}\hat{\boldsymbol{V}}_s\right), \tag{6}$$

$$\text{C-BIC}(s) = -2c\ell_s(\hat{\boldsymbol{\psi}}_s; \boldsymbol{X}) + \log N\text{tr}\left(\hat{\boldsymbol{H}}_s^{-1}\hat{\boldsymbol{V}}_s\right), \tag{7}$$

respectively. $\mathbf{H}$ is the sensitivity matrix, $\mathbf{H} = E(-\nabla^2 c\ell(\boldsymbol{\psi}; \mathbf{X}))$, while $\mathbf{V}$ is the variability matrix (the covariance matrix of the score vector), $\mathbf{V} = Var(\nabla c\ell(\boldsymbol{\psi}; \mathbf{X}))$. They have the same structure of AIC and BIC; the only difference is the way used to measure the complexity of the model. In the former ones, it is simply the number of parameters; in the composite version, it is given by the $\text{tr}(\boldsymbol{H}^{-1}\boldsymbol{V})$. It is interesting to note that the former ones are a particular case of the latter ones. When the log-likelihood is used, then $\mathbf{H} = -\mathbf{V}$ and $\text{tr}(\boldsymbol{H}^{-1}\boldsymbol{V})$ is the number of parameters; it follows that C-BIC and C-AIC coincide with BIC and AIC, respectively. Sample estimates of $\mathbf{H}$ and $\mathbf{V}$ could be

$$\hat{\boldsymbol{H}} = -\frac{1}{N}\sum_{n=1}^{N} \nabla^2 c\ell(\hat{\boldsymbol{\psi}}; \mathbf{x}_n)$$

and

$$\hat{\boldsymbol{V}} = \frac{1}{N}\sum_{n=1}^{N} (\nabla c\ell(\hat{\boldsymbol{\psi}}; \mathbf{x}_n))(\nabla c\ell(\hat{\boldsymbol{\psi}}; \mathbf{x}_n))'.$$

Numerical approximation techniques to compute first and second derivatives can be used. However, from a computational point of view, these criteria result to be computationally time-consuming. Comparing $2\text{tr}\left(\hat{\boldsymbol{H}}^{-1}\hat{\boldsymbol{V}}\right)$ with $\log(N)\text{tr}\left(\hat{\boldsymbol{H}}^{-1}\hat{\boldsymbol{V}}\right)$, we note that C-BIC penalizes the model complexity more heavily than C-AIC, if $N \geq 8$.

3.2 Cross-Validation

The K-fold cross-validation procedure here considered can be summarized as follows. This approach involves randomly dividing the set of observations into K groups, or folds, of approximately equal size. The k-th fold $\mathbf{X}^{(k)}$ is treated as a validation set, and the model with a given number of components, say s, is fit on the remaining $K - 1$ folds by obtaining the estimate $\hat{\boldsymbol{\psi}}_s^{(k)}$. The composite log-likelihood is then computed on the observations in the held-out fold $\mathbf{X}^{(k)}$ by obtainnig the value $c\ell_s(\hat{\boldsymbol{\psi}}_s^{(k)}, \mathbf{X}^{(k)})$. This procedure is repeated for $k = 1, 2, \ldots, K$; each time, a different fold of observations is treated as a validation set. This process results in K estimates of the composite log-likelihood. The cross-validated composite log-likelihood is defined as

$$CV(s) = \sum_{k=1}^{K} c\ell_s(\hat{\boldsymbol{\psi}}_s^{(k)}, \mathbf{X}^{(k)})$$

that is, as the sum of the contributions of all K test sets. We select the number of components s maximizing the cross-validation composite log-likelihood.

3.3 *Elbow Method*

All the previous methods imply an important computational burden. To avoid it, we explore a different strategy. This is based on the elbow method that is frequently used in cluster analysis heuristically to select the number of clusters. The criterion is based on choosing G at the elbow of the curve. In order to make it more objective and automatic, we follow the approach of [2] that fits a piecewise linear regression with one breakpoint to the elbow curve. This method has been used for the mixture models, although in different contexts and for different purposes (see, e.g., [6] or [3]). The method can be algorithmically described in the following way. First, for $G = 1, 2, \ldots, G_{\max}$, the model is estimated by maximizing the composite log-likelihood with G components and the maximum value l_G recorded. Then, once again for $G = 1, 2, \ldots, G_{\max}$, a piecewise linear regression with a single breakpoint in G is fitted to the data $(1, l_1), (2, l_2), \ldots, (G_{\max}, l_{G_{\max}})$ and the corresponding residual sum of squares recorded. The optimal G is where the residual sum of squares takes a minimum. Differently from [2] and [3], we require the breakpoint to be exactly on G.

Unlike C-AIC and C-BIC that are based on a strong statistical theory, the elbow method is heuristic. However, the logic is the same: we increase by 1 the number of components if the corresponding increment in the composite log-likelihood is greater than a certain threshold. The difference among the criteria lies in how the threshold is computed. In C-AIC, it is a function of the number of parameters; in C-BIC, it is a function of the parameters and the sample size; in the elbow method, it depends on the antecedent and subsequent increments. Compared to C-AIC, C-BIC, and cross-validation, the elbow method is the most simple from a computationally point of view.

4 Simulation Study

In order to investigate the performance of C-AIC, C-BIC, K-fold cross-validation, and elbow method, we have simulated samples of four ordinal variables with five categories from a mixture model with three components, considering two experimental factors: different sample sizes ($N = 1000, 5000$) and the separation between clusters (well separated or non-well separated). In the last case, the separation between clusters is given by distance between mixture component means and cluster dispersion. In each scenario, we simulated 250 samples. The mixture weights are $p_1 = 0.25$, $p_2 = 0.35$, and $p_3 = 0.40$. The mean vectors and the covariance matrices are generated randomly for each sample, while the thresholds are the quantiles of the latent mixture of Gaussians. For each dataset, we have fitted different number of components, $G = 2, \ldots, 5$. The EM-like algorithm was stopped when the relative increase of the composite log-likelihood between two consecutive steps was less than $\epsilon = 10^{-5}$. For the cross-validation, the number of folds used was $K = 5$. To initialize the model parameters, we worked out the output of the parsimonious Gaussian mixture model (using the R package `mclust`) where we treated all variables as they were continuous with initial partition of `mclust` given by the fitted group membership of k-means. The initial values for the thresholds have been computed as follows: for each variable, we have considered the empirical relative frequency of each category, and then, we have minimized the quadratic difference between this frequency and the corresponding probability of the mixture. We suggest to use a rational start because, in our experience, a purely random initialization is generally extremely worse than our rational start in terms of local optima and computational time. However, our rational start does not guarantee to reach the global optimum, and further studies are needed to improve it.

Looking at Table 1, it possible to note that as the sample size increases, all methods perform better, while as the degree of separation between clusters decreases, the performance get worse. C-AIC and C-BIC behave similarly over all the scenarios. The cross-validation method seems the method is more affected by the sample size; its lowest performances are due to the higher chance of reaching local spurious maxima. Finally, as regards the elbow method, we note encouraging results. Even if we cannot trust on this method completely, we think it could be used to have an idea about the possible number of components. The method is very fast and computationally stable. For these reasons, it totally deserves further research to explore more in depth its behavior.

To better understand the behavior of the elbow method, we also conducted a simulation study under a simplified setting. We considered $P = 8$ continuous

Table 1 Percentages of times out of the 250 samples for which C-AIC, C-BIC, CV, and the elbow method select the true number of components ($G = 3$) by different scenarios. Samples are generated from a mixture model for continuous data 4

Scenario	C-AIC	C-BIC	CV	Elbow method
Well separated				
$N = 1000$	88	86	74	90
Well separated				
$N = 5000$	96	98.8	86.8	92
Non-well separated				
$N = 1000$	78	80	67.2	70
Non-well separated				
$N = 5000$	86.4	85.2	80	90

variables sampled from a mixture model with $G = 3, 4$ components, considering two experimental factors: different sample sizes ($N = 250, 500$) and the separation between clusters (well separated or non-well separated). The mixture weights are $p_1 = 0.25$, $p_2 = 0.35$, and $p_3 = 0.40$ when $G = 3$; $p_g = 0.25$, with $g = 1, 2, \ldots, G$ when $G = 4$. The mean vectors and the covariance matrices are generated randomly for each sample. For each dataset, we have fitted different number of components, $G = 1, 2, \ldots, 5$. The EM algorithm was stopped when the increase of the log-likelihood between two consecutive steps was less than $\epsilon = 10^{-6}$, using the R package `mclust`, with initial partition of `mclust` given by the fitted group membership of k-means. We compared the standard criteria based on the full log-likelihood, AIC and BIC, with the CV and elbow methods. Of course, the case of ordinal data is more complicated. The choice of the starting points for the model parameters is crucial, and the final results of CV and elbow methods are highly affected by the presence of local maxima. Further research is needed to reduce the computational burden as well as to try different sets of starting points.

Looking at Table 2, it possible to note that overall, the best performance is given by the AIC. When the number of groups increases and/or the groups are less separated, BIC do not always select the true number of components. However, as the sample size increases, the performance improves, selecting all the times the right number of components. The cross-validation method shows very similar results, although in more complicated settings, it could become very time-consuming (from a computational point of view), and it could converge to local maxima. Finally, as regards the elbow method, we note that it is able to select the right number of components, performing very well. We note that a very small number of times it do not select the right number of components (when $G = 4$, the groups are not well separated and the sample size is small). The method is very fast and computationally stable. For these reasons, it totally deserves further research to explore more in depth its behavior within more complicated settings.

5 Conclusions

In this paper, we have briefly surveyed some model selection criteria (AIC anf BIC) and the corresponding versions under a composite likelihood approach, along with some methods, cross-validation and elbow method, used within a mixture framework. The simulation study gave interesting results. The elbow method leaves some open points and needs further research. For the continuous data, it is very precise. As regards the ordinal data, it could be used to restrict the number of components, and then, C-AIC or C-BIC could be used to select the final G among the restricted set.

Table 2 Percentages of times out of the 250 samples for which AIC, BIC, CV, and the elbow method select the true number of components ($G = 3, 4$) by different scenarios. Samples are generated from a mixture model for continuous data 1

Scenario	C-AIC	C-BIC	CV	Elbow method
$G = 3$ Well separated $N = 250$	100	100	100	100
$G = 3$ Well separated $N = 500$	100	100	99.2	100
$G = 3$ Non-well separated $N = 250$	100	98.4	100	100
$G = 3$ Non-well separated $N = 500$	100	100	100	100
$G = 4$ Well separated $N = 250$	100	95.6	99.6	100
$G = 4$ Well separated $N = 500$	100	99.6	100	100
$G = 4$ Non-well separated $N = 250$	100	80.4	97.2	99.6
$G = 4$ Non-well separated $N = 500$	100	100	100	100

References

1. Arlot, S., Celisse, A.: A survey of cross-validation procedures for model selection. Stat. Surv. **4**, 40 – 79 (2010)
2. Baudry, J.P., Raftery, A.E., Celeux, G., Lo, K., Gottardo, R.: Combining mixture components for clustering. J. Comput. Graph. Stat. **19**(2), 332–353 (2010)
3. Demidenko, E.: The next-generation k-means algorithm. Stat. Anal. Data Mining: ASA Data Sci. J. **11**(4), 153–166 (2018)
4. Everitt, B.: A finite mixture model for the clustering of mixed-mode data. Stat. Probab. Lett. **6**(5), 305–309 (1988)
5. Gao, X., Song, P.X.K.: Composite likelihood Bayesian information criteria for model selection in high-dimensional data. J. Am. Stat. Assoc. **105**(492), 1531–1540 (2010)
6. Gattone, S., Rocci, R.: Clustering curves on a reduced subspace. J. Comput. Graph. Stat. **21**, 361–379 (2012)
7. Hennig, C., Meila, M., Murtagh, F., Rocci, R.: Handbook of Cluster Analysis. CRC Press, Boca Raton (2015)

8. Lubke, G., Neale, M.: Distinguishing between latent classes and continuous factors with categorical outcomes: class invariance of parameters of factor mixture models. Multivariate Behav. Res. **43**(4), 592–620 (2008)
9. Muthén, B.: A general structural equation model with dichotomous, ordered categorical, and continuous latent variable indicators. Psychometrika **49**(1), 115–132 (1984)
10. Ranalli, M., Rocci, R.: Mixture models for ordinal data: a pairwise likelihood approach. Stat. Comput. **26**, 529–547 (2016)
11. Ranalli, M., Rocci, R.: Mixture models for mixed-type data through a composite likelihood approach. Comput. Stat. Data Anal. **110**(C), 87–102 (2017). https://doi.org/10.1016/j.csda.2016.12.01
12. Ranalli, M., Rocci, R.: A model-based approach to simultaneous clustering and dimensional reduction of ordinal data. Psychometrika (2017). https://doi.org/10.1007/s11336-017-9578-5
13. Varin, C., Reid, N., Firth, D.: An overview of composite likelihood methods. Statistica Sinica **21**(1), 1–41 (2011)
14. Varin, C., Vidoni, P.: A note on composite likelihood inference and model selection. Biometrika **92**(3), 519–528 (2005)

Clustering of Italian Higher Education Institutions Based on a Destination-Specific Approach

Luca Scaffidi Domianello

Abstract Student mobility flows are typically examined by using gravity models. These models generally assume the same relationship for each origin-destination pair. This assumption may not hold for the Italian higher education system, where, as a consequence of the decentralization process, some universities have a national vocation, while others primarily target local populations. For this reason, we adopt a destination-specific approach to gather detailed insights for each university. Indeed, our empirical analysis reveals distinct interaction dynamics among Italian universities, corroborating our hypothesis of spatial heterogeneity. Furthermore, we propose a fuzzy clustering technique based on the estimated distance parameters that measure the deterrence effect. Through this clustering procedure, we identify two distinct groups: one comprising universities with a national vocation and the other comprising those with a local focus.

1 Introduction

Mobility of students across a country in higher education has gained increasing attention in the last years, due to the economic development of the area where the university is located. Specularly, student migration, especially remarkable for bachelor degree programs, has a relevant socioeconomic impact on those areas suffering the so called *brain drain*. Flows between an origin and a destination are generally examined by using gravity models, which rely on Newton's law of universal gravitation: the interactions among two areas are proportional to the product of their masses (attraction effect) and inversely proportional to their distance (deterrence effect). Generally, these models, in order to explain the flows between the area of origin of the student and the university of destination, assume the same relationship for each origin-destination pair—see, e.g., Sa et al. [14] for the Dutch

L. Scaffidi Domianello (✉)
Department of Economics and Business, University of Catania, Catania, Italy
e-mail: luca.scaffidi@unict.it

G. Giordano et al. (eds.), *Statistical Models and Learning Methods for Complex Data*, Studies in Classification, Data Analysis, and Knowledge Organization,
https://doi.org/10.1007/978-3-031-84702-8_19

universities, while Bacci and Bertaccini [2] for the Italian ones—and they do not consider possibly different interaction dynamics that is spatial heterogeneity.

In the present contribution, we consider a destination-specific gravity model to gather detailed insights for each university [11]: allowing the model parameters to vary across the space is a crucial task for the Italian higher education system. Indeed, in the last decade of the twentieth century, a decentralization process took place, through which new types of faculties and degree courses were created, and, as a consequence, there are some universities with a national vocation, while others target local territory. For this purpose, we allow the distance parameter, reflecting the deterrence effect, to assume different values whether the university is more or less 250 km away from the student province of residence. The idea is that if we have a huge increase in the value of the parameters as the distance increases, the university has a wider catchment area, rather than confined to the neighborhood.

To gather deeper insights, we conduct also a clustering procedure based on the similarity of the estimated distance parameters, aiming at identifying groups of universities sharing a similar deterrence effect, and then, allowing us to infer whether they are characterized by a local or a national focus.[1] Indeed, identifying the catchment area is of interest to university administrators for their marketing strategies.

The data employed for this work comes from the Italian National Student Registry (in Italian, Anagrafe Nazionale Studenti, ANS), the Italian administrative database that records the students, by their province of residence, enrolled in any degree program in a certain university located in Italy. In this work, we focus on *first level* mobility students—that is, students enrolled in a bachelor or five-year degree program [1]—and more specifically on the subset of students enrolled in Science & Technologies (S&T) courses: they are of particular interest for policymakers because they are directly related to the technological development of the area where the university is located [6].

The work is structured as follows: Sect. 2 describes the model and the clustering procedure we employ for the empirical analysis, Sect. 3 analyzes the data and comments on the estimation results, while Sect. 4 offers some concluding remarks.

2 Theoretical Framework

In order to analyze student mobility in higher education, we adopt gravity models, a useful tool for describing people flows over a geographic area [10]. By considering the flow T_{ij}, denoting the number of students moving from the province of residence i $(i = 1, \ldots, I)$ to the university of destination located in the province j $(j = 1, \ldots, J)$, as an outcome of a Poisson process, its conditional mean, λ_{ij}, can be expressed as follows [8]:

[1] Nakaya [13] proposed to group origin-destination pairs based on the coefficients estimated by using the geographically weighted regression approach.

$$\lambda_{ij} = exp\left(k + \sum_{p=1}^{P} \alpha_p \log x_{ip} + \sum_{q=1}^{Q} \beta_q \log z_{jq} + \gamma \log d_{ij}\right) \tag{1}$$

where $x_{i1}, \ldots, x_{iP}$ are P covariates measuring the origin propulsiveness, while $z_{j1}, \ldots, z_{jQ}$ are Q covariates related to destination attractiveness, d_{ij} is the road distance, expressed in kilometers, for each origin-destination pair (that we expect has a negative influence on the student flow, since larger distances should deter students from enrolling in those universities), k is a constant of proportionality, while α_p, β_q, and γ are other parameters to be estimated.

In the model specified above, we assume the same relationship for each origin-destination pair. Then, with the purpose to get disaggregated information, we estimate the model in (1) [11]:

$$\lambda_{ij} = exp\left(k_j + \sum_{p=1}^{P} \alpha_{jp} \log x_{ip} + \gamma_j \log d_{ij}\right) \quad \forall j \tag{2}$$

Notice that, differently from the global model in (1), $\sum_{q=1}^{Q} \beta_q \log z_{jq}$ is constant, as the destination is unique for each j, then it is enclosed in the parameter k_j, while the parameters related to the origin, $\alpha_{j1}, \ldots, \alpha_{jP}$ are specific to destination (see [11]).

As opposed to the log-normal models, the Poisson regression allows us to deal with the problem of zero-valued flows, while this is not the case when we have to use the log transformation of the dependent variable. To circumvent this issue, it has been proposed to add a small positive constant to each observation. However, when there are a lot of zeros, this leads to biased estimates of the coefficients. Furthermore, through the Poisson regression, we directly estimate the expected value, λ_{ij}, not the expected value of its logarithm, $\log \lambda_{ij}$, whose antilogarithm is a biased estimate of λ_{ij}.

Additionally, we estimate the parameters through the Poisson-pseudo-maximum-likelihood (PPML) estimator [15], to obtain consistent estimates of the parameters even if the assumed distribution is no more valid, except the correct specification of the conditional mean, as it could be the case when dealing with a large amount of zero-valued flows. Consequently, we have to estimate a robust specification of the variance covariance matrix [16].

Once we estimate the destination-specific model, we propose to group the universities through a clustering approach based on the similarity of the estimated parameters related to the deterrence effect. Generally, clustering approaches can be divided into two classes: hard clustering (e.g., *k-means*), in which units belong to one cluster, and soft clustering (e.g., fuzzy *k-means*), in which units can belong to more than one cluster according to a membership degree, whose value ranges between 0 and 1, and their sum must be equal to 1. We adopt a fuzzy clustering approach, due to its flexibility, that minimizes the following objective function:

$$\begin{cases} \min : \sum_{i=1}^{N} \sum_{c=1}^{C} u_{i,c}^{m} \left[\left(\gamma_i^{(1)} - \gamma_c^{(1)} \right)^2 + \left(\gamma_i^{(2)} - \gamma_c^{(2)} \right)^2 \right] \\ \text{s.t.} \quad \sum_{c=1}^{C} u_{i,c} = 1 \quad \text{and} \quad u_{i,c} \geq 0, \quad \forall i, \forall c \end{cases} \tag{3}$$

where $\gamma_c^{(1)}$ and $\gamma_c^{(2)}$ are the centroids of the c-th cluster, $m > 1$ is a tuning fuzziness parameter (the higher its value, the higher the degree of fuzziness), and $u_{i,c}$ the membership degree of the i-th unit to the c-th cluster. The above minimization problem can be solved iteratively, whose conditionally optimal iterative solutions are given by Bezdek [3]:

$$\begin{aligned} u_{i,c} &= \frac{1}{\sum_{c'=1}^{C} \left[\frac{\left(\gamma_i^{(1)} - \gamma_c^{(1)}\right)^2 + \left(\gamma_i^{(2)} - \gamma_c^{(2)}\right)^2}{\left(\gamma_i^{(1)} - \gamma_{c'}^{(1)}\right)^2 + \left(\gamma_i^{(2)} - \gamma_{c'}^{(2)}\right)^2} \right]^{1/m-1}}, \ \gamma_c^{(1)} = \frac{\sum_{i=1}^{N} \gamma_i^{(1)} u_{i,c}^m}{\sum_{i=1}^{N} u_{i,c}^m}, \ \gamma_c^{(2)} \\ &= \frac{\sum_{i=1}^{N} \gamma_i^{(2)} u_{i,c}^m}{\sum_{i=1}^{N} u_{i,c}^m} \end{aligned} \tag{4}$$

Nevertheless, noise data can influence the outcome of the clustering process. To face this issue, Dave [4] proposed to consider a c-*th* a noise cluster in addition to the $c-1$ good clusters. The idea is that the noisy observations belong to the *bad* (or noise) cluster. The objective function to be minimized is the following:

$$\begin{cases} \min : \sum_{i=1}^{N} \sum_{c=1}^{C-1} u_{i,c}^{m} \left[\left(\gamma_i^{(1)} - \gamma_c^{(1)} \right)^2 + \left(\gamma_i^{(2)} - \gamma_c^{(2)} \right)^2 \right] + \left(1 - \sum_{c=1}^{C-1} u_{i,c} \right)^m \delta^2 \\ \text{s.t.} \quad \sum_{c=1}^{C-1} u_{i,c} \leq 1 \quad \text{and} \quad u_{i,c} \geq 0, \quad \forall i, \forall c \end{cases} \tag{5}$$

with δ the constant distance, to be specified in advance [4], among each point and the noise prototype. Notice that, according to the constraints in Eq. (5), the membership degree does not need to sum to one for the *good* clusters. Furthermore, for $i = 1$ to $C - 1$, the iterative solutions for the centroids are the same as in Eq. (4), while those for the membership grade matrix are the following:

$$u_{i,c} = \frac{1}{\sum_{c'=1}^{C-1} \left[\frac{\left(\gamma_i^{(1)} - \gamma_c^{(1)}\right)^2 + \left(\gamma_i^{(2)} - \gamma_c^{(2)}\right)^2}{\left(\gamma_i^{(1)} - \gamma_{c'}^{(1)}\right)^2 + \left(\gamma_i^{(2)} - \gamma_{c'}^{(2)}\right)^2} \right]^{1/m-1} + \left[\frac{\left(\gamma_i^{(1)} - \gamma_c^{(1)}\right)^2 + \left(\gamma_i^{(2)} - \gamma_c^{(2)}\right)^2}{\delta^2} \right]^{1/m-1}} \tag{6}$$

3 Empirical Analysis

The data of this work come from ANS, the Italian administrative database that records students' enrollment in Italian universities, by their province of residence. We focus on the cohort of students enrolled in a bachelor's or five-year degree programs (first-level mobility) for the academic year 2011–2012 and more specifically on the subgroup of students attending Science & Technologies (S&T) courses (ISCED 5, 6, and 7), due to their impact on the technological development of the area where the university is located [6]. The resulting 65,812 students are distributed across 62 universities and 107 provinces. For what concerns the universities with more than one campus, we decide to take into account the province where the main campus is located, while if the campuses have similar relevance, we take into account the specific provinces (see [2]). As a proxy for the origin propulsiveness (O Mass), we use the total number of students by their province of residence as in Dotti et al. [6], while for the road distance, expressed in kilometers, we allow its parameter to vary whether it is more or less than 250 kilometers.

Table 1 reports the summary results of the estimated global and destination-specific gravity models.[2] The results of the global model (top part of Table 1) show the expected effect on interactions: the parameters related to the distance, measuring the deterrence effect, have a negative influence on the flows, while the parameter related to the mass, measuring the attraction effect, has a positive impact. For what concerns the destination-specific model, as we can see in the bottom part of Table 1,

Table 1 Summary results of the estimated global Poisson gravity model (top) and of the Poisson destination-specific gravity model (bottom) for student attending S&T courses in Italian universities for the academic year 2011–2012

	Global model		
Coefficients	Estimates	Std. Errors	*p*-value
Constant	2.312	0.513	0.000
$d < 250$	−0.612	0.020	0.000
$d > 250$	−0.863	0.015	0.000
O Mass	0.554	0.083	0.000
Pseudo-R^2	0.712		

	Destination-specific model					
Coefficients	Min	Mean	Max	1Q	2Q	3Q
Constant	−6.537	3.373	12.593	0.824	3.409	5.437
$d < 250$	−2.036	−0.682	−0.261	−0.783	−0.657	−0.533
$d > 250$	−4.252	−1.159	−0.524	−1.331	−1.067	−0.861
O Mass	−0.866	0.359	1.823	0.081	0.310	0.643
Pseudo-R^2	0.291	0.802	0.972	0.718	0.834	0.917

[2] The global and destination-specific gravity models are estimated in R (`http://www.r-project.org/`) through the function "`glm`".

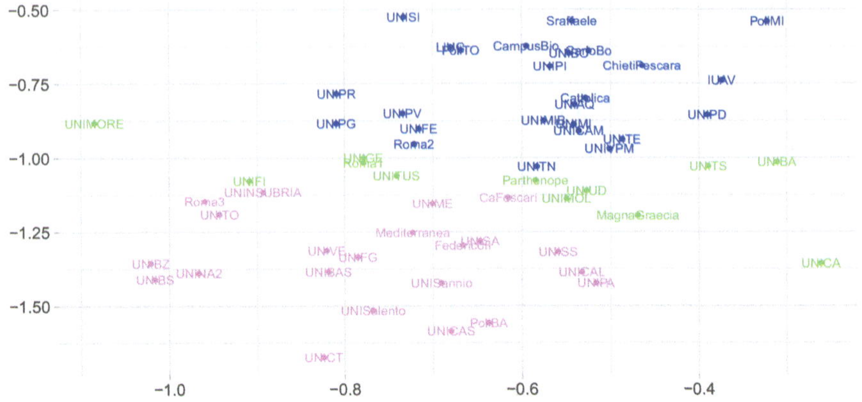

Fig. 1 Obtained clustering through fuzzy k-means with noise cluster. $\gamma_i^{(1)}$ (x-axis), $\gamma_i^{(2)}$ (y-axis). 1st group (blue points), 2nd group (violet points), fuzzy units (green points)

the value of the distance parameter related to the first class is higher, and there is a slight difference among the quartiles, while the other class has a difference among the first and the third quartile more pronounced. As expected, the population proxy shows a positive impact, and overall, we can say that there is a lot of variation in the value of the coefficients among the universities, thus supporting our hypothesis of spatial heterogeneity. For the lake of space, we do not report the estimates for each destination, but they are available on request.

To gather further insights, we conduct a fuzzy clustering procedure based on the similarity of the estimated university-specific distance parameters. More specifically, we adopt a fuzzy clustering approach due to its greater flexibility with respect to hard clustering (e.g., *k-means*). Additionally, in order to avoid misleading results in the clustering process, we adopt the robust procedure to the noise data of Dave [4].[3] We select two clusters, that is, the number of clusters chosen by minimizing the Xie-Beni index [17]. The University of Bergamo, Piemonte Orientale, and Kore belong to the noise cluster. We conduct also a plain fuzzy *k-means*, which, by optimizing the Xie-Beni index, suggests we should choose three clusters and one of them comprises the three universities belonging to the noise cluster provided by the robust approach described above. By inspecting the composition of the two *good* clusters, see Fig. 1, we can say that, generally, they are different in terms of the values of both distance parameters. The first group, characterized by a lower value of both distance parameters (blue points), includes mainly universities offering very specialized degree programs (e.g., Polytechnic of Milan and Turin), and we can conclude that these universities have a national vocation. The second group, characterized by a higher value of distance parameters (violet points), comprises most of the universities of the South, which according to

[3] We used the library *fclust* [7] to perform the fuzzy clustering procedure.

our interpretation have a local focus. There are other universities, like the University of Modena and Reggio Emilia, whose belonging to one of the two clusters is ambiguous: indeed, they are almost equidistant to centroids of the two clusters, and they represent the so-called fuzzy units.

4 Conclusion

This work analyzes student mobility flows in Italian higher education through a Poisson gravity model. More specifically, we concentrate on the subset of students attending bachelor or five-year degree programs related to the S&T courses due to their impact on the technological development of the area where the university is located. For what concerns the specification of the adopted model, we rely on a destination-specific approach, thus allowing the model parameters to vary across different institutions to detect some heterogeneity in the interaction behaviors. Empirical analysis supports our hypothesis of spatial heterogeneity by showing a high degree of variation between the values of the estimated parameters for each university. Furthermore, we propose to group the universities according to the pattern of the parameters measuring the deterrence effect: according to the results of the employed clustering procedure, we identify two groups of universities, one associated with a national vocation, while the other with a local focus. The identification of the catchment area of higher education institutions is of relevant interest to university administrators in implementing their orientation strategies aimed at high school students. As future research, it could be interesting to obtain deeper information by allowing the parameters to vary also according to the origin, e.g., through the geographically weighted regression (GWR) technique [9, 12]. Furthermore, it could be interesting to cluster universities through a mixture of Poisson regressions,[4] allowing, through the EM algorithm [5], iterative learning of the weights and the coefficients related to the components.

References

1. Attanasio, M., Enea, M.: La mobilità degli studenti universitari nell'ultimo decennio in italia. In De Santis, G., Pirani, E., Porcu, M. (eds.) Rapporto sulla popolazione. L'istruzione in Italia, pp. 43–58. Il Mulino (2019)
2. Bacci, S., Bertaccini, B.: Assessment of the university reputation through the analysis of the student mobility. Soc. Indicators Res. **156**, 363–388 (2021)
3. Bezdek, J.C.: Pattern Recognition with Fuzzy Objective Function Algorithms. Plenum Press, New York (1981)

[4] See, e.g., Xiong and Yeung [18] for time series clustering through autoregressive moving average (ARMA) mixture models.

4. Dave, R.N.: Characterization and detection of noise in clustering. Pattern Recognit. Lett. **12**(11), 657–664 (1991)
5. Dempster, A., Laird, N., Rubin, D.: Maximimum likelihood from incomplete data via the EM algorithm. J. Roy. Stat. Soc. Ser. B **39**, 1–38 (1977)
6. Dotti, N.F., Fratesi, U., Lenzi, C., Percoco, M.: Local labour market conditions and the spatial mobility of science and technology university students: evidence from Italy. Rev. Regional Res. **34**, 119–137 (2014)
7. Ferraro, M.B., Giordani, P., Serafini, A., et al.: fclust: an R package for fuzzy clustering. R J. 1–18
8. Flowerdew, R., Aitkin, M.: A method of fitting the gravity model based on the poisson distribution. J. Regional Sci. **22**(2), 191–202 (1982)
9. Fotheringham, A.S., Charlton, M.E., Brunsdon, C.: Geographically weighted regression: a natural evolution of the expansion method for spatial data analysis. Environ. Plan. A **30**(11), 1905–1927 (1998)
10. Fotheringham, A.S., O'Kelly, M.E.: Spatial Interaction Models: Formulations and Applications. Kluwer Academic Publishers, Dordrecht (1989)
11. Haynes, K.E., Fotheringham, A.S.: Gravity and Spatial Interaction Models. Sage Publications, Thousand Oaks (1984)
12. Kordi, M., Fotheringham, A.S.: Spatially weighted interaction models (SWIM). Ann. Am. Assoc. Geographers **106**(5), 990–1012 (2016)
13. Nakaya, T.: Local spatial interaction modelling based on the geographically weighted regression approach. GeoJournal **53**(4), 347–358 (2001)
14. Sa, C., Florax, R.J., Rietveld, P.: Determinants of the regional demand for higher education in the Netherlands: a gravity model approach. Regional Stud. **38**(4), 375–392 (2004)
15. Silva, J.S., Tenreyro, S.: The log of gravity. Rev. Econ. Stat. **88**(4), 641–658 (2006)
16. White, H.: Maximum likelihood estimation of misspecified models. Econometrica **50**, 1–25 (1982)
17. Xie, X.L., Beni, G.: A validity measure for fuzzy clustering. IEEE Trans. Pattern Anal. Mach. Intell. **13**(8), 841–847 (1991)
18. Xiong, Y., Yeung, D.-Y.: Time series clustering with ARMA mixtures. Pattern Recognit. **37**(8), 1675–1689 (2004)

Analyzing Italian Crime Data Using Matrix-Variate Hidden Markov Models

Salvatore D. Tomarchio, Antonio Punzo, and Antonello Maruotti

Abstract Hidden Markov models (HMMs) emerge as a powerful tool for examining various time-dependent data structures. Within the context of matrix-variate longitudinal data, the literature has seen notable progress in recent years. In this study, we apply a set of parsimonious matrix-variate HMMs to a dataset that records crime indicators across 106 provinces in Italy. Utilizing the flexibility of HMMs and matrix-variate structure, various states are identified, and the transitions of provinces between these states are examined over time.

1 Introduction

The statistical literature concerning hidden Markov models (HMMs) is extensive and evolving; for an in-depth survey, see [16]. Characterized by their dependent mixture structures, HMMs offer more flexibility compared to single distributions, and their mathematical framework renders them applicable to many real-world scenarios.

An interesting sphere of implementation involves the use of HMMs for the analysis of univariate or multivariate longitudinal data (see, e.g., [5, 7, 8]). However, in recent years, efforts have been made to generalize HMMs for modeling matrix-variate longitudinal data both in unconditional [12, 13] and regression settings [14, 15]. In the unconditional framework, which is the focus of this paper, these models are specifically designed to handle datasets where, for each statistical unit and at each time frame, a $P \times R$ matrix of variables is observed. Such data can be

S. D. Tomarchio (✉) · A. Punzo
Department of Economics and Business, University of Catania, Catania, Italy
e-mail: daniele.tomarchio@unict.it; antonio.punzo@unict.it

A. Maruotti
Department of Law, Economics, Politics and Modern Languages, LUMSA University, Rome, Italy
e-mail: a.maruotti@lumsa.it

G. Giordano et al. (eds.), *Statistical Models and Learning Methods for Complex Data*, Studies in Classification, Data Analysis, and Knowledge Organization,
https://doi.org/10.1007/978-3-031-84702-8_20

conveniently organized into four-way arrays with dimensions $P \times R \times I \times T$, where I represents the sample size and T denotes the time length.

In this manuscript, we leverage a family of parsimonious matrix-variate HMMs (MV-HMMs) to evaluate crime across 106 Italian provinces over the last 4 years of available data. Understanding crime is a multifaceted task that requires exploration at appropriate disaggregated temporal and territorial levels. With a specific focus on the Italian context, some empirical research has been conducted about crime at the provincial level (NUTS3, according to the European Nomenclature of Territorial Units for Statistics); see, for example, [3, 4, 10, 11]. They generally focused on the socioeconomic determinants of criminality, and their economic consequences, and provided different patterns of the territorial diffusion of crime across the country.

The aim of this paper is not to ascertain such determinants or pretend to provide a unique representation of the crime diffusion in Italy. Conversely, we aim to illustrate how the use of MV-HMMs can (i) highlight the presence of different states (or clusters) sharing similar crime characteristics, (ii) allow the monitoring of how the provinces switch from one state to another over time, and (iii) benefit of the interpretability and parsimony granted by the matrix-variate structure.

The manuscript is organized as follows. In Sect. 2, we summarize the methodological foundation of the family of parsimonious MV-HMMs. In Section 3, we provide specific details about the dataset and results when the considered models are fitted. Finally, some concluding remarks are given in Sect. 4.

2 Methodology

In a general sense, MV-HMMs represent a specific type of time-dependent mixture encompassing two processes:

1. A latent first-order Markov chain $\{S_{it}; i = 1, \ldots, I, t = 1, \ldots, T\}$ with a state space $\{1, \ldots, k, \ldots, K\}$ that adheres to the Markov property:

$$\Pr\left(S_{it} = s_{it} | S_{i1} = s_{i1}, \ldots, S_{it-1} = s_{it-1}\right) = \Pr\left(S_{it} = s_{it} | S_{it-1} = s_{it-1}\right),$$

2. A state-dependent observation process $\{\mathcal{X}_{it}; i = 1, \ldots, I, t = 1, \ldots, T\}$, where $\mathcal{X}_{it} \in \mathbb{R}^{P \times R}$, adhering to the conditional independence property:

$$\begin{aligned} & f\left(\mathcal{X}_{it} = \mathbf{X}_{it} | \mathcal{X}_{i1} = \mathbf{X}_{i1}, \ldots, \mathcal{X}_{it-1} = \mathbf{X}_{it-1}, S_{i1} = s_{i1} \ldots, S_{it} = s_{it}\right) \\ =& f\left(\mathcal{X}_{it} = \mathbf{X}_{it} | S_{it} = s_{it}\right), \end{aligned}$$

where $f(\cdot)$ represents a matrix-variate probability density function (pdf).

The functional form of $f(\cdot)$ adopted herein corresponds to the matrix-variate normal distribution, expressed as

$$f\left(\mathbf{X}_{it}|S_{it}=k;\boldsymbol{\Theta}_k\right)=\frac{\exp\left\{-\frac{1}{2}\operatorname{tr}\left[\boldsymbol{\Sigma}_k^{-1}(\mathbf{X}-\mathbf{M}_k)\boldsymbol{\Psi}_k^{-1}(\mathbf{X}-\mathbf{M}_k)'\right]\right\}}{(2\pi)^{\frac{PR}{2}}|\boldsymbol{\Sigma}_k|^{\frac{R}{2}}|\boldsymbol{\Psi}_k|^{\frac{P}{2}}}, \tag{1}$$

where $\mathbf{M}_k$ represents the $P \times R$ matrix of means and $\boldsymbol{\Sigma}_k$ and $\boldsymbol{\Psi}_k$ are the $P \times P$ and $R \times R$ covariance matrices associated with the P rows and R columns, respectively. Note that $\operatorname{tr}(\cdot)$ denotes the trace operator, and $\boldsymbol{\Theta}_k = \{\mathbf{M}_k, \boldsymbol{\Sigma}_k, \boldsymbol{\Psi}_k\}$.

In addition to $\boldsymbol{\Theta}_k$, we also have the parameters governing the hidden Markov chain. Specifically, these include:

1. The initial probabilities $\pi_{ik} = \Pr(S_{i1} = k)$, where $k = 1, \ldots, K$, consolidated into the K-dimensional vector $\boldsymbol{\pi}$
2. The probabilities that regulate the state transitions

$$\pi_{ik|j} = \Pr\left(S_{it} = k|S_{it-1} = j\right), \quad t = 2, \ldots, T \text{ and } j, k = 1, \ldots, K. \tag{2}$$

These probabilities are organized in the $K \times K$ transition matrix denoted as $\boldsymbol{\Pi}$. In Equation (2), the symbol k denotes the current state, while j refers to the previously visited one. For simplicity, we consider homogeneous MV-HMMs, implying that $\pi_{ik|j} = \pi_{k|j}$ and $\pi_{ik} = \pi_k$ for all $i = 1, \ldots, I$.

To achieve parsimony, the eigen-decomposition of a covariance matrix [2] is applied to $\boldsymbol{\Sigma}_k$ and $\boldsymbol{\Psi}_k$ in (1). This process yields 14 parsimonious structures for $\boldsymbol{\Sigma}_k$ (EII, VII, EEI, VEI, EVI, VVI, EEE, VEE, EVE, VVE, EEV, VEV, EVV, VVV) and 7 for $\boldsymbol{\Psi}_k$ (II, EI, VI, EE, VE, EV, VV). The different number of structures arises from identifiability considerations, as elaborated in [12]. In the acronyms for $\boldsymbol{\Sigma}_k$, the positions of the letters signify the volume, shape, and orientation of the k-state, while in the acronyms for $\boldsymbol{\Psi}_k$, the volume letter is omitted. Furthermore, for both covariance matrices, "E" denotes equal, "V" signifies varying, and "I" represents the identity matrix. The combination of these parsimonious structures results in a family of 98 MV-HMMs. To enhance notational clarity, each MV-HMM is identified by hyphenating the acronyms used for the row and column covariance matrices.

Parameter estimation is accomplished, through a maximum likelihood approach, utilizing the expectation conditional maximization (ECM) algorithm [6]. In this framework, the complete data are $\mathcal{S}_c = \{\mathbf{X}_{it}, z_{it}, zz_{it}; i = 1, \ldots, I, t = 1, \ldots, T\}$, where $z_{it} = (z_{it1}, \ldots, z_{itk}, \ldots, z_{itK})'$ refers to the latent state membership, and

$$zz_{it} = \begin{bmatrix} zz_{it11} & \cdots & zz_{it1k} & \cdots & zz_{it1K} \\ \vdots & & \vdots & & \vdots \\ zz_{itj1} & \cdots & zz_{itjk} & \cdots & zz_{itjK} \\ \vdots & & \vdots & & \vdots \\ zz_{itK1} & \cdots & zz_{itKk} & \cdots & zz_{itKK} \end{bmatrix},$$

with

$$z_{itk} = \begin{cases} 1 & \text{if } S_{it} = k \\ 0 & \text{otherwise} \end{cases} \quad \text{and} \quad zz_{itjk} = \begin{cases} 1 & \text{if } S_{it-1} = j \text{ and } S_{it} = k \\ 0 & \text{otherwise} \end{cases},$$

refers to the unobserved states transition. The corresponding complete-data log-likelihood is

$$l_c\left(\boldsymbol{\Omega}|\mathcal{S}_c\right) = \sum_{i=1}^{I}\sum_{k=1}^{K} z_{i1k}\log\left(\pi_k\right) + \sum_{i=1}^{I}\sum_{t=2}^{T}\sum_{k=1}^{K}\sum_{j=1}^{K} zz_{itjk}\log\left(\pi_{k|j}\right) + l_{c_1}\left(\boldsymbol{\Theta}|\mathcal{S}_c\right), \tag{3}$$

where $\boldsymbol{\Omega}$ contains all the parameters of the model, $\boldsymbol{\Theta} = \{\boldsymbol{\Theta}_k; k = 1, \ldots, K\}$ and

$$l_{c_1}\left(\boldsymbol{\Theta}|\mathcal{S}_c\right) = \sum_{i=1}^{I}\sum_{t=1}^{T}\sum_{k=1}^{K} z_{itk}\Bigg[-\frac{PR}{2}\log\left(2\pi\right) - \frac{R}{2}\log|\boldsymbol{\Sigma}_k| - \frac{P}{2}\log|\boldsymbol{\Psi}_k| \\ - \frac{1}{2}\,\mathrm{tr}\left[\boldsymbol{\Sigma}_k^{-1}(\mathbf{X}_{it} - \mathbf{M}_k)\boldsymbol{\Psi}_k^{-1}(\mathbf{X}_{it} - \mathbf{M}_k)'\right]\Bigg].$$

From a computational point of view, in the E-step, a forward recursion approach commonly used in HMM literature is employed for computing the conditional expectation of (3) [1]. Then, in the CM-steps, the model parameters are updated. Because of space constraints, we remand to [12] for a detailed discussion.

3 Application to Italian Crime Data

As introduced in Sect. 1, the object of our analysis is a dataset concerning crime indicators for $I = 106$ Italian provinces. Notably, [4] pointed out that the provincial level provides more detailed insights than the regional level. This granularity allows the identification of diverse patterns within Italian regions (NUTS2), particularly in regions without a marked historical presence of criminality.

Our data comes from the online repository (accessible at https://lab24.ilsole24ore.com/) of an Italian financial newspaper, *Il Sole 24 Ore*, which annually assesses the quality of life in the Italian provinces via a battery of indicators collected from various thematic areas. By focusing on the criminality topic, we directed our attention to the only $P = 2$ indicators that are separately evaluated according to $R = 2$ execution methods, thus providing deeper and more precise information on the crime nature. Particularly, thefts (T) and robberies (R) are measured (as the number of complaints per 100 thousand inhabitants) at commercial businesses (C) and houses (H); thus, for each province, we have a 2×2 matrix. Though these measurements don't encompass extremely violent crimes, they still furnish valuable insights into safety levels across diverse geographical areas. The information spans the years 2020–2023 (i.e., $T = 4$), resulting in a final dataset presented as a four-way array with dimensions $2 \times 2 \times 106 \times 4$.

In analyzing this dataset, several aspects are worth to be investigated. The first concerns the existence of areas with similar crime levels across the country. Another noteworthy aspect is evaluating the temporal dependence strength, quantified by the transition probability matrix, and understanding how provinces transition between hidden states. This latter aspect can be particularly useful in identifying broad shifts in crime levels. Finally, by using the matrix-variate form factor, useful insights can be gained by analyzing the within-and-between variability as measured by the two covariance matrices.

3.1 Results and Discussion

We fit the family of parsimonious MV-HMMs for $K \in \{1, \ldots, 7\}$. Model selection, using the Bayesian information criterion (BIC; [9]), identified the best-fitting model as the VVV-EE MV-HMM with $K = 5$ states, yielding a BIC value of 12304.25.

Initially, we can attempt to understand the hidden states identified by this model by examining its estimated mean matrices:

$$\mathbf{M}_1 = \begin{array}{c} \\ \mathrm{T} \\ \mathrm{R} \end{array} \!\!\begin{array}{c} \begin{array}{cc} \mathrm{C} & \mathrm{H} \end{array} \\ \begin{bmatrix} 42.26 & 99.60 \\ 1.75 & 2.11 \end{bmatrix} \end{array}, \quad \mathbf{M}_2 = \begin{array}{c} \\ \mathrm{T} \\ \mathrm{R} \end{array} \!\!\begin{array}{c} \begin{array}{cc} \mathrm{C} & \mathrm{H} \end{array} \\ \begin{bmatrix} 65.56 & 190.76 \\ 1.94 & 2.18 \end{bmatrix} \end{array},$$

$$\mathbf{M}_3 = \begin{array}{c} \\ \mathrm{T} \\ \mathrm{R} \end{array} \!\!\begin{array}{c} \begin{array}{cc} \mathrm{C} & \mathrm{H} \end{array} \\ \begin{bmatrix} 73.05 & 144.72 \\ 6.39 & 2.87 \end{bmatrix} \end{array},$$

$$\mathbf{M}_4 = \begin{array}{c} \\ \mathrm{T} \\ \mathrm{R} \end{array} \!\!\begin{array}{c} \begin{array}{cc} \mathrm{C} & \mathrm{H} \end{array} \\ \begin{bmatrix} 110.14 & 299.41 \\ 4.32 & 2.91 \end{bmatrix} \end{array}, \quad \mathbf{M}_5 = \begin{array}{c} \\ \mathrm{T} \\ \mathrm{R} \end{array} \!\!\begin{array}{c} \begin{array}{cc} \mathrm{C} & \mathrm{H} \end{array} \\ \begin{bmatrix} 179.77 & 270.77 \\ 9.87 & 3.22 \end{bmatrix} \end{array}.$$

The first message we can draw is that there seems to be a relative variability among the states; hence, different levels of crimes are observed across the country. Moreover, it is intriguing to note that the five states can be roughly arranged in ascending order based on their crime rates. Consequently, the first state encompasses the safer provinces, while the fifth state is associated with the riskier ones.

Another interesting aspect is that, in all the states, the average thefts in houses are greater than those in commercial businesses. However, although this trend also holds for robberies in the first two states, a distinct pattern emerges in the remaining three states, where incidents of robberies in commercial businesses exceed those in houses.

To visualize the detected states, we report in Fig. 1 the maps of the Italian province colored according to the estimated state memberships at each of the four

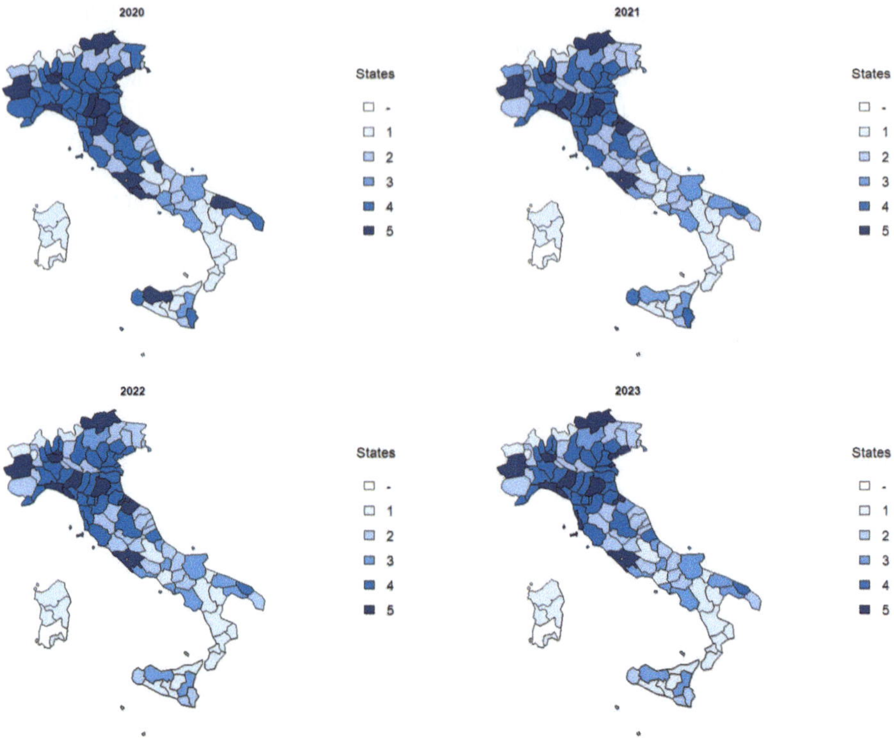

Fig. 1 Political map of Italy with provinces colored according to the estimated state memberships

years. Note that the province colored in white (Sud Sardinia) is not included in the analysis because of not available data.

First of all, it is possible to observe that, over the years, there have been a certain number of transitions among the states. Specifically, 25 provinces underwent state changes from 2020 to 2021. However, this number significantly decreased to 5 and 4 in the subsequent two years, respectively. A predominant trend in these transitions is the movement of provinces from riskier to safer states, particularly evident in the shift from 2020 to 2021. One plausible explanation for this phenomenon is linked to the stringent restrictions imposed in Italy during the COVID-19 pandemic, which significantly impacted movements and activities (and consequently these kinds of crimes).

In terms of province clustering, we start by analyzing the fifth state. Despite some changes over the years, this state generally aggregates the biggest metropolitan areas of the country. As such, the criminality levels are inevitably higher than in other areas. Provinces such as Bologna, Genoa, Milan, Rome, and Turin are constantly present in this state.

The fourth state is the most regularly visited, given that, for all the years, it has the highest number of provinces. Several industrial districts belong to this state, which mostly contains provinces located in the central-northern part of the country.

The third state is the least visited instead. Here, we mainly find large metropolitan areas located in the southern part of Italy. For example, provinces such as Bari, Catania, Naples, and Palermo are present in this state. However, the provinces of Trento and Trieste, located in North Italy, are herein almost regularly classified.

The second state has a varied composition. Indeed, here we found relatively small provinces spanning all over the country, such as Ragusa, Siena, Udine, and Vercelli.

Lastly, the first state mostly considers provinces located in the southern-insular part of Italy, with only a few exceptions (Aosta, Biella, Sondrio, and Verbano-Cusio-Ossola). It is interesting to report that all the provinces of the Sardinia, Calabria, and Basilicata regions are herein assigned. As previously discussed, this state can be considered the safer one, in terms of the two crime indexes considered.

To recap the results discussed thus far, it appears that there are disparities in the diffusion of crime across Italian provinces. A distinct North-South divide is not always easily discernible, differently from many other socioeconomic phenomena.

We now report the estimated transition probability matrix, i.e.,

$$\boldsymbol{\Pi} = \begin{bmatrix} 0.98 & 0.00 & 0.02 & 0.00 & 0.00 \\ 0.00 & 0.96 & 0.04 & 0.00 & 0.00 \\ 0.09 & 0.00 & 0.91 & 0.00 & 0.00 \\ 0.00 & 0.11 & 0.00 & 0.85 & 0.04 \\ 0.00 & 0.00 & 0.14 & 0.12 & 0.74 \end{bmatrix}.$$

We note that as we move from the first to the fifth state, i.e., from the safer to the riskier, the transitions among the states increase. This suggests that when a province attains a safer risk state, it tends to exhibit increased resistance to criminal activities. Additionally, state transitions frequently occur between neighboring states, while they are often null among distant states.

Lastly, insights can be obtained from the analysis of the estimated correlation matrices. Because of the 2×2 dimension of the matrices, we simply report the correlation coefficient between each pair of variables, over the five states. In detail

$$\sigma_1^{\text{TR}} = 0.37, \ \sigma_2^{\text{TR}} = 0.35, \ \sigma_3^{\text{TR}} = 0.21, \ \sigma_4^{\text{TR}} = 0.33, \ \sigma_5^{\text{TR}} = 0.37, \ \psi_{1,\dots,5}^{\text{CH}} = 0.26.$$

We note that the correlations among the T and R variables are positive, not relatively high, and only moderately different across the states. Conversely, due to the EE parsimonious structure, we have a single correlation for the states concerning the C and H crime execution methods, which is positive but relatively low. Hence, there is a positive but low correlation between crimes and among criminal tactics.

4 Conclusions

Hidden Markov models (HMMs) stand at the forefront of analyzing time-dependent data. Here, a family of matrix-variate HMMs has been applied to a dataset measuring crime levels among 106 Italian provinces over the most four recent years.

Our results disclose, at first, the presence of five hidden states across Italy having quite different levels of criminality. Such information is gathered by the analysis of the estimated mean matrices, which allow us to sort the states from the safer (the first) to the riskier (the fifth). The estimated transition matrices reported that as we move from the first to the fifth state the transitions among the states increase and that transitions between the states mostly occur when they are adjacent. From the analysis of the switches of the provinces, we noted that most of them occurred from 2020 to 2021 and from riskier to safer states. By looking at the Italian political maps, we identified peculiarities in the memberships of the provinces in the five states, with disparities in the diffusion of crime across Italy. Lastly, correlations among the crimes and between the criminal tactics are positive but low correlated in each state.

Acknowledgments The authors have been partially supported by MUR, grant no. 2022XRHT8R (CUP: E53D23005950006)—The SMILE project: Statistical Modelling and Inference to Live the Environment, funded by the European Union, Next Generation EU.

References

1. Baum, L.E., Petrie, T., Soules, G., Weiss, N.: A maximization technique occurring in the statistical analysis of probabilistic functions of Markov chains. Ann. Math. Stat. **41**(1), 164–171 (1970)
2. Celeux, G., Govaert, G.: Gaussian parsimonious clustering models. Pattern Recogn. **28**(5), 781–793 (1995)
3. Cracolici, M.F., Uberti, T.E.: Geographical distribution of crime in Italian provinces: a spatial econometric analysis. Rev. Reg. Res. **1**(29), 1–28 (2009)
4. Lombardo, R.: Is there also a north-south divide in the diffusion of crime? A cluster analysis of Italian provinces. Rev. Dev. Econ. **20**(2), 443–455 (2016)
5. Maruotti, A., Punzo, A., Bagnato, L.: Hidden Markov and semi-Markov models with multivariate leptokurtic-normal components for robust modeling of daily returns series. J. Financ. Econ. **17**(1), 91–117 (2019)
6. Meng, X.L., Rubin, D.B.: Maximum likelihood estimation via the ECM algorithm: a general framework. Biometrika **80**(2), 267–278 (1993)
7. Punzo, A., Maruotti, A.: Clustering multivariate longitudinal observations: the contaminated Gaussian hidden Markov model. J. Comput. Graph. Stat. **25**(4), 1097–1098 (2016)
8. Punzo, A., Ingrassia, S., Maruotti, A.: Multivariate hidden Markov regression models: random covariates and heavy-tailed distributions. Stat. Pap. **62**(3), 1519–1555 (2021)
9. Schwarz, G.: Estimating the dimension of a model. Ann. Stat. **6**(2), 461–464 (1978)
10. Solivetti, L.M.: Crime patterns between tradition and change: a territorial analysis of the Italian Provinces. Soc. Indic. Res. **128**(2), 531–558 (2016)
11. Viroli, C.: Model based clustering for three-way data structures. Bayesian Anal. **6**(4), 573–602 (2011)

12. Tomarchio, S.D., Punzo, A., Maruotti, A.: Parsimonious hidden Markov models for matrix-variate longitudinal data. Stat. Comput. **32**(3), 1–18 (2022)
13. Tomarchio, S.D., Punzo, A., Maruotti, A.: Hidden Markov models for four-way data. In: Balzanella, A., Bini, M., Cavicchia, C., Verde, R. (eds.) Book of the Short Papers, pp. 127–132 (2022)
14. Tomarchio, S.D., Punzo, A., Maruotti, A.: Matrix-variate hidden Markov regression models: fixed and random covariates. J. Classif. **41**(3), 429–454 (2023)
15. Tomarchio, S.D., Punzo, A., Maruotti, A.: Matrix-variate hidden Markov regressions. In: Rampichini, C., La Rocca, M., Coretto, P., Giordano, G., Parrella, M.L. (eds.) CLADAG 2023 Book of Abstracts and Short Papers, pp. 648–651 (2023)
16. Zucchini, W., MacDonald, I.L., Langrock, R.: Hidden Markov Models for Time Series: An Introduction Using R. CRC Press, Boca Raton (2017)

MIX
Papier aus verantwortungsvollen Quellen
Paper from responsible sources
FSC® C105338

If you have any concerns about our products,
you can contact us on
ProductSafety@springernature.com

In case Publisher is established outside the EU,
the EU authorized representative is:
Springer Nature Customer Service Center GmbH
Europaplatz 3, 69115 Heidelberg, Germany

Printed by Libri Plureos GmbH
in Hamburg, Germany

SpringerBriefs in Environment, Security, Development and Peace

Volume 20

Series editor

Hans Günter Brauch, Mosbach, Germany

More information about this series at http://www.springer.com/series/10357
http://www.afes-press-books.de/html/SpringerBriefs_ESDP.htm

John W. Tai

Building Civil Society in Authoritarian China

Importance of Leadership Connections for Establishing Effective Nongovernmental Organizations in a Non-Democracy

Springer

John W. Tai
Elliott School of International Affairs
George Washington University
Washington, DC
USA

ISSN 2193-3162 ISSN 2193-3170 (electronic)
ISBN 978-3-319-03664-9 ISBN 978-3-319-03665-6 (eBook)
DOI 10.1007/978-3-319-03665-6

Library of Congress Control Number: 2014945261

Springer Cham Heidelberg New York Dordrecht London

Cover photo: The famous terracotta warriors stand guard at the tomb of China's first emperor on the outskirt of the ancient capital of Xi'an (photo by author).

Copyediting: PD Dr. Hans Günter Brauch, AFES-PRESS e.V., Mosbach, Germany

Printed on acid-free paper

Springer is part of Springer Science+Business Media (www.springer.com)

To Dad, Mom, Rebecca

Acknowledgments

The situation of China's civil society is fluid. Much of that has to do with the confusing signals that the state has been sending about its attitude toward civil society organizations. In the past few years, the state has cracked down on migrant worker NGOs in Guangdong, closed a Beijing-based NGO devoted to political reform, and convicted and imprisoned a former NGO leader who has been at the forefront of advocacy for legal and political reforms. Yet, at the same time, the state has rolled out a pilot program to ease the registration requirements for NGOs. Before this initiative, the state had begun experimentation with government procurement of NGO services. At the micro level, a few private organizations have achieved official nonprofit registration status, including Rural Women, which is featured in this book.

Amidst this uncertainty, China's NGO sector has been growing and it appears that NGOs are here to stay. However, there is no question that China's NGOs remain at a nascent stage of development, which reflects the fact that China's civil society is still under construction. While the public legitimacy of China's NGO is increasing, it still cannot rival that of the state. In addition, they continue to be resource-challenged while the state has only grown more resource-rich. In this state of affairs, the leaders of China's NGOs are critical for navigating their respective organizations through these uncertain terrains and, just as importantly, as *resource agents* for their respective organizations. Therefore, in a sense, China's NGO leaders are both planners and engineers of this great civil society-building project.

The critical importance of China's NGO leaders for their respective organizations reflects the under-institutionalized nature of China's civil society and its constituent parts. As time progresses, China's NGOs may one day achieve the organizational legitimacy that will sustain them even after their founders pass from the scene. However, until that day arrives, we must not ignore the men and women who are working daily to sustain this great construction effort. They serve as the collective inspiration for this study.

I have had the good fortune of learning about the building of China's civil society in general and NGOs in particular directly from some of these individuals. This book would not have been possible without the willingness of Chinese NGO leaders, volunteers, and experts to take the time to share with me their travails as well as their hopes and dreams. Here I would like to especially express my gratitude to Chen Jian, Chen Yuying, Gao Guanglin, Ma Yang, Wang Fang, Wei Wei, Xie Lihua, and Zhang Wenjuan, who opened their organizations to me. However, they are not to be held accountable for my portrayal and assessment of their organizations.

At the George Washington University, I benefited immensely from the guidance of outstanding advisors. Bruce Dickson, the chairman of my dissertation committee, was unfailingly patient in guiding me through every major step of the dissertation process. I could not have asked for a better mentor. Nathan Brown, the senior member of my dissertation committee, was equally kind and generous with his time. Gina Lambright, the third member of the committee, was also encouraging. Together, Nathan and Gina helped me to learn that NGOs in other countries and regions, notably the Middle East and Africa, faced similar opportunities and challenges as those in China. In addition, I am grateful to Edward McCord and Steven Balla for serving as external reviewers on my dissertation committee. Moreover, I express my appreciation to other scholars who have read portions of my dissertation, upon which this book is built: Jianling Liao, Thomas Lum, Catherine Sim, and Karla Simon.

Michael Zhao deserves special thanks for playing a critical role in funding my research in China. By offering me the opportunity to become the first visiting scholar at the Beijing Study Center of the Institute for the International Education of Students, he allowed my family and me to live in China together for an extended period of time so that I could begin the field research that has now culminated in this book. At the Beijing Study Center, I benefited from insights and interview referrals by my colleagues, Yuan Yinghua and Wendy Kang, whose knowledge of China's NGO sector was invaluable. Through the Beijing Study Center, I was also introduced to He Ping, whose experience and insights about China's NGO sector in general and environmental NGOs in particular contributed a great deal to my understanding. Also there, I had the pleasure of meeting Yang Jingyong, who introduced me to the world of the physically disabled persons in China. It was also at the Beijing Study Center that I met Wang Kan, to whom I am greatly indebted for his interview referrals and insights about China's migrant workers, state of labor relations, and NGOs.

I must extend my appreciation to Hans Günter Brauch, editor of the SpringerBriefs in Environment, Security, Development and Peace, for his willingness to consider how this study might fit into his series and for making the arrangement to review the manuscript and, ultimately, to usher it to its final form. Similarly, I benefited greatly from the comments and suggestions of the two anonymous reviewers. I have tried my best to incorporate their insights and I alone must bear the responsibility for any remaining deficiencies.

Finally, this book would not have been possible without the support of my family. They have encouraged me with their words and with their own achievements. In particular, my brother, a highly accomplished creative artist, has inspired me by his example. Similarly, my father-in-law has encouraged me with his endless pursuit of scholarship, even after over three decades in an army uniform that resulted in reaching the pinnacle of his profession. A special note about my sons, who were born just before I began this study. They have witnessed my journey, even though they were most likely oblivious to most of what had happened, which is perhaps a good thing. In any case, they have been a source of joy and they are definitely gifts from God.

I reserve my greatest love and most profound gratitude to three people. My parents have made this book a possibility a long time ago when they decided to bring my brother and me to the United States, where they have allowed me and my brother to pursue our interests and dreams without interference. My wife has not only supported my decision to return to school at the cost of a significant reduction to our household income; she has also endured my emotional ups and downs during the journey. There is no question that she is the better half in our relationship. It is to these three wonderful people that I dedicate this book.

Contents

Abbreviations

ACLA	All-China Lawyers' Association
ACWF	All-China Women's Federation
AMO	Administered Mass Organization
CDPF	China Disabled Persons' Federation
CJCRA	China Juvenile Crime Research Association
CMWSCC	China Migrant Worker Social Service Center
CPC	Communist Party of China
CPPCC	Chinese People's Political Consultative Conference
CYL	Communist Youth League
ENGO	Environmental NGO
EU	European Union
FON	Friends of Nature
GONGO	Government-Organized NGO
INGO	International NGO
NGO	Nongovernmental Organization
NPC	National People's Congress
NPO	Nonprofit Organization
PRC	People's Republic of China
SEPA	State Environmental Protection Administration
UNDP	United Nations Development Program

Chapter 1
Building Civil Society Under the Shadows of Authoritarianism

Abstract Writings about civil society in general and nongovernmental organizations (NGOs) in particular have downplayed the value of state linkage for the development of non-state social actors in non-democratic settings. Those writings have generally emphasized the state's constraints on the development of NGOs in authoritarian countries. However, an increasing number of scholars have discovered that NGOs in non-democratic countries can accrue real benefits by establishing strong linkages to the authoritarian state, which control critical organizational resources. Similarly, in existing studies about NGOs in non-democratic states, the importance of NGO leadership has been overshadowed by the attention to structural factors. Yet, NGO leaders are valuable resources for their respective organizations. Among the critical roles that NGO leaders play in advancing organizational effectiveness is their role as resource agents. The NGO's dependence on external resources means that the leader's social ties are among the most valuable leadership skills. Given the state's control of critical organizational resources in contemporary China, state linkages constitute the most valuable form of leadership ties for China's NGOs.

Keywords Nongovernmental organization (NGO) · Civil society · Leadership ties · State linkage · Media connections · International ties · Resource dependence · Organizational effectiveness · Accountability · Legitimacy · Organizational capacity

Sitting in a small meeting room on top of a storefront on the edge of the Yangzi River, I sat listening to a young, disabled woman talk about her life. It was a day in the month of May in Chongqing, one of China's famed "furnaces." The temperature was still at a reasonable 80 °F, albeit the humidity was clearly felt. Throughout the one-hour session, the young woman, who was approximately 30 years old, smiled even as she recounted her tragic past. She spoke of how she worked as a migrant worker in sweatshop conditions in Shenzhen, the driver of China's economic boom that sits immediately north of Hong Kong. She spoke about how she survived a major factory fire that claimed the lives of 100 workers. Although she lived through the disaster, she suffered severe burn to 75 % of her body. It would take several

J.W. Tai, *Building Civil Society in Authoritarian China*,
SpringerBriefs in Environment, Security, Development and Peace 20,
DOI 10.1007/978-3-319-03665-6_1

years of physical and psychological healing before she regained her independence of movement and self-esteem. During her recuperation, she found the love of her life and eventually became the mother of a young girl.

If her story ended there, it would have been an inspiring tale of individual courage. However, for those involved in building China's civil society, the story thankfully did not end there. Eventually, Chen Yuying, who is affectionately known as Xiao Ying or "Little Ying," founded the Self-Empowerment Service Center for Disabled Persons, a nongovernmental organization (NGO) that has been committed to educating and assisting local residents who are either preparing to work or are working in other parts of China as rural migrant workers. The organization also provides assistance to residents who have returned to their hometown due to occupational injuries.

This book is about men and women like Xiao Ying, people who are working to build China's civil society. In particular, it seeks to examine how leaders of China's NGOs have built those organizations. Therefore, it is unlike many other books on China's civil society, which emphasize the institutional context, notably the obstacles, to the establishment of NGOs and, ultimately, the development of a modern civil society that is familiar to Westerners. Instead, by examining individual Chinese NGOs, this book attempts to uncover the means by which modern civil society is being created in a political environment that is known for its intolerance of independent social actors.

With this approach, I am not suggesting that institutional factors are insignificant. However, the emphasis on them has cast a long shadow over analysis about the fast-developing civil society landscape in China. Similarly, this emphasis suggests that NGOs in non-democracies cannot be effective organizations because they have not brought about dramatic political changes. However, the absence of dramatic political changes does not mean that these organizations have failed to meet organizational objectives and in the process have an impact on state policies and practices while contributing to societal well-being. In fact, many NGOs in China have been effective in precisely those terms.

Another drawback to the focus on institutions is that it has drawn our attention away from the agents of those changes, i.e., the men and women who have committed their lives to the building of a modern civil society in China. A fuller understanding of China's civil society in general and its NGOs in particular cannot be achieved without examining these individuals and, most importantly, how they are leveraging their social ties to create effective organizations.

The basic premise of this book is that relations with state institutions are by far the social ties with the greatest value for China's NGOs. By extension, the value of other social ties is best judged by whether they can facilitate linkages with the state. In this respect, aside from state linkages, media connections have proven to constitute a valuable set of social ties. By contrast, international ties have been significantly less impactful. In fact, the value of international ties seems to be contingent on the presence of state linkages, media connections, or both. However, as indicated in the following section, experts have generally dismissed the importance of state linkages for civil society building in non-democracies in general and in China in particular due to their perception about the proper nature of state-civil society relationship, a perception that can impede the civil society-building project.

1.1 NGOs in Non-Democracies: The Dilemma of External Dependence

A common conceptualization of civil society is one that is labeled by some scholars as the "political" definition (White et al. 1996; Simon 2013). Those who subscribe to this conceptualization argue that civil society should be "autonomous from the state" (Diamond 1994: 5) and even in "permanent tension" with the state (Keane 1998: 6). Therefore, experts have argued that NGOs, as principal units of a modern civil society, should adopt the strategy of maintaining organizational autonomy from the state.[1]

This prescribed strategy is connected to the presumed function of civil society in general and NGOs in particular, which is to check the state: "A robust civil society, [relatively autonomous from the state], with the capacity to generate political alternatives and to monitor government and state can help [democratic] transitions get started, help resist reversals, help push [democratic] transitions to their completion, help [democracies] consolidate, and help deepen democracy" (Linz and Stepan 1996: 7, 9). This statement also suggests that in non-democratic states, NGOs and other civil society actors are expected to bring about dramatic political change, notably the overturn of non-democratic states. Given the expected adversarial relationship between civil society organizations and the authoritarian state, it is no wonder that organizational autonomy has been perceived to be of utmost importance.

This zero-sum notion has influenced even those who are not necessarily advocating that NGOs undertake programs to overturn non-democratic regimes. Julie Fisher, who has written widely on NGOs in developing countries, stated: "[The] key factor associated with the political clout of NGOs appears to be organizational autonomy…. [Organizations] that strengthen their own identities and autonomy before seeking to influence policy makers are likely to have greater latitude in initiating their own political strategies or in responding to government policies within a wide range of political contexts" (Fisher 1998: 76). According to Michael Edwards and David Hulme, the lack of organizational autonomy vis-à-vis the state means that NGOs "run the risk of being co-opted into [government] agendas" and

[1] One notable exception is Andrew Wells-Dang, who argued that networks of organizations and individuals are the most important constituents of civil society in non-democracies like China and Vietnam. See Wells-Dang (2012).

thereby lose their legitimacy as nongovernmental, grassroots-oriented institutions (Edwards and Hulme 1996: 4).

Some scholars who study China's NGO sector have likewise emphasized the value of organizational autonomy in their analyses (Shue 1994; Howell 1998; He 2008). For instance, Yiyi Lu argued that dependence on the state, or the lack of autonomy vis-à-vis the state, has been the main, indeed if not the only, cause of the ineffectiveness of Chinese NGOs: "[There] is no question that [the NGOs'] dependence on the state limits their usefulness as champions of interests and values that are different from those of the state" (Lu 2009: 116).

As these statements show, many experts have defined NGOs as "anti-government" organizations. In this conceptualization, state ties are viewed as liabilities because those relationships are perceived to act as obstacles to NGOs carrying out their functions to either check the state or overturn it. However, the political functions of NGOs are not limited to these two. Another function, as detailed in the welfare state literature, is participation in societal governance. In carrying out this function, the relationship between the state and NGOs must be more complementary, even collaborative, than adversarial. James Douglas' analysis has illustrated this idea very well (Douglas 1987: 46–7). In fact, an NGO's effectiveness in performing this function is premised on the presence of a positive state-NGO relationship.

In recent years, an increasing number of studies on China's civil society have also highlighted the complementary nature of state-civil society relations. In his study of China's response to the severe acute respiratory syndrome (SARS), Jonathan Schwartz revealed that civil society actors worked closely with the state, often at the state's instigation, to combat the crisis (Schwartz 2009). Similarly, Joan Kaufman found that the state has worked closely with NGOs to respond to China's AIDS crisis (Kaufman 2009). Jennifer Hsu and Reza Hasmath have also made similar observations about the situation pertaining to China's migrant population (Hsu and Hasmath 2013).

In short, despite the fact that many experts advocate the value of organizational autonomy for NGOs, there are indications which suggest that achieving organizational autonomy should not, and cannot, be interpreted to mean that NGOs should and are able to avoid the state without negative consequences for organizational effectiveness. This is the case even in non-democracies, especially powerful states like the one found in contemporary China. For instance, for scholars interested in the effects of civil society actors on the political system, notably NGOs involved in policy advocacy, the notion of NGO autonomy vis-à-vis the state means that autonomous groups are at risk of losing the opportunity to change the state. As Mary Gallagher wrote: "[State]-civil society interaction is important because it is changing the nature of the Chinese state" (Gallagher 2004: 422). This is a useful reminder that, for NGOs wanting to influence existing state policies and practices, they should be most interested in engaging the state. The alternative could be marginalization by the state, which are likely to result in organizational ineffectiveness.

Gallagher's statement also highlights the value of state linkage as a means to access state *resources*, which in this case is the state's policy apparatus. This idea is in contrast to the perception that an NGO must maintain resource autonomy vis-à-vis the state out of concern that the price of obtaining state resources is the loss of "decisional competence, recruitment of leaders, and control of important economic and managerial resources" (Hyden 1997). According to this perspective, a reliance on "state funding and support can also be dangerous to [an NGO] in that it may compromise the group's ability to criticize state policies…" (Sperling 2006: 164). The implication of this perspective on organizational autonomy is that NGOs should diversify the origins of organizational resources, with the clear goal of maintaining a distance from the state.

However, the challenging resource environment has often meant that efforts to achieve resource autonomy from the state have led NGOs in non-democracies to develop dependence on foreign resources. Such resource dependence has led to the criticism that NGOs have succumbed to foreign agenda, become elitist, and lost touch with both the state and the people, thereby becoming illegitimate (Hudock 1999; Hemment 2007). Even more damning has been the charge that foreign funding has had the effect of "fueling" the very socio-political system that NGOs were supposed to transform (Chabal and Daloz 1999: 23). Just as importantly, foreign funding is very limited and difficult to access. In other words, a reliance on foreign resources can simply result in substituting state dependence with a reliance on foreign donors, with negative consequences for organizational effectiveness.

Ultimately, the preference for resource autonomy vis-à-vis the state is likely to be an unsustainable proposition. In a resource-poor environment in which the powerful state controls all key organizational resources, including policy access and material resources, the standard of organizational autonomy vis-à-vis the state would place NGOs in non-democracies in a bind. Resource attainment, regardless of the sources, should be the priority for NGOs. Similarly, policy advocacy NGOs must do all it can to gain access to the state. In this sense, it seems counterproductive for NGOs to maintain autonomy from the state. In addition, it is important to be reminded that the establishment of state linkages does not automatically imply that NGOs lose the ability to maintain organizational control. In sum, in order for NGOs in non-democracies to achieve their objectives, they need to engage the state.

1.2 Disaggregated State: Opportunities for Accessing Resources and Achieving Effectiveness

Similar to Gallagher's finding in the context of the Chinese NGOs, Kate Thomson, in her study of Russian disability NGOs, found that close ties to government agencies have helped some Russian organizations access state and non-state resources (Thomson 2006: 237). Likewise, Amaney Jamal found that in Palestine organizations linked to the Palestinian National Authority have enjoyed access to important organizational resources (Jamal 2007: 51–58). These and other authors pointed out that resources of importance to NGOs have not been limited to material and financial goods. For example, Jamal identified "a sense of reassurance and security" as an important organizational resource (Jamal 2007: 56). Lisa Sundstrom and Laura Henry suggested that public legitimacy is another key organizational resource for NGOs in non-democratic countries because in those places "many citizens continue to believe that the state's authority should be largely unquestioned, and that the state's imposition of order is more important than freedom or democracy" (Sundstrom and Henry 2006: 318). Collectively, these findings suggest that the rewards for engaging the state can be quite significant for NGOs. But why would a non-democratic state be willing to allow NGOs, which often draw state suspicion, to access valuable state resources?

An answer may be found in the fact that authoritarian states behave differently with respect to different policy issue areas. For instance, Suzaina Kadir found that in Singapore, NGOs advocating women's issues have had greater success than those addressing ethnic Malay issues (Kadir 2004). She concluded that this phenomenon had to do with how the state viewed the two sets of issues. According to Kadir, ethnic issues were perceived by the state to touch on the sensitive area of identity politics, which the Singaporean state has always been careful to control out of the fear that these matters, if improperly handled, could lead to political instability. However, as important as issue area is for explaining variations in state-NGO relations, the more fundamental cause in Kadir's study was the fact that the Singaporean state felt "the need to engage with civil society," although it wanted "to do so at a pace and scope" that it found "agreeable" (Kadir 2004: 349).

Kadir's study makes it clear that there are opportunities for NGOs to survive and thrive in non-democracies, and those opportunities have been the products of state policies. Similarly, in his study of environmental, HIV/AIDS, and gay and lesbian NGOs in China, Timothy Hildebrandt credited government policies for creating a demand for NGOs to work in these issue areas (Hildebrandt 2013: 26–31). However, he also pointed out that the state's program of decentralization has given more responsibilities and powers to sub-national governments, which have played a critical role in determining the nature of government relations with NGOs (Hildebrandt 2013: 44–52).

Hildebrandt's findings highlight the fact that states, even non-democratic ones, are not monolithic. As Jennifer Coston observed, "governments are not monolithic: [regimes] of all types may incorporate agencies and actors that are more cooperative or repressive than the overall regime" (Coston 1998: 363). The model of "fragmented authoritarianism" developed by Kenneth Lieberthal and Michel Oksenberg shows that this general notion about the disaggregated nature of the state is also applicable to the situation in China (Lieberthal and Oksenberg 1988; Lieberthal 1995).

This understanding about the state is important because it demonstrates that there are many state actors and those actors may harbor different interests. On the one hand, this has created an uncertain political environment for the NGOs to engage the state. On the other hand, it also indicates that there are opportunities for NGOs to develop beneficial relations with state actors. The right NGO leader will be able to uncover and exploit those opportunities to the organization's advantages.

1.3 NGO Leader as a Resource Agent

As mentioned at the beginning of this chapter, the growing literature on China's civil society remains fixated on structural factors. The importance of institutions notwithstanding, a focus on structural constraints yields an incomplete picture of the relationship between the state and societal actors because it privileges the state side of the equation and by extension favors the state's perspective. Consequently, the NGO's perspective has not been given sufficient attention. With a few exceptions (Economy 2004; Ma 2006), NGOs have either been viewed in a negative light (Lu 2009), overshadowed by structural constraints (Hildebrandt 2013), or analyzed as a component of other units of analysis (Wells-Dang 2012). In addition, this conceptual emphasis has overshadowed the importance of organizational leadership for China's NGOs.

In contrast to the writings on China's NGOs, organizational leadership has been a subject of considerable interest among the large body of works on nonprofit organizations (NPOs). These works offer some observations about the qualities and skills of successful leaders. For instance, one identified quality is charisma. A charismatic leader is valuable because he or she "inspires staff and members to put aside personal self-interest for the common good of the organization and to have confidence in their ability to achieve the 'extraordinary' challenges before them" (Anheier 2005: 163). According to some NPO experts, a successful leader is also often an entrepreneur who fosters innovation (Young 1987: 168). A third quality is that NPO leaders, in contrast to their business and political peers, are motivated neither by political ambitions nor by financial gains (James 1987: 404–405). Lastly, in keeping with these observations, David Bornstein indicated that as "social entrepreneurs" who aim to provide solutions to social problems, NGO leaders have often displayed a willingness to "self-correct," "share credit," "break free of established structures," "cross disciplinary boundaries," "work quietly," as well as demonstrating a "strong ethical impetus" (Bornstein 2007: 238–246).

These leadership qualities are surely important. However, we must not overlook another important leadership quality, or skill, which is resource acquisition. Resource attainment is important for any type of organization. The importance of acquiring resources thus makes "resource dependence" an important organizational attribute. According to Jeffrey Pfeffer and Gerald Salancik, resource dependence demonstrates that every organization is affected by its environment because organizational resources often originate outside the organization (Pfeffer and Salancik 2003). In this situation, an organization's leaders/managers are important because they manage the organization's resource environment. Put simply, these individuals are central to resource acquisition.

As indicated earlier, the powerful non-democratic state is an important object of resource attainment for NGOs because the former controls, in fact monopolizes, critical resources. Therefore, arguably a key responsibility of NGO leaders should be to play a decisive role in facilitating access to state resources, including whether to engage the state. In fact, scholars have identified this role among NGO leaders in non-democracies around the world. For instance, leaders of environmental NGOs (ENGOs) in Russia make the "decision of whether to oppose, cooperate with, or avoid the state…" (Henry 2006: 224). Highlighting the linkage between NGO leaders and access to state resources, Maha Abdelrahman has found that a disproportionate number of NGO leaders in Egypt were government employees (Abdelrahman 2004: 154). Scholars who study NGOs in China have made similar observations. According to Kang Xiaoguang and his colleagues, NGO leaders in China have had a positive effect on the efforts of Chinese NGOs to obtain virtually all forms of state resources (Kang et al. 2010: 26–30). Likewise, Jiang Ru and Leonard Ortolano have found that successful Chinese ENGOs have often been led by leaders with connections to the state (Ru and Ortolano 2008: 62).

Put simply, from the perspective of resource dependence, the NGO leader's function as a *resource agent* stands out as the most important leadership role. His or her ability to help the organization access resources, notably state resources, goes a long way to define the extent to which an NGO leader is a successful one, and the extent to which the organization can achieve effectiveness.

1.4 Concepts: NGO Effectiveness and Leadership Ties

The resource-dependent nature of NGOs and their significant reliance on external resources make it very clear that the most important skill of the leaders of Chinese NGOs is their social ties. However, before analyzing how the social ties of NGO leaders in China can contribute to organizational effectiveness, it is necessary to define the concept of *NGO effectiveness* and determine the *leadership ties* that will be the focus of this study. As will be seen in the remainder of this section, the concept of NGO effectiveness is a heavily-contested one. In comparison, the leadership ties identified are less controversial. However, there is still a need to explain why some leadership ties are removed from more detailed examination.

1.4.1 NGO Effectiveness

The focus of this study is to examine how NGO leaders contribute to organizational success. Hence, I do not emphasize the capacity of NGOs to achieve dramatic political change, which does not correspond to reality in China because many, if not most, NGO leaders do not perceive this to be their aim. Instead, organizational success is defined in this book as *organizational effectiveness*. In turn, organizational effectiveness is defined as "goal effectiveness": "the extent to which [an NGO] is able to achieve the goals it sets for itself" (Calnan 2008: 231).

One reason for choosing this definition is to avoid the endless debate about the best definition for organizational effectiveness. Indeed, it is nearly impossible for experts to reach a consensus on the definition of effectiveness. One challenge is that depending on the type of work in which an NGO engages, its specific performance indicators may be different from another NGO that conducts a different type of work. For example, the performance indicators for an NGO that provides services to autistic children may include ones that measure the number of autistic children under its care, the number of staff devoted to each child, the number of beds in the facilities, and other similar indicators. The same performance indicators obviously cannot be applied to a policy research NGO, the effectiveness of which is probably best assessed in terms of the extent to which its policy recommendations have been adopted by government agencies.

The lack of consensus about the measurements for effective NGOs can be readily observed by the comments made by the staff of international donors who work in China. One former program officer for an international donor in China told me that donors have generally agreed that effectiveness should be measured in terms of "process" rather than "outcome." A process-based indicator of effectiveness is one that assesses the extent to which an NGO effectively engages the organization's multiple stakeholders (Fowler 1997: 177–179). In a nutshell, this type of measurement is associated with a focus on the extent to which NGOs are accountable. In contrast to this line of reasoning, a former research director for a U.S.-based donor shared with me his opinion that organizational capacity should be the fundamental criterion for NGO effectiveness.

In addition to the notion that effectiveness should be assessed based on stakeholder relationships, some experts and practitioners have indicated that there are some common indicators of effectiveness that would apply to all NGOs because there are certain organizational characteristics that distinguish NGOs from other types of organizations. The shared NGO characteristics include the prominence of values in NGO work and the public-interest nature of NGO missions and goals. Therefore, based on existing studies, many writers advocate that NGO effectiveness be measured in terms of the NGO's organizational capacity, accountability, and public legitimacy.

Although these three characteristics have been distinguished for analytical purposes, they are in reality inter-connected. The reason is that NGOs are widely perceived to be value-driven entities, as indicated above, especially in comparison to for-profit firms (Anheier 2005: 160). Therefore, many experts have argued that the effectiveness of any NGO cannot be judged solely on capacity indicators, as important as they are (Fowler 1997). Moreover, because NGOs engage in public-interest activities, they are also perceived to be accountable to multiple stakeholders, both within and outside the NGOs. Therefore, the effectiveness of any NGO must also be assessed in terms of whether it can maintain accountability and establish public legitimacy.

In situations such as China, where grassroots NGOs face a significantly constrained resource environment, there is considerable validity to the observation that NGO effectiveness should be synonymous with NGO survival (McMahon 2006: 49). Even so, some NGO leaders have clearly been interested in surviving "better" than others. An indication of this desire has been the comments made by some

Chinese NGO leaders that their priority has been the building up of their respective organizational "brand names" (Interview BJ040109; Wei Wei Interview 2009).

In any case, the issue of organizational survival is essentially one about organizational capacity, or the ability of the organization to carry out its stated objectives. Sarah Michael, in her study of local African development NGOs, has made a direct connection between organizational effectiveness and organizational capacity (Michael 2004). In her study, Michael defined effectiveness in terms of power. She observed that four features have characterized the most powerful development NGOs in the world: size, wealth, reputation, and the ability to write and rewrite the rules (Michael 2004: 24–37). The Bangladesh Rural Advancement Committee (BRAC) was cited by her as an illustration of an effective, or powerful, NGO. As Michael described it, BRAC had employed 60,000 people, not including volunteers, to administer projects involving services to over 4 million people. The organization's 1999 budget was over $130 million.

However, there are reasons to believe that organizational capacity may not be the best indicator of NGO effectiveness, especially with respect to NGOs operating in contemporary China. First, no Chinese grassroots NGO has achieved the level of effectiveness that has been demonstrated by BRAC. In fact, Chinese experts on the Chinese NGO sector have observed consistently that the Chinese NGOs, irrespective of their registration status, have lacked strong organizational capacity (Deng 2007). A more important reason, however, is that we cannot presuppose that organizations with strong capacity will put those resources into good use, just as we can no more assume that organizations lacking capacity will be ineffective. The case of the Self-Empowerment Service Center for Disabled Persons will testify to this point. The NGO's organizational capacity pales in comparison to those of many other Chinese NGOs; but the NGO has been quite effective in servicing its constituents.

Experts have also identified accountability to be another important criterion to evaluate NGO effectiveness (Edwards and Hulme 1996; Fowler 1997: 180–183; Jordan and van Tuijl 2006). At the core of the issue of accountability is the notion that NGOs, as public organizations with multiple stakeholders, should be held accountable for how they use the resources at their command. Transparency is therefore directly related to the notion of accountability. And there are internal and external dimensions of accountability. Internal accountability often involves the presence of a board of directors (Tandon 1996). External accountability, on the other hand, may involve the extent to which the NGO is transparent to stakeholders outside the organization about its activities through regular reports and audits that are made available publicly in an easily accessible manner (Deng 2007: 175).

Organizational accountability or lack thereof can certainly have implications for NGO effectiveness. An accountable NGO is arguably more likely to use its resources wisely and efficiently, thereby contributing to optimal organizational performance. Moreover, an NGO that enjoys a reputation as one that upholds the principles of accountability is more likely to attract more resources to support its efforts to meet the organization's goals. However, accountability by itself may not be the best indicator of organizational effectiveness. One reason is that any assessment of accountability has to take into account multiple objects of accountability, or stakeholders, and stakeholders do not and should not exercise equal influence upon the organization. The employment of accountability as an indicator of organizational effectiveness is therefore likely to require the analyst to engage in somewhat arbitrary selection and prioritization of the various objects of accountability. In this vein, it will be difficult to come up with a set of indicators that can objectively measure accountability. Furthermore, as is the case with organizational capacity, we cannot assume that an accountable NGO will perform effectively.

Related to the issue of accountability is the notion of organizational legitimacy. For Chinese NGOs, legitimacy has important meanings because the public has generally displayed very little trust in these organizations. Nonetheless, some Chinese NGOs have clearly enjoyed greater legitimacy than others, as reflected in the positive evaluations by government officials, donors, media agencies, outside experts, and beneficiaries (Deng 2007: 152, 177). Indeed, strong organizational legitimacy may be a means for the NGO to achieve effectiveness because it facilitates access to organizational resources and protects the organization from state predation. But many of the same problems associated with the use of accountability as an indicator for effectiveness are also applicable to the use of public legitimacy as an indicator of effectiveness.

In sum, a review of the existing literature makes it quite clear that despite the plethora of offerings by experts about indicators for NGO effectiveness, most of the indicators are either difficult to measure or have serious deficiencies. Therefore, the definition that I have chosen appears to be the best by default. However, there are more valid reasons to adopt goal effectiveness as the definition for NGO effectiveness. One important reason is that by using the organization's own set of standards, it eliminates the inevitable biases that come with imposing outside standards on the organization. Another important reason is that this definition of NGO effectiveness best facilitates comparisons among NGOs, regardless of any differences with respect to issue area and focus.

1.4.2 Leadership Ties

As a resource agent, the NGO leader must possess the skills to acquire resources for her organization. Among those skills, her social ties arguably exercise the most decisive influence on her ability to acquire resources. Given the significance of state resources, those social ties must first and foremost help her access state resources. In this study, I will examine the extent to which the state linkages, media connections, and international ties of NGO leaders have impacted organizational performance.

The choice of these three sets of social ties has been informed by the writings of other analysts and by my interactions with NGO leaders in China. However, while mentioning the value of these three types of social ties, writers have not examined in detail how those ties contribute to China's NGOs. They also have not examined the interactions among the ties and how those interactions affect China's NGOs. Therefore, in this book, I will not only examine how each type of social tie is translated into benefits for the NGOs in question; I will also examine how they interact with each other and, ultimately, how they stack up against each other in terms of contributions to NGO effectiveness.

In emphasizing these ties, I am clearly leaving out others. Therefore, it is important to address potential criticisms about those ties that I have left out, notably the relationship with other NGOs.

Inter-NGO linkages may be valuable for NGOs. In recent years, we have witnessed the presence of various types of networks involving Chinese NGOs. However, based on my research, we can only speak about the benefits of inter-NGO linkages in China in potential terms. There are at least two reasons for this observation. First, substantive inter-NGO linkages, through which NGOs interact with each other on a regular basis and which involve substantial transfers of benefits, remain limited in scope. The most developed networks in this respect have been found among ENGOs, and these networks have generally revolved around China's most prominent ENGOs (Ru and Ortolano 2008; Wells-Dang 2012). More importantly, the NGOs involved in those networks constitute only a small fraction of the Chinese NGO population. Second, due to the limited resources available to NGOs, most Chinese NGO leaders, even the leaders of the most prominent and long-standing NGOs, have made it very clear that their priority is to build their respective organizational brand (*pinpai*). This means that they have jealously guarded their hard-earned resources, which has resulted in a minimal transfer of resources among NGOs. Furthermore, given the Chinese NGO sector's lack of political and popular legitimacy, and given the fact that organizational resources needed by the NGOs are either owned or regulated by the state, building linkages among NGOs can yield very little, if any, direct and immediate benefits on individual NGOs. In fact, inter-NGO linkages may be detrimental to NGOs' relations with the state (Hildebrandt 2013: 79). All of this is not to deny the fact that NGO alliances have had noticeable effects on major advocacy campaigns and, most recently, post-disaster relief efforts. But the effects of those alliances have required the presence of friendly state officials (Mertha 2008; Wells-Dang 2012). Moreover, the effects of such alliances on the effectiveness of individual NGOs have been inconsistent at best.

Therefore, it should be unsurprising that in non-democracies like China, NGOs have generally sought linkages with non-NGO entities that could best advance the organization's interests. The importance of the state in China means that it is important for the NGO to establish ties with the state. And the degree to which the leadership possesses state linkages is correlated with the relative effectiveness of the NGO.

Although it is not necessary to be state employees in order to have state linkages, membership in the officialdom is certainly a useful means to acquire state linkages. Scholars have indeed observed that many effective NGOs in non-democracies have been led by current and former government employees. In some cases, such as Egypt, this phenomenon has reflected the enormous scale of that country's state bureaucracy and public sector (Abdelrahman 2004: 155). Like Egypt, China is also home to a huge state bureaucracy and public sector (Pei 2006). But other factors may also account for the value of being state officials in China. One of those factors may be historical. Throughout Chinese history, membership in the officialdom has usually conferred that individual with a degree of respect and legitimacy that virtually no other social position could rival.

Whatever the nature of state linkages, NGO leaders who possess this type of social ties may be more likely to benefit their organizations mainly because in comparison with those leaders who do not possess similar linkages, NGO leaders who possess them are more likely to have access to state resources, including financial capital and political networks. Similarly, leaders with state linkages are presumably more likely to have access to the policymaking process. In this manner, NGOs led by such individuals will more likely to have opportunities to impact policies.

Interestingly, NGOs that are led by leaders with state linkages have also been successful in accessing non-state resources. One reason is that state linkages have often been accepted by ordinary citizens in those countries as a sign of legitimacy and reliability. Abdelrahman observed that in Egypt, local people would trust NGOs more if they were led by government officials because people expected that "respectable civil servants [would] not misuse public money or their position in the NGO for any personal gain" (Abdelrahman 2004: 155). In China, there has also been a perception that NGO leaders with state linkages are more trustworthy because they are more likely to manage their organizations in accordance with rules and regulations (Interview BJZX042209).

Another type of social ties is media connections. A significant number of Chinese NGO leaders are either current or former journalists. The media's impact on NGO effectiveness cannot be underestimated, especially in China where the media have historically served as a powerful instrument of the authoritarian state. By virtue of the media's political association with the state, there has been a popular perception among both government officials and ordinary people that media workers have access to the most important and updated information about political and social issues. This also suggests that media professionals also have access to policymakers and the policy process. In recent years, as China's media organizations have attempted to carve out more autonomous spaces, Chinese journalists have increasingly written news reports that are critical of state agencies and officials. As a result, Chinese media organizations in general and Chinese journalists in particular have earned increasing popular legitimacy. One consequence has been the fact that ordinary Chinese people have sought out journalists to provide information about official wrongdoing (Economy 2004). The increasing popular legitimacy of the Chinese media has also meant that media reports about NGOs have the potential to do much to enhance "the visibility, reputation, and influence of the NGOs" (Shieh 2009: 36). In fact, Chinese journalists and media outlets have become more outspoken and critical of state policies and practices despite the common perception that China remains a "propaganda state."

However, just as it would be simplistic to assume that collectively the Chinese media are simply the mouthpiece of the state, we should not assume that a journalist profession is the only ticket to the media for an NGO leader. In other words, an NGO leader may possess media connections without ever having worked as a media professional. As media reporting has enhanced the NGOs' organizational visibility and reputation, media reporting has also raised the profile of NGO leaders who were not professional journalists. As "media darlings," these NGO leaders have enjoyed regular media exposure. Access to media has become an important vehicle for these individuals to establish state connections, both for them as individuals and for the organizations that they lead. Wei Wei, the founder of the Little Bird Migrant Workers Mutual Support Hotline, which is one of the four core NGO cases featured in this book, is a well-known representative of NGO leaders who have benefited from tremendous media support without owning a journalist pedigree.

The third and final type of leadership ties is international ties. It is a well-known fact that NGOs in non-democracies and developing countries have relied on foreign donors, notably those from Western countries, for the necessary organizational resources. The reasons for this phenomenon are many and they have to do with the situations in non-democracies and developing countries. One reason has been that in those parts of the world, states have not been supportive of NGOs, either because the states have lacked the capacity to do so, lacked the will to do so, or, in most cases, both. Another reason has been that the fundraising environment outside the state apparatus in those countries has been equally unsupportive. In those cases, the resource-poor environment might be either an outcome of state regulations, the underdevelopment of an indigenous culture of philanthropy, or the NGO's lack of legitimacy as a social institution.

However, in contrast to the NGO leader's political ties and his media connections, international ties may not facilitate positive connections with the state in a direct manner. In fact, some scholars have argued that in China, international ties have generally served as a means by which NGOs could achieve greater organizational autonomy vis-à-vis the state (Shieh 2009: 35). Indeed, there is evidence to suggest that state linkages and media connections may be stronger predictors of international ties than the other way around.

1.5 Organization of the Book

The remainder of this book is structured around a presentation of each of the three types of leadership ties, namely state linkages, media connections, and international ties. In Chap. 2 I will provide a description of the NGO landscape in contemporary China, with an emphasis on the political challenges facing NGOs in China, which highlights the importance of the state for NGOs operating there. I will also examine the state of China's rural migrants, with a look at the NGOs servicing this large population. Chapter 2 will end with a formal presentation of the four core cases and three shadow cases that constitute the key analytical elements of this book.

The empirical findings will be presented in the three subsequent chapters. Each chapter is structured around one of the three types of leadership ties. The aim of each chapter is to analyze the processes through which each type of ties has exerted positive impact on NGO effectiveness. As a part of the comparative framework, the discussions in Chaps. 4 and 5 will include analyses about how the different types of leadership ties interact with each other. Taken together, the three chapters should make clear the relative causal weight of each type of ties through an examination of the interactions among the three types of ties and the effects those interactions have on NGO effectiveness.

In Chap. 6 and the concluding chapter, I will review the central argument of this thesis as well as its main findings and their implications. I will also suggest new avenues for further research. In addition, I will make some recommendations to international donors and domestic Chinese NGOs about how best to advance the civil society-building project in contemporary China.

The last section of the book is the appendix, which provides details about the scope of the study, the data sources, the research methodology, as well as the study's limitations.

References

Abdelrahman, Maha M. 2004. *Civil Society Exposed: The Politics of NGOs in Egypt*. London: Tauris Academic Studies.

Anheier, Helmut K. 2005. *Nonprofit Organizations: Theory, management, policy*. New York: Routledge.

Bornstein, David. 2007. *How to Change the World: Social Entrepreneurs and the Power of New Ideas*. New York: Oxford University Press.

Calnan, Scott. 2008. *The Effectiveness of Domestic Human Rights NGOs: A Comparative Study*. Leiden: Martinus Nijhoff Publishers.

Chabal, Patric and Jean-Pasacal Daloz. 1999. *Africa Works: Disorder as Political Instrument*. Bloomington: Indiana University Press.

Coston, Jennifer M. 1998. A Model and Typology of Government-NGO Relationships. *Nonprofit and Voluntary Sector Quarterly* 27, 3: 358-382.

Deng Guosheng. 2007. *Mingjian Zuzhi Pinggu Tixi: Lilun, Fangfa Yu Zhibiao Tixi* [*NGO Evaluation: Theory, Methods & Indicator System*]. Beijing: Peking University Press.

Diamond, Larry. 1994. Toward Democratic Consolidation. *Journal of Democracy* 5, 3: 4-17.

Douglas, James. 1987. Political Theories of Nonprofit Organization. In *The Nonprofit Sector: A Research Handbook*, ed. Walter W. Powell, 43-54. New Haven: Yale University Press.

Economy, Elizabeth C. 2004. *The River Runs Black: the Environmental Challenge to China's Future*. Ithaca: Cornell University Press.

Edwards, Michael and David Hulme. 1996. Introduction: NGO Performance and Accountability. In *Beyond the Magic Bullet: NGO Performance and Accountability in the Post-Cold War World*, eds. Michael Edwards and David Hulme, 1-20. West Hartford: Kumarian Press.

Fisher, Julie. 1998. *Nongovernments: NGOs and the Political Development of the Third World*. West Hartford: Kumarian Press.

Fowler, Alan. 1997. *Striking a Balance: A Guide to Enhancing the Effectiveness of Non-Governmental Organizations in International Development*. London: Earthscan.
Gallagher, Mary E. 2004. China: The Limits of Civil Society in a Late Leninist State. In *Civil Society and Political Change in Asia: Expanding and Contracting Democratic Space*, ed. Muthiah Alagappa, 419–452. Stanford: Stanford University Press.
He, Zengke. 2008. Institutional barriers to the development of civil society in China. In *China's Opening Society: The non-state sector and governance*, eds. Zheng Yongnian and Joseph Fewsmith, 161–173. London: Routledge.
Hemment, Julie. 2007. *Empowering Women in Russia: Activism, AID, and NGOs*. Bloomington: Indiana University Press.
Henry, Laura A. 2006. Russian Environmentalists and Civil Society. In *Russian Civil Society: A Critical Assessment*, eds. Alfred B. Evans, Jr. et al, 211–228. Armonk: M.E. Sharpe.
Hildebrandt, Timothy. 2013. *Social Organizations and the Authoritarian State in China*. New York: Cambridge University Press.
Howell, Jude. 1998. An Unholy Trinity? Civil Society, Economic Liberalization and Democratization in Post-Mao China. *Government and Opposition* 33, 1: 56-80.
Hsu, Jennifer Y.J. and Reza Hasmath. 2013. The Rise and Impact of the Local State on the NGO Sector. In *The Chinese Corporatist State: Adaptation, Survival and Resistance*, eds. Jennifer Y. J. Hsu and Reza Hasmath, 120-135. New York: Routledge.
Hudock, Ann C. 1999. *NGOs and Civil Society: Democracy by Proxy*? Cambridge: Polity Press.
Hyden, Goran. 1997. Civil Society, Social Capital, and Development: Dissection of a Complex Discourse. *Studies in Comparative International Development* 32, 1: 3-30.
Jamal, Amaney. 2007. *Barriers to Democracy: The Other Side of Social Capital in Palestine and the Arab World*. Princeton: Princeton University Press.
James, Estelle. 1987. The Nonprofit Sector in Comparative Perspective. In *The Nonprofit Sector: A Research Handbook*, ed. Walter W. Powell, 397-415. New Haven: Yale University Press.
Jordan, Lisa and Peter van Tuijl. 2006. *NGO Accountability: Politics, Principles & Innovations*. London: Earthscan.
Kadir, Suzaina. 2004. Singapore: Engagement and Autonomy Within the Political Status Quo. In *Civil Society and Political Change in Asia: Expanding and Contracting Democratic Space*, ed. Muthiah Alagappa, 324-354. Stanford: Stanford University Press.
Kang et al. 2010. *NGO Yu zZhengfu Hezuo Celue* [*Strategic Study on NGO-Government Collaborations*]. Beijing: Social Sciences Academic Press.
Kaufman, Joan. 2009. The role of NGOs in China's AIDS Crisis: Challenges and possibilities. In *State and Society Responses to Social Welfare Needs in China: Serving the people*, eds. Jonathan Schwartz and Shawn Shieh, 156-173. New York: Routledge.
Keane, John. 1998. *Civil Society*. Cambridge: Polity Press.
Lieberthal, Kenneth. 1995. *Governing China: From Revolution Through Reform*. New York: W.W. Norton.
Lieberthal, Kenneth and Michel Oksenberg. 1988. *Policy Making in China: Leaders, Structures, and Processes*. Princeton: Princeton University Press.
Linz, Juan J. and Alfred Stepan. 1996. *Problems of Democratic Consolidation: Southern Europe, South America, and Post-Communist Europe. Baltimore*: The Johns Hopkins University Press.
Lu, Yiyi. 2009. *Non-Governmental Organizations in China: The rise of dependent autonomy*. New York: Routledge.
McMahon, Patrice C. 2006. International Actors and Women's NGOs in Poland and Hungary. In *The Power and Limits of NGOs: A Critical Look at Building Democracy in Eastern Europe and Eurasia*, eds. Sarah E. Mendelson and John K. Glenn, 29-53. New York: Columbia University Press.
Ma, Qiusha. 2006. *Non-Governmental Organizations in Contemporary China: Paving the way to civil society*? New York: Routledge.
Mertha, Andrew C. 2008. *China's Water Warriors: Citizen Action and Policy Change*. Ithaca: Cornell University Press.

Michael, Sarah. 2004. *Undermining Development: The Absence of Power Among Local NGOs in Africa*. Bloomington: Indiana University Press.

Pei, Minxin. 2006. *China's Trapped Transition: The Limits of Developmental Autocracy*. Cambridge: Harvard University Press.

Pfeffer, Jeffrey and Gerald Salancik. 2003. *The External Control of Organizations: A Resource Dependence Perspective*. Stanford: Stanford University Press.

Ru, Jiang and Leonard Ortolano. 2008. Corporatist Control of Environmental Nongovernmental Organizations: a State Perspective. In *China's Embedded Activism: Opportunities and Constraints of a Social Movement*, eds. Peter Ho and Richard Edmonds, 44-68. London: Routledge.

Schwartz, Jonathan. 2009. The Impact of Crises on Social Service Provision in China: The State and Society Respond to SARS. In *State and Society Responses to Social Welfare Needs in China: Serving the people*, 133-155. New York: Routledge.

Shieh, Shawn. 2009. Beyond Corporatism and Civil Society: Three Modes of State-NGO Interaction in China. In *State and Society Responses to Social Welfare Needs in China: Serving the people*, 22-41. New York: Routledge.

Shue, Vivienne. 1994. State Power and Social Organization in China. In *State Power and Social Forces: Domination and Transformation in the Third World*, eds. Joel S. Migdal et al, 65-88. New York: Cambridge University Press.

Simon, Karla W. 2013. *Civil Society in China: The Legal Framework from Ancient Times to the "New Reform Era."* New York: Oxford University Press.

Sperling, Valerie. 2006. Women's Organizations: Institutionalized Interest Groups or Vulnerable Dissidents? In *Russian Civil Society: A Critical Assessment*, eds. Alfred B. Evans, Jr. et al, 161-177. Armonk: M.E. Sharpe.

Sundstrom, Lisa McIntosh and Laura A. Henry. 2006. Russian Civil Society: Tensions and Trajectories. In *Russian Civil Society: A Critical Assessment*, eds. Alfred B. Evans, Jr. et al, 305-322. Armonk: M.E. Sharpe.

Tandon, Rajesh. 1996. Board Games: Governance and Accountability in NGOs. In *Beyond the Magic Bullet: NGO Performance and Accountability in the Post-Cold War World*, eds. Michael Edwards and David Hulme, 53063. West Hartford: Kumarian Press.

Thomson, Kate. 2006. Disability Organizations in the Regions. In *Russian Civil Society: A Critical Assessment*, eds. Alfred B. Evans, Jr. et al, 229-245. Armonk: M.E. Sharpe.

Wells-Dang. 2012. *Civil Society Networks in China and Vietnam: Informal Pathbreakers in Health and the Environment*. London: Palgrave Macmillan.

White, Gordon et al. 1996. *In Search of Civil Society: Market Reform and Social Change in Contemporary China*. New York: Oxford University Press.

Young, Dennis R. 1987. Executive Leadership in Nonprofit Organizations. In *The Nonprofit Sector: A Research Handbook*, ed. Walter W. Powell, 167-179. New Haven: Yale University Press.

Chapter 2
Chinese NGOs: Thriving Amidst Adversity

Abstract Notwithstanding media reports about how the authoritarian state in China has repressed its civil society, China's NGO sector has grown significantly in number over the past two decades. In fact, the state has played a decisive role in the emergence of the NGO sector in China, which is also highly diverse. It includes both organizations that are officially registered as nonprofit organizations and those that do not possess this status; it also includes organizations that should be more properly viewed as state institutions and those organizations that meet the conventional criteria for NGOs. However, despite the state's positive contributions to the development of China's NGO sector, it continues to view NGOs with suspicion and has adopted legal, political, and practical measures to control China's NGOs. A significant consequence of these control measures is that many Chinese NGOs are unable to attain official nonprofit registration status, thereby negatively impacting their social and political legitimacy. This fact, in combination with the state's overwhelming legitimacy, makes it all the more important for individual Chinese NGOs to reach out to the state, thereby affirming the value of state linkages. In recent years, the rising demand for rural migrant workers to contribute to China's modernization and urbanization has highlighted the institutionalized discrimination against this population group. Consequently, migrant NGOs have emerged to advocate on behalf of migrant workers and their dependents as well as to serve their needs. These NGOs are keenly aware of the value of state linkages. Some of these organizations have also been quite effective. Therefore, a close look at these organizations can help us understand how China's NGOs have been able to achieve their goals while allowing us to gain an in-depth understanding of migrant NGOs in contemporary China.

Keywords Government-organized NGO (GONGO) · Mass organization · State control · Registration status · Organizational legitimacy · Rural migrant worker · *Hukou* · Shadow case

J.W. Tai, *Building Civil Society in Authoritarian China*,
SpringerBriefs in Environment, Security, Development and Peace 20,
DOI 10.1007/978-3-319-03665-6_2

2.1 Rapid Growth of China's NGO Sector

China has been undergoing an "association revolution" over the last 20 years, particularly since the second half of the 1990s. By the end of 2013, there were over 500,000 registered NGOs in China. Between 1988 and 2013, the number of registered NGOs had increased over 100 times, from just under 4,500 in 1988 to over 540,000 in 2013 (Fig. 2.1). In addition, during the same time period, the categories of registered NGOs have also increased from one in 1988 (social groups) to three by 2004, with the introduction of non-state non-commercial groups and foundations. Most of these registered organizations were, and are, former state institutions like hospitals and schools, as well as state-sponsored trade and professional associations. However, there are also genuine NGOs involved in other sectors and engaged in other types of activities within this growing group.

According to many experts, the aforementioned number of NGOs does not even come close to representing the size of the NGO sector in China. For example, He Jianyu and Wang Shaoguang estimated that there might be over 8 million NGOs in China (He and Wang 2008: 162). Like the aforementioned population of registered groups, a majority of the 8 million groups are not genuine NGOs because they were created by the state, which continues to exercise direct control over the groups' personnel, financial, and general management decisions. Moreover, a significant number of those groups are exempted from the registration requirement. Examples of these groups include the Communist Youth League (CYL) at all administrative levels, which are essentially the administrative arms of the state.

Still, there is a growing number of grassroots NGOs that have been created by private citizens, including individual state officials who have funded their own NGO operations and who have retained the powers to make important organizational decisions without state interference. This type of NGO is characterized by a

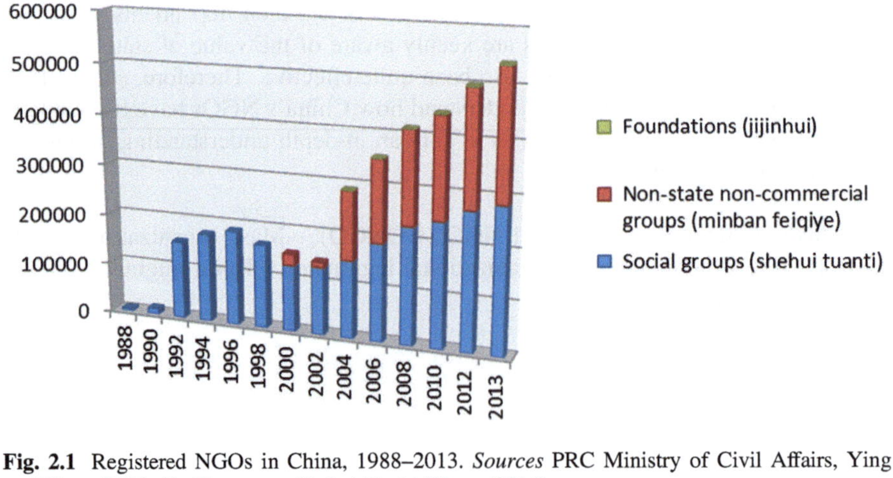

Fig. 2.1 Registered NGOs in China, 1988–2013. *Sources* PRC Ministry of Civil Affairs, Ying and Zhao (2010: 9), Zhongguo Shehui Zuzhi Wang (2014)

wide range of legal status. Based on surveys conducted by Wang and He, there might be as many as 1 million such grassroots NGOs in China (Ibid).

As suggested by the preceding discussion, the Chinese NGO sector is characterized by diversity in terms of their linkages to the state. Based on this observation, He and Wang developed four categories of domestic Chinese NGOs. In the first category are the "mass organizations," which number eight at the national level. These mass organizations are what Gregory Kasza referred to as "administered mass organizations" (AMOs) (Kasza 1995). Like AMOs elsewhere, China's mass organizations were created by the "incumbent regime…to augment [its] control over the rest of society" (Ibid: 8). Indeed, China's mass organizations were established to "execute the various policies and orders of the [Communist Party of China (CPC)] and the state" as well as to "understand and collect the opinions, perspectives, and interests of the masses" for the regime to consider during policy formulation (He and Wang 2008: 135). In this sense, China's mass organizations are not truly NGOs. In fact, they most resemble state agencies, but they may also fall under the broader category of government-organized NGOs (GONGOs). A prominent example of the mass organizations in China is the aforementioned CYL, which was headed by Hu Jintao before he became the head of the CPC and president of the People's Republic of China (PRC).

In He and Wang's conceptualization, "quasi-governmental organizations" constitute the second category of NGOs. Like the mass organizations, quasi-governmental organizations were founded by the government. They also have the function of mobilizing individuals and organizations within their respective industries or sectors and, in this manner, assist relevant government agencies (Ibid: 154). According to He and Wang, the quasi-governmental organizations may possess more limited political autonomy than the mass organizations. However, they may function more like NGOs due to their representative role. One example of quasi-governmental organizations is the China Disabled Persons' Federation (CDPF), which was founded in 1988 by Deng Pufang, a son of former CPC leader Deng Xiaoping. According to He and Wang, the CDPF is subsidized as an item in the state budget and its personnel are recruited through the civil service selection mechanism (Ibid). The CDPF indicated that in addition to "representing the interests of disabled persons" and "defending the legal rights of disabled persons," the organization is also "commissioned by the government to manage and develop the affairs of the disabled persons" (CDPF 2011).

The last two categories according to He and Wang are the "registered social groups" and the unregistered "grassroots social groups." The registered groups are made up of three types of registered NGOs: the social groups, the non-state non-commercial groups, and the foundations. Based on their 2003 national survey of over 3,000 registered social groups, He and Wang noted that the great majority of these registered groups were started up by government or CPC agencies (Ibid: 158). Yet, an equally large number of the registered social groups have retained organizational autonomy vis-à-vis the state in daily operations. In other words, many of the registered social groups operate as "hybrid" NGOs.

Altogether, He and Wang observed that the various types of NGOs in China numbered approximately 8.2 million by 2007 (Ibid: 162). In all likelihood China's NGO sector has grown significantly since the publication of He and Wang's report. For instance, at the time of their report, there were just over 200,000 registered social groups. In 2012, however, there were over 270,000 registered social groups. Similarly, the number of unregistered grassroots social groups has most likely exceeded He and Wang's estimate of 1 million. Therefore, there is reason to believe that China's NGO sector has moved even closer to the 10 million mark.

When examining China's NGO sector, it is clear that the state has been instrumental in its creation. In this respect, although the aforementioned grassroots groups are unregistered, state involvement is often observed in the founding and operation of those groups (Ibid: 161). Therefore, the development of China's NGO sector closely tracks global trends, which indicate that the expansion of the NGO sector has been closely related to government support, notably monetary support (Anheier 2005: 282).

The growth of China's NGO sector also reflects state policy. In particular, the emergence of the NGO sector in China in the 1990s was an outcome of the state's policy of promoting "small government, big society." This policy reflected the state's intention to divest its social responsibilities as a means to facilitate economic reform as well as the state's admission that its capacity was limited (Saich 2003). In this context, the state viewed NGOs as useful for performing many important functions of governance. In particular, NGOs could reduce the state's burden of social service provision (Wong 1998). In addition, NGOs could serve as vehicles for mobilizing societal resources to supplement state spending (Lu 2005). Moreover, NGOs could be created to accommodate excess government personnel (Gallagher 2004; Ma 2006). Therefore, recognizing the value of NGOs as a governance vehicle, the Chinese government has continued the call for all state agencies to support NGOs as a part of the effort to encourage society to become more involved in welfare provision (People's Daily 2010).

2.2 State Control and the Politics of Registration

Given the state's prominent role in the development of China's NGO sector, one may forget that China remains a one-party authoritarian regime. As such, the party-state has remained vigilant about potentially subversive forces even as it tried to promote the idea that more NGOs should be involved in the provision of social welfare. This ambivalence toward the NGOs helps to explain continued state restrictions and even repression against Chinese NGOs since various color revolutions swept across parts of the former Soviet Union between 2003 and 2005. In fact, senior PRC officials have made it clear that the color revolutions constituted the source of the government's suspicion about NGOs (Wang 2012).

For instance, before and during the 2008 Beijing Olympic Games, the 2009 celebration of the 60th anniversary of the founding of the People's Republic of China, and the 2010 Shanghai World Exposition, the state had maintained strict vigilance over the Chinese NGOs. Chinese NGO leaders were told not to host foreign visitors during these important national events and NGO leaders have either been placed under house arrest and extensive surveillance, or encouraged to go abroad.[1]

However, state sanctions against NGOs have not been limited to periods during which events of national significance were held. In fact, they have occurred quite regularly. For instance, in 2007 the original English version of the *China Development Brief*, which had been an authoritative source for the latest news and analyses about Chinese NGOs for over 10 years, was closed down by the government and the founder of the publication, who was not a PRC citizen, was eventually forced to leave China. In 2008, in the aftermath of the Wenchuan Earthquake in Sichuan Province, the government took control of the disaster area and effectively restricted the activities of Chinese NGOs. In 2009, Gongmeng, a Beijing-based NGO focused on providing legal aid to the common people and supporting constitutional reform, was closed down by the Beijing government on charges of tax evasion. In 2010, Peking University dissociated itself from its Women's Legal Aid Center, allegedly due to government pressure. In the same year, Wan Yanhai, China's leading AIDS activist and head of the Aizixing NGO, announced that he and his family had left China for the United States due to government harassment. Most recently, in July 2013, the Transition Institute, which has been committed to research and advocacy on governance issues and institutional reforms, was officially shut down by the Beijing government.

The state's behavior reflects its fundamental desire to control civil society actors. Enforcing the legal requirement that all NGOs must register with the government is a means to exercise that control. In particular, civil affairs agencies, which are tasked with registering and managing NGOs, are legally empowered with a range of broad justifications to deny registration. For instance, civil affairs agencies can deny registration if the work and mission of the applicant NGO are determined to contravene stipulations in the PRC Constitution, harm the interests of the state and the "unity of the nationalities," and go against the "society's moral customs." In addition, the government can deny registration to NGO applicants if registered NGOs engaged in similar work are already present within the same geographical administrative area. This stipulation ensures that existing GONGOs will continue to enjoy sectoral monopoly (Saich 2000: 131). In a related manner, it may provide the legal basis for local government officials to create GONGOs to prevent NGOs from obtaining government registration (Simon 2013: 244). Furthermore, registration may be denied if the leader(s) of the applicant NGO is found to lack the "capacity for civil conduct."

---

[1] I had intended to take a group of American students to visit a local NGO in Shanghai in May 2010, but was told by the NGO leader that local government officials had informed the organization that it should not host any foreign visitors during the period that the World Exposition was held in Shanghai.

A certificate of non-state non-commercial group registration. *Source* The author

The most important legal stipulation that allows the state to deny registration is a well-known clause in both the "Regulations for the Management and Registration of Social Organizations" and the "Temporary Regulations for the Management and Registration of Non-State Non-Commercial Groups," which stipulates that a key condition for attaining the registration status is that the applicant NGO must secure the sponsorship of a governmental or quasi-governmental agency that works in the relevant sector. In practice, many, if not most, NGOs are unable to become registered in large part because they could not meet the requirement of sponsorship (Deng 2010). Why?

Karla Simon identified two major reasons that account for this phenomenon. First, the potential sponsors may simply be unwilling to bear the responsibility of overseeing NGOs, especially since the sponsor is legally and politically liable for the activities and finances of the organizations under their charge (Simon 2013: 242). Second, the "unwillingness" of the potential sponsors may be a result of the fact that the law does not explicitly require the potential sponsoring entities to act on an application, even though there is an implicit understanding that they should take action to determine whether the applicant NGOs are qualified for sponsorship. Moreover, the lack of clarity with respect to the documents needed for approval further encourages the potential sponsors to drag their feet (Ibid). Ultimately, since potential sponsors have nearly absolute discretion in determining whether to sponsor any NGO, these legal considerations may simply serve as useful justifications for de facto denial of registration to NGOs.

In 2012, the government of Guangdong province launched a pilot program in which NGOs were permitted to acquire the nonprofit registration status without a sponsor (Caixin 2012). By the end of 2013, similar pilot programs have reportedly been launched in 26 provinces and five cities, enabling "more than 19,000 [NGOs] to register directly" with the government (China Daily 2013). Although many experts have expressed optimism about this development, it is still too early to know for certain the extent to which this pilot program will be institutionalized across the country. In addition, the pilot program has a limited scope in terms of the fields to which it applies: only NGOs providing social services and engage in non-political activities have been explicitly permitted to register under the pilot program (The Economist 2014). Labor advocacy is not included in the list of approved fields, even though some labor NGOs have apparently been approached about direct registration (Ibid).

There are also practical reasons for government agencies to deny registration to NGOs. These reasons again highlight the tremendous discretion that government agencies wield in determining whether and which NGOs may be registered. One reason is that local governments recognize that most NGOs exercise positive influences on society. However, registration cannot be granted to many NGOs because they are unable to meet the stringent legal requirements. Yet, banning those organizations are likely to result in some degree of resistance, not including the financial cost of implementing the ban. Therefore, local government agencies have instituted the unwritten rule of "no banning, no recognition, and no intervention" (Deng 2010: 195). In this case, the denial of registration has been motivated in part by pragmatism and expediency.

Lastly, government officials may also decide against registering NGOs because they believe that unregistered NGOs are easier to control since they will be more dependent on good relations with the state than are registered groups (Hildebrandt 2013: 72). In a related manner, since unregistered NGOs cannot access certain financial resources in a direct manner due to the organization's illegal status, they must rely on informal ties with state institutions and officials to obtain those resources. This type of situation offers government agencies and officials an opportunity to enrich their organizational coffers and individual pockets. This is especially the case for NGOs working in the HIV/AIDS sector, in which significant foreign funding is administered through the state (Ibid). Therefore, there are financial incentives for government officials to deny registration to NGOs.

2.3 Crisis of Legitimacy

The inability to obtain government approval has serious long-term consequences for China's NGOs. One important consequence is that these organizations possess weak legitimacy, both politically and socially. This is in contrast to the state's overwhelming legitimacy. In this context, it is not difficult to see how state linkages can be valuable assets for China's NGOs.

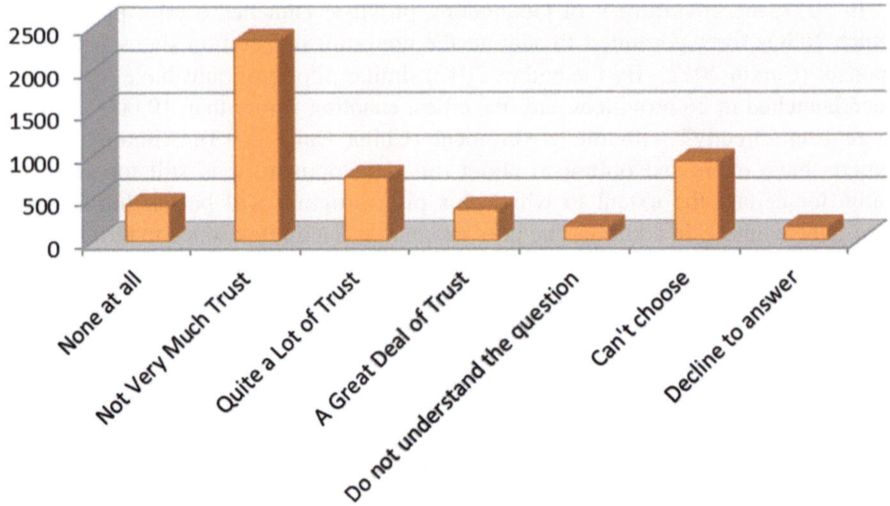

Fig. 2.2 Public trust in NGOs. *Source* Asian Barometer (2008); $n = 5098$

Recent research suggests that China's NGOs are enjoying greater social legitimacy even in the absence of state linkages (Brown and Hu 2012; Spires et al. 2014). As China's NGOs make important contributions to society by delivering valuable services and engaging in policy advocacy in an increasing manner, it is very likely that they will enjoy increasing popular support. However, that day has not yet arrived.

Indeed, the majority of the ordinary Chinese people have no knowledge of NGOs and certainly no understanding of its operations.[2] Some studies even show that in general people in China are highly suspicious of NGOs because these organizations are perceived to be anti-government and anarchical (He 2012: 170–171). In 2008, the investigators of the Asian Barometer survey project found that over 45 % of the respondents in China indicated that they did not harbor very much trust in NGOs and over 8 % of the respondents stated that they had no trust in NGOs at all (Fig. 2.2). This means that over half of the respondents expressed little trust in NGOs. At the same time, <15 % of the respondents said that they had "quite a lot of trust" in NGOs and only about 7 % of the respondents said that they had a "great deal of trust" in NGOs (Ibid). Altogether, <25 % of the respondents expressed confidence in NGOs.

The NGOs' state of public legitimacy is in significant contrast to that of the government. Today, over 20 years after the top Chinese leaders decided to use

[2] In my field interviews, a theme that recurred on a regular basis is that the ordinary Chinese people do not understand why nonprofit organizations would need to solicit outside funding. Therefore, in the mindset of many, if not most, Chinese persons, organizations that claim to be nonprofit in nature and yet openly seek funding must have ulterior motives and thus cannot be trusted.

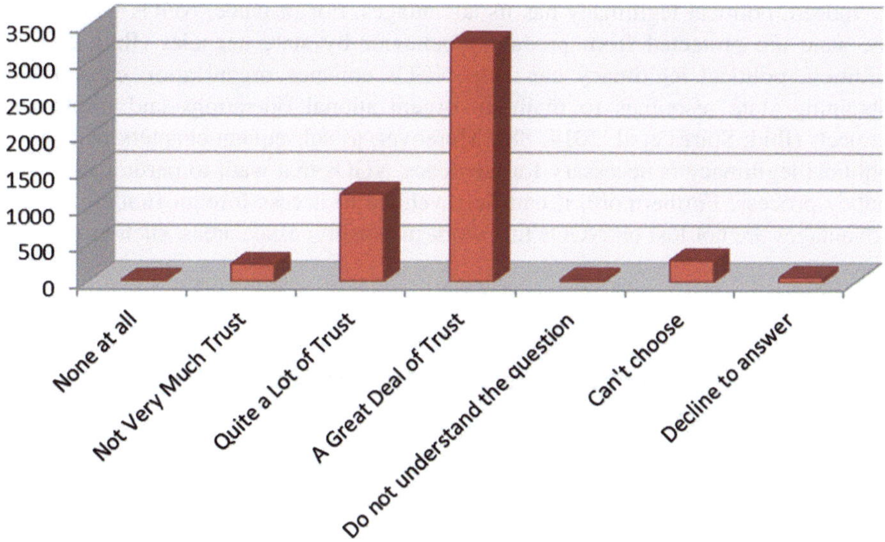

Fig. 2.3 Public trust in central government. *Source* Asian Barometer (2008); $n = 5098$

military force against protesting students assembling in Beijing's Tiananmen Square, the state seems securely ensconced in its position as the most legitimate entity in China. As experts have noted, the state has successfully employed a range of incentives and coercive measures to ensure general political compliance (Tong 2009; Chen and Dickson 2010; Michelson and Liu 2010; Wright 2010). Notwithstanding the regular grumblings of individual Chinese persons and the rise of local protests about official corruption, the unresponsiveness of government agencies, and even the inappropriate use of force against Chinese citizens, the reality is that the communist regime is the only political game in town. In survey after survey, popular support for the current political system has remained high (Chen 2004).

Again, the findings of the Asian Barometer are instructive. According to the survey's 2008 findings, 64 % of the respondents in China expressed a "great deal of trust" in China's central government while another 23.5 % expressed "quite a lot of trust" in the central government (Fig. 2.3). This means that nearly 90 % of the respondents expressed considerable trust in the central government. Even China's local governments, which have been the main objects of popular grievances, received better overall ratings than the NGOs, with over 50 % of the respondents expressing confidence in them.

These findings suggest that China's NGOs must look to the state in their search for social legitimacy. Indeed, research has shown that China's NGOs acquire public legitimacy by displaying their ties to the state (Lu 2009: 53). However, as valuable as public trust may be for the long-term development of China's NGOs, the above findings also make it clear that gaining legitimacy in the eyes of the state, i.e., political legitimacy, should be even more important to the NGOs.

Indeed, political legitimacy has its advantages. For instance, NGOs with ties to the state are protected from predatory behavior by state agencies (Ibid: 56). In addition, political legitimacy can help NGOs enhance organization capacity by obtaining state resources to maintain organizational operations and implement projects (Ibid; Spires et al. 2014: 88). Moreover, as subsequent chapters will show, political legitimacy is necessary for advocacy NGOs that want to participate in the policy process. Furthermore, it can be a vehicle to access foreign funding. These advantages are not lost on NGOs that work in sensitive issue areas, such as migrant labor.

2.4 Migrant Workers: Objects of Institutionalized Discrimination

Labor-related issues have always been high on the agenda of any government because the state of labor concerns the state of the economy, which ultimately has political implications. For non-democratic states, there is an added dimension to the general concern about labor, which is the fear that labor could act as a force to effect political change. Particularly for non-democratic states with a socialist past, labor must be treated with special care, i.e., controlled, given the states' historical claim to represent the interests of the workers. Therefore, for the Chinese state, the fact that hundreds of millions of rural migrant workers (*nongmingong*) have been running around the country, seemingly out of control, has understandably been a source of concern even as the state has recognized that this "floating population" is necessary for the state to achieve its ultimate modernization goals.

According to the 1990 PRC government census, there were approximately 30 million rural migrant workers in China who engaged regularly in work outside their official places of residence (Liang 2004: 251). Ten years later, the 2000 census reported that the total number of rural migrant workers had reached 79 million (cited in Liang and Ma 2004: 470). At the end of the first decade of the 21st century, there were over 220 million rural migrant workers in China (Xinhua 2011). In 2010, the All-China Federation of Trade Unions released a report about the state of migrant workers who were between the ages of 16 and 30, who constituted nearly 62 % of the overall rural migrant population (All-China Federation of Trade Unions 2010). The report indicated that as of 2009, the average age of this cohort was 23 years old. Most of them began their lives as migrant workers before they turned 20. Eighty percent were unmarried and approximately 70 % had obtained a middle-school education or less. In a nutshell, Chinese rural migrant workers are young, single, and have low educational attainment.

Deng Xiaoping's policy of economic reform and opening has produced an economic miracle in China that has arguably been unprecedented in terms of the speed, scale, and nature of the transformation. The social consequences of this economic transformation are multiple and are still being felt today. The emergence of the rural migrant workers is one of those consequences. By definition, rural migrant workers originated from non-urban areas where they had been expected to engage in agricultural work.[3] Migrant workers have flocked to urban areas, especially the first-tier cities along China's eastern coast—Beijing, Shanghai, and Guangzhou/Shenzhen, in increasing numbers at an accelerated pace over the past 25 years.

Individually and collectively, rural migrant workers have contributed to China's modernization, urbanization, and growing affluence. China's economic miracle has essentially been a product of their sweat and blood. Yet, migrant workers and their dependents have been living on the margins of the Chinese society, both literally and figuratively. In fact, scholars have observed that the Chinese rural migrant workers are at best second-class citizens. Some experts have even argued that the rural migrant workers are really third-class and fourth-class citizens (Fan 2002).

There is a consensus that the fundamental reason for the low socio-political status of the rural migrants is institutional. The household registration (*hukou*) system has been identified as the source of institutionalized discrimination against the rural migrants. Every Chinese citizen is required to maintain a *hukou*, which identifies his or her official place of residence and the type as well as quantity of social benefits to which he or she is entitled. In essence, the *hukou* system is a system for social control and resource allocation or subsidization. It channels critical resources to urban residents as a part of the state-led program to industrialize and modernize China. In this respect, it has served as the political-legal means to justify, not without popular resistance, the fact that persons with rural *hukou* status receive significantly less social benefits and of poorer quality than their urban peers.[4] It is also a mechanism for regulating internal migration. More specifically, it is a means for limiting the flow of rural residents to urban areas. Therefore, the system has not only institutionalized the wide disparities between the rural areas and the urban areas in China; it has in fact exacerbated the urban-rural gap.

[3] Although there seems to be a consensus among researchers, including Chinese government researchers, about the rural origins of the migrant workers, this understanding is not reflected in the government census, which uses the category of "floating population" to describe all persons who "lived in places other than the towns (townships or streets) of their household registration where they had left for over 6 months" National Bureau of Statistics of China (2011).

[4] Fei-ling Wang, a leading expert on the *hukou* system, wrote that until 2006, "victims with different *hukou* types suffering the same wrongful injury or death were compensated very differently….[In] death, the family of a victim with a rural *hukou* generally brought about half to one-third of that awarded to a victim with an urban *hukou*." Wang (2010), 92–93.

Courtyard/playground of Beijing's Mingyuan School for migrant children. *Source* The author

The *hukou* system, because it has institutionalized the disparity in social status between rural residents and their urban peers, has thus given license to official and public discriminatory views and practices about rural migrants. For example, for a long time, the general perception has been that rural migrants have contributed significantly to rising crime rates in major urban areas. Experts, however, have generally found the widespread perception to be inconsistent with reality (Zhao and Kipnis 2000; Han 2010). Notwithstanding these expert findings, it has proven difficult to change popular perception.[5]

Similarly, the *hukou* system has exerted direct effects on the ability of the rural migrants to access the necessary social welfare resources while they live and work in urban areas. Because rural migrants are not legal urban residents, they cannot access the basic health services that are available to those who have legal urban residencies (Jacka 2006: 111–113). Likewise, the lack of legal residency has prevented children of migrant workers from accessing local public schools (Kwong 2004). Although health services and public education can be purchased, rural migrants' low incomes have effectively barred them and their children from

[5] During the 4 years that I lived in Beijing and Shanghai (2007–2011), just before the annual lunar New Year holidays, local residents consistently reminded each other to be cautious of cash-strapped rural migrant workers who might perpetrate crimes against local residents.

accessing social resources, notably public education, that have been guaranteed by law. As a result, migrants have had to establish schools, often unregistered, for their children, which pale in comparison to public schools in terms of the quality of school facilities and instructional resources. These problems notwithstanding, the main long-standing complaint among migrant workers has been the inability to collect wage arrears from employers. A related problem has been the lack of knowledge among the migrant workers about their legal rights, especially before they leave for their eventual destinations.

In general, the state's response to the problems that migrant workers have faced can be characterized as negative, slow, and inconsistent. Only in the last 10 years have individual central government leaders come out publicly to speak favorably about migrant workers (Associated Press 2010). And that only happened after media reports about the dire consequences of the existing institutions for individual migrants have caused public uproar.[6]

2.5 Migrant NGOs: Introducing the Core Cases

In this environment, NGOs that engage in migrant worker advocacy and service provisions to migrant workers and their families have been established across China, especially in China's major urban centers along the country's eastern coast (Hsu 2012a, b). This book focuses on four of these NGOs, all of which are based in Beijing, but the leaders of all four organizations have made the development of nationwide reach an important organizational goal. In this respect, all four organizations have established physical presence beyond China's capital city. The four organizations, however, vary in organizational effectiveness. In the remainder of this section, I will describe the general conditions of the four NGOs, with a particular focus on the extent to which each organization has achieved its stated objectives, i.e., effectiveness.

2.5.1 Zhicheng Public Interest Law NGO

The first NGO is the Zhicheng Public Interest Law NGO (hereafter referred to as Zhicheng). The organization has been committed to the protection of the rights of minors and migrant workers, and the establishment of appropriate laws and regulations in these areas. This organization began operation in 1999 as a part of a Beijing law firm under the jurisdiction of a sub-municipal justice bureau. No

[6] The most well-known case here is the 2003 death of Sun Zhigang, who was a college-educated migrant worker in Guangdong who was beaten to death by the local police while in detention for failing to produce documents that proved he was legally permitted to work in Guangdong.

member of the NGO staff appears to be on the payroll of any state agency.[7] This NGO has two major organizational arms, reflecting the organization's interest in two areas of law: the rights of migrant workers and the rights of minors. The two organizational arms have attained the nonprofit registration status separately. To support the NGO's mission, its leader and staff have engaged in the provision of free legal aid, the establishment of nationwide networks of public interest lawyers, conducting research and publishing writings about the two areas of the law, and participating in the lawmaking processes at the national and sub-national levels. Meanwhile, Zhicheng and its founder, Tong Lihua, have enjoyed a strong reputation in the field of public interest law in China. Tong Lihua, in particular, has been widely considered one of China's top 10 public interest lawyers.[8] Tong has been featured in media reports along with other giants in the field of public interest law, such as Guo Jianmei, the founder of the former Women's Legal Aid Center at the Peking University.[9]

Zhicheng has been quite successful in meeting its own goals. Since its establishment, the organization's staff has handled numerous legal cases concerning children and migrant workers. For example, in 2010, on average, each month the NGO provided free legal advice involving nearly 600 cases concerning the welfare of minors (Zhicheng Beijing Children's Legal Aid and Research Center 2010). During the same time period, the NGO directly handled over 300 cases concerning the rights of minors and provided support to lawyers who were involved in over 500 cases throughout China (People's Daily 2011).

Although the migrant worker arm of the NGO had a later beginning, its record has been even more impressive than the record of the children's arm. As of the end of April 2011, the organization has provided free legal advice involving over 35,000 cases throughout China (People's Daily 2011). Over 100,000 migrant workers have reportedly benefited from the NGO's work. Meanwhile, the organization had directly handled and completed nearly 16,000 cases that involved nearly 26,000 migrant workers. The sum of money associated with these cases totaled over

[7] According to Tong's biographical sketch, the original Zhicheng Law Office was administratively attached to Fengtai district's justice bureau in Beijing. When Tong became the head of that office in 1998, the office still held this administrative status. According to Wang Fang, a Zhicheng staff lawyer who I interviewed in 2008, the law office had been privatized since the establishment of the NGO, and all the lawyers who were employed at the law office when the founder was first employed there had since then resigned from the law office. All current members of the NGO staff were hired much later. For example, the majority of the staff lawyers in the Zhicheng Migrant Center joined the NGO after 2004. Wang's claim that the law firm had been privatized corresponds to the understanding that by 2001, all state-owned law firms in Beijing had been privatized. Michelson (2007), 373.

[8] LawFirm50 (2009). LawFirm50 is reportedly an NGO that focuses on collecting and reporting about the quality of lawyers and law firms in China. The organization reported that the ranking was based on the survey responses of nearly 2,500 lawyers, government officials, and other legal practitioners in China.

[9] Feng (2007). In this news report, which was re-published on the website of Guo Jianmei's NGO, Guo and Tong were the only two public interest lawyers who were featured.

300 million Chinese yuan (approximately $50 million), of which nearly 200 million were already distributed to the victims (People's Daily 2011).

As of the end of 2011, Zhicheng coordinated a national network of 20 migrant worker legal aid stations, involving approximately 60 lawyers and 40 administrative staff working in different Chinese provinces and other provincial-level administrative areas. Similarly, it has created a network of 8,900 lawyers working in over 20 provinces and provincial-level administrative areas to protect the interests and rights of minors. As of 2014, the network of lawyers working on issues pertaining to the rights of minors has expanded to include 9,000 lawyers. In addition, the organization coordinates the work of 30 public interest law organizations on the rights of migrant workers (US-Asia Law Institute 2014).

The NGO has also been quite prolific with respect to publishing reports concerning the rights of minors and the rights of migrant workers. As of 2010, Zhicheng's staff produced over 50 book-length reports. The most recent report was published in 2010 and dealt with legal issues in China's rural areas (Tong 2010). The volume includes a chapter based on the NGO's research about corruption among village cadres.

The most impressive aspect of Zhicheng's history, however, has been the NGO staff's repeated participation in law-making processes at both the national and the municipal levels. Between 2002 and 2011, Tong and his colleagues were involved in the modification of nearly a dozen national laws and municipal regulations concerning minors and migrant workers. The NGO has also been commissioned by central government and municipal agencies to conduct original policy research and the crafting of legal handbooks.

2.5.2 Rural Women

The Rural Women NGO, which was established in 1993 as a magazine, has been the leading Chinese NGO that addresses issues concerning Chinese women from the rural parts of the country. The organization is committed to the promotion and protection of the interests and rights of women in China's rural areas, which has included helping female migrant workers in urban areas to acquire the necessary skills and awareness to transition successfully to urban life. Rural Women is made up of three main components: the Cultural Development Center for Rural Women; the Migrant Women's Club, which is administratively part of the Cultural Development Center; and the Practical Skills Training Center. Up until 2012, the NGO also published a monthly magazine entitled *Rural Women*. Between 1993 and 2012, the magazine was operating legally through the flagship newspaper of the All-China Women's Federation (ACWF), which is the leading mass organization on women's issues. The Cultural Development Center was registered as a commercial enterprise while the Practical Skills Training Center was registered as a private, nonprofit education provider. To support its mission of developing the potential of rural women, defending the rights of rural women, and cultivating rural women talents, the NGO has provided opportunities for ordinary rural women to educate themselves and speak out, provided skills training to rural women, and, to an increasing degree, published research reports about the situations concerning rural women in general and female migrants in particular as well as engage in legal advocacy. Since 2007, Rural Women has also been involved in training and promoting female public officials in China's villages.

The NGO's publications have constituted an important avenue through which ordinary rural women could educate and express themselves. In addition to the *Rural Women* magazine, the NGO has also published a newsletter through the Migrant Women's Club and a publication through the book club project. A few years ago, the NGO introduced a publication targeting female village cadres entitled *Women Village Head*. However, this publication appears to have gone out of public circulation. Through these publications, rural women have been able to obtain advice about marriage, child-rearing, and their legal rights. Moreover, they could submit essays to have their voices heard. And the publications were distributed throughout the country, essentially via the nationwide networks of sub-national ACWFs. In this manner, the NGO's impact was felt by hundreds of thousands of Chinese rural women.

The primary organizational network of Rural Women has been the 66 village book clubs around the country. The central aim of the book club project is to combat illiteracy among rural women. And the project has involved nearly 10,000 rural women as of 2010. As mentioned in the preceding paragraph, a publication was associated with the project. The publication primarily served as a platform for rural women to express themselves through short written pieces.

Perhaps the most invested component of Rural Women's efforts to empower rural women in general and female migrant workers in particular has been its training center, which is located on the outskirts of Beijing municipality. Since the center's establishment in 1998, it has trained over 10,000 persons, most of whom until recently have been female migrant workers in the Beijing area. At the center, female migrant workers, most of whom were quite young and who hailed from different parts of China, have received vocational and basic skill training, mostly in preparation for jobs in the domestic help industry and kinder care. Training has also included post-training job placement. In recent years, the center has included an entrepreneurship program to encourage rural women to become business owners.

Logos of migrant worker NGOs and schools for migrant children. *Source* The author

Before 2010, the NGO had a very small legal aid project (Milwertz 2002; Shang 2007).[10] However since early 2010, it has devoted, on an annual basis, nearly one million Chinese yuan (approximately $154,000) to legal aid-related projects. According to Rural Women's 2010 Annual Report, it expected that its legal aid program would affect several thousand people on an annual basis (Rural Women 2010). Likewise, before 2010, the NGO had published no book-length research report.[11] However, in 2010, under the direction of the organization's newly-appointed chief executive officer, who is a well-known and published expert on

[10] When I first visited the NGO in June 2008, I met with one of the two lawyers on the staff. It was clear from our conversation that the NGO handled very few cases and was more involved in referring cases to other NGOs. It is also noteworthy that the NGO's legal aid program has not been mentioned by other writers who have studied this NGO.

[11] One of the two co-founders of Rural Women, however, did co-author a book about the history of China's women's movement in the reform era, which profiled the key Chinese NGOs, including Rural Women.

rural women in China, the NGO published a survey report about the land rights of migrant women.[12]

2.5.3 *Little Bird Migrant Workers Mutual Support Hotline*

The Little Bird Migrant Workers Mutual Support Hotline (hereafter referred to as Little Bird) was established in 1999 as a mutual support group for migrant workers in China's capital. Very soon after the organization's establishment, the founder of the NGO committed the organization to a focus on the defense of the legal rights of the migrant workers. Although it has not attained nonprofit registration status, since 2004 Little Bird has been in a partnership with a local justice bureau in Beijing and has functioned as an authorized labor dispute resolution agency.

Since 1999 and ending in December 2012, by the organization's own documented account, Little Bird has helped migrant workers around the country obtain over 180 million Chinese yuan (approximately $30 million) in wage arrears (Little Bird 2013). The NGO's telephone hotline, which was established in 2000, had accounted for a majority of the legal cases that the NGO has handled. Since its inception, the telephone hotline has received over 166,000 telephone calls (Little Bird 2013). In 2010 alone, the NGO received nearly 10,000 telephone calls and helped migrant workers obtain over 14 million Chinese yuan ($2.15 million) in wage arrears (Little Bird 2010).

Even more impressively, as the first state-approved non-state dispute resolution committee (*renmin tiaojie weiyuanhui*) in China that addresses the labor dispute needs of migrant workers, Little Bird has successfully settled 70 % of the labor dispute cases that it managed. Just as impressively, Wei claimed that his organization was able to resolve 50 % of the legal dispute cases over the telephone (Interview with Wei Wei 2009).

Little Bird has expanded its operations beyond Beijing. It has offices in Beijing, Shenzhen, Shenyang (center of China's northeast rust belt), and Shanghai. This means that the NGO has a presence in the most important labor centers in China. Wei Wei, the founder of Little Bird, had plans to establish branches in Xi'an, Zhengzhou, and Chongqing, which represent the three largest inland Chinese cities (Little Bird 2011).

---

[12] The book was by Wu Zhiping and entitled *Zhongguo Liudong Funu Tudi Quanyi Zhuangkuang Diaocha* [*An Investigation on the State of Chinese Migrant Women's Land Rights*]. It was published by the Social Sciences Academic Press of China. Since then, a book on female participation in public affairs in China by the deputy CEO of the NGO was also published by the same publisher. In addition, the NGO published a handbook on protecting the land rights of female migrant workers.

2.5.4 On Action International Cultural Center

The On Action International Cultural Center (hereafter referred to as On Action) was founded in 2006 and was registered as a commercial enterprise. At the establishment of the NGO, its founder had set as the goals of the organization (1) the provision of legal aid; (2) the promotion of corporate social responsibility; (3) the holding of activities to build up the "knowledge, physical, and mental capacity" of individual migrant workers; (4) policy advocacy through the publication of original research reports; (5) the encouragement of migrant workers to engage in self-empowerment and mutual understanding through the performing arts; and (6) the sponsorship of activities to promote the building of civil society (On Action). Unfortunately, On Action has yet to deliver on its founder's promises. In this study, On Action is clearly the least effective among the four featured cases.

With respect to the area of policy advocacy through the publication of original research, since the founding of the NGO in late 2006, the organization has conducted small-sample surveys about the state of migrant workers in Beijing and Suzhou. The surveys were conducted years apart and none offered new insights with expert analyses about potentially new issues facing the migrant working population. The survey also did not garner significant media coverage. Needless to say, the impact of both surveys has been minimal to nonexistent.

On Action has committed more resources to the dissemination of legal information. For example, On Action's 2010 annual report has shown that well over 80 % of the organization's activities in that year appear to have involved the dissemination of information about laws and regulations, notably information about the 2008 Labor Contract Law (On Action 2010). However, the impact of this effort is not obvious, perhaps due to the fact that the activity was seasonal and indirect, meaning that information dissemination did not generally involve visits to factories and construction sites, although the organization's volunteers have disseminated information in schools for the children of migrant workers and in areas where migrants lived.

The remaining few non-legal aid activities that On Action had participated in 2010, as reported in the NGO's 2010 annual report, involved citizenship education and NGO capacity building. Virtually all of the listed activities were not sponsored by the NGO. The NGO staff members were essentially attendees. Although the NGO's website has posted information about lawyers who may be able to provide legal aid, there is no indication of any significant legal aid activity being sponsored on a continuous basis by the organization. A foreign observer had lamented how Chinese NGOs have spent too much time and efforts on holding activities (Interview BJ032209). It appears that On Action has exemplified this problem and has not been particularly successful even in this respect.

In 2009, the organization suffered a serious financial scandal, which was reported in the chief Chinese publication about NGOs. The financial scandal involved accusations by several organizational staff and members of the board of directors that the founder of the NGO had engaged in deliberate false reporting about the performance of the NGO and its accounting. The immediate consequence was that all the members of the board resigned.

Through a detailed treatment of each of these core NGO cases, I intend to explore the relative significance of the three types of leadership ties and examine the process through which each type of leadership ties contributes to NGO effectiveness. At the same time, I hope that we will be able to gain a more profound understanding of how NGOs have served rural migrant workers in contemporary China.

2.6 The Shadow Cases

Based on the information provided about each core case, one may understandably point to the possibility that On Action's organizational youthfulness may be a factor in determining its level of effectiveness. One may also suggest that the cases are applicable only to NGOs in Beijing that service migrant workers. To address these concerns, the following shadow cases will be examined with several objectives in mind. First, they will corroborate the central argument of this book, which is that leadership ties that facilitate state linkages for the organization have constituted the main contributor to NGO effectiveness in China. Second, the shadow cases will show that similar leadership mechanisms can be found in NGOs working in different issue areas and in different geographical locations throughout China. Third, they will illustrate that leadership ties are able to account for organizational effectiveness regardless of organizational age.

2.6.1 *Friends of Nature*

The Beijing-based Friends of Nature (FON) is arguably China's most well-known NGO. It is certainly the most prominent Chinese environmental NGO. Since its establishment in 1994, FON has been committed to the promotion of environmental protection in China. The NGO, through its founder, had led or coordinated several well-publicized environmental campaigns involving government officials and other NGOs, the most famous of which was the 2003 campaign to halt the Nu River dam project in the southwestern province of Yunnan. It has published an annual volume, with essays written by journalists, government officials, and academic researchers, which reports on the state of environment in China. The NGO has also developed a nationwide network of individual and organizational members, and FON has contributed directly to the establishment of other environmental NGOs in China.[13]

[13] The founder of the Beijing-based Brooks NGO was a project manager at FON before she established her own organization. Interview BJ121207. Likewise, the founder of the Chongqing Green League has been a long-time FON member. Interview CQ040908.

2.6.2 *Yirenping*

Although the organization was only established in 2006, the Beijing-based Yirenping has become China's most successful advocate of the rights of individuals and groups that have been the objects of discrimination in the workplace and in schools. Yirenping's main constituents have been the approximately 120 million Hepatitis-B carriers and patients in China. However, it has expanded its constituencies to include people who are physically handicapped as well as mental health patients. In the short period of its existence, Yirenping has already successfully contributed to changes in the government's policies and practices and those of multinational corporations.

2.6.3 *Self-Empowerment Service Center for Disabled Persons*

Established in 2002 in a rural county of the Chongqing municipality in south-central China, the Self-Empowerment Service Center for Disabled Persons (hereafter referred to as the Self-Empowerment Center) has committed its resources to service the 230,000 local residents, approximately one-third of them either are or have been migrant workers in other parts of China. The NGO's primary activities have been educating outbound migrant workers and providing physical and psychological examinations and counseling to residents who have returned to the county after suffering debilitating injuries while working as migrant workers in China's coastal regions. In this manner, the NGO has reached out to tens of thousands of local residents over the years. Lastly, the NGO has provided limited input to local governments with respect to policies concerning handicapped persons.

2.7 Conclusion

China's NGOs are born in the midst of tremendous political uncertainty. The state is at once their benefactor and their oppressor. NGOs that work in the field of migrant labor face additional challenges due to the sensitivity of the topic that they deal with. Yet, many of these organizations have not only survived; they have thrived. The main factor for success has been the ability of the NGOs to become beneficiaries of the state. And the leaders of those organizations play the critical role in facilitating access to state resources. In the following chapters, we will see how the leaders of China's NGOs, notably those that serve migrant workers, have been able to do just that through their various social ties. We will begin our story in the next chapter with a look at how China's NGO leaders have used their ties to the state to advance the missions of the organizations that they lead.

References

All-China Federation of Trade Unions. 2010. Guanyu Xingshengdai Nongmingong Wenti De Yanjiu Baogao [The Research Report Concerning the Issue of the New Generation of Rural Migrant Workers]. http://www.acftu.org/template/10004/file.jsp?cid=222&aid=83614. Accessed 5 May 2011.

Anheier, Helmut K. 2005. *Nonprofit Organizations: Theory, management, policy*. New York: Routledge.

Associated Press. 2010. China Premier Wen Jiabao Urges Improved Treatment of Migrant Workers Amid Labor Unrest. http://www.foxnews.com/world/2010/06/15/china-premier-wen-jiabao-urges-improved-treatment-migrant-workers-amid-labor/. Accessed 5 May 2011.

Zhicheng Beijing Children's Legal Aid and Research Center. 2010. Statistical Summaries of Cases for January – December 2010. http://www.chinachild.org/zhi/3fyzx/ajtj/index.asp. Accessed 7 June 2011.

Brown, L. David and Xing Hu. 2012. Building Local Support for Chinese Civil Society with International Resources. *Voluntas* 23: 711-713.

Chen, Jie. 2004. Popular Political Support in Urban China. Stanford: Stanford University Press.

Chen, Jie and Bruce J. Dickson. 2010. *Allies of the State: China's Private Entrepreneurs and Democratic Change*. Cambridge: Harvard University Press.

China Disabled Persons' Federation. 2011. Zhongguo Canlian Zhuyao Zhineng [Main Functions of the China Disabled Persons' Federation]. http://www.cdpf.org.cn/clgk/ content/2011-04/13/content_30318407.htm. Accessed 15 March 2014.

Asian Barometer. 2008. http://www.asianbarometer.org/newenglish/surveys/. Accessed 12 April 2014.

Deng Guosheng. 2010. The Hidden Rules Governing China's Unregistered NGOs: Management and Consequences. *The China Review* 10, 1: 183-206.

Fan C. Cindy. 2002. The Elite, the Natives, and the Outsiders: Migration and Labor Market Segmentation in Urban China. *Annals of the Association of American Geographers* 92, 1: 103-124.

Feng Jian. 13 April 2007. Minjian Falu Yuanzhu: Jiannanzhong Chengzhang [Nongovernmental Legal Aid: Growing in the Midst of Trials]. *Beijing Zhoubao* [*Beijing Weekly*]. http://www.beijingreview.com.cn/2009news/renwu/2007-04/13/content_201223.htm. Accessed 27 July 2011.

Gallagher, Mary E. 2004. China: The Limits of Civil Society in a Late Leninist State. In *Civil Society and Political Change in Asia: Expanding and Contracting Democratic Space*, ed. Muthiah Alagappa, 419-452. Stanford: Stanford University Press.

Han, Dong. 2010. Policing and Racialization of Rural Migrant Workers in Chinese Cities. *Ethnic and Racial Studies* 33, 4: 593–610.

He Dan. 29 December 2013. NPO Rules Expected in 2014. *China Daily*. http://usa.chinadaily.com.cn/china/2013-12/29/content_17203671.htm. Accessed 15 April 2014.

He Jianyu and Wang Shaoguang. 2008. Zhongguoshi De Shetuan Geming – Dui Shetuan Quanjingtu De Dingliang Miaoshu [China's Association Revolution – A Quantitative Description of the Association Panorama]. In *Zhongguo Gongmin Shehui Fazhan Lanpishu* [*Bluebook on Civil Society Development in China*], eds. Gao Bingzhong and Yuan Ruijun, 133-163. Beijing, China: Peking University Press.

He Zengke. 2012. Institutional Barriers for the Development of Civil Society in Current China. In *Civil Society and Governance in China*, eds. Jianxing Yu and Sujian Guo, 157-173. New York: Palgrave Macmillan.

Hildebrandt, Timothy. 2013. *Social Organizations and the Authoritarian State in China*. New York: Cambridge University Press.

Hsu, Jennifer Y.J. 2012a. Spaces of Civil Society: the Role of Migrant Non-Governmental Organizations in Beijing and Shanghai. *Progress in Development Studies* 12, 1: 63-76.

Hsu, Jennifer. 2012b. Layers of the Urban State: Migrant Organisations and the Chinese State. *Urban Studies* 49, 16: 3513-3530.

Hu Shuli et al. March 2012. Minzhengbu Tan 'Jieshe' [Ministry of Civil Affairs Discuss 'Association'] *Caixin Xinshiji*. http://08charterbbs.blogspot.com/2012/03/blog-post_5182.html . Accessed 15 April 2014.

Jacka, Tamara. 2006. *Rural Women in Urban China: Gender, Migration, and Social Change*. Armonk: M.E. Sharpe.

Kasza, Gregory J. 1995. *The Conscription Society: administered mass organizations*. New Haven: Yale University Press.

Kwong, Julia. 2004. Educating Migrant Children: Negotiations between State and Civil Society. *The China Quarterly* 180: 1073-1088.

LawFirm50. 2009 Nian Zhongguo Shida Gongyi Lushi [China's Top 10 Public Interest Lawyers in 2009]. http://www.lawfirm50.com/index/index.php/2010-05-22-02-39-43/2010-05-22-02-59-04/141-lawfirm502009. Accessed 27 July 2011.

Liang, Zai. 2004. Patterns of Migration and Occupational Attainment in Contemporary China: 1985-1990. *Development and Society* 33, 2: 251–274.

Liang, Zai and Zhongdong Ma. 2004. China's Floating Population: New Evidence from the 2000 Census. *Population and Development Review* 30, 3: 467–488.

Little Bird. 2010. Work Report in 2010. http://www.xiaoxiaoniao.org.cn/html/aboutus/2011-04-06/174639.html. Accessed 29 June 2011.

Little Bird. 2011. Little Bird Hotline – Henan Television Broadcast. http://www.xiaoxiaoniao.org.cn/html/lm_93/2011-03-28/140246.html. Accessed 8 July 2011.

Little Bird. 2013. Organization Introduction. http://www.xiaoxiaoniao.org.cn/Article/Show.asp?id=3106. Accessed 23 March 2014.

Lu, Yiyi. 2005. The Growth of Civil Society in China: Key Challenges for NGOs. *Briefing Paper* ASP BP 05/01, Chatham House, Asia Programme. http://www.chathamhouse.org/sites/default/files/public/Research/Asia/china.pdf. Accessed 14 November 2011.

Lu, Yiyi. 2009. *Non-Governmental Organisations in China*. London: Routledge.

Ma, Qiusha. 2006. *Non-Governmental Organizations in Contemporary China: Paving the way to civil society?* New York: Routledge.

Michelson, Ethan. 2007. Lawyers, Political Embeddedness, and Institutional Continuity in China's Transition from Socialism. *American Journal of Sociology* 113, 2: 352–414.

Michelson, Ethan and Sida Liu. 2010. What do Chinese Lawyers Want? Political Values and Legal Practice. In *China's Emerging Middle Class: Beyond Economic Transformation*, ed. Cheng Li, 310-333. Washington, DC: Brookings Institution Press.

Milwertz, Cecilia. 2002. *Beijing Women Organizing for Change: A New Wave of the Chinese Women's Movement*. Copenhagen, Denmark: NIAS Press.

Ministry of Civil Affairs. 1990-2012. Tongji Gongbao [Statistical Communiqués]. http://cws.mca.gov.cn/article/tjbg/?. Accessed 15 March 2014.

National Bureau of Statistics of China. 28 April 2011. Press Release on Major Figures of the 2010 National Population Census. http://www.stats.gov.cn/english/newsandcomingevents/t20110428_402722237.htm. Accessed 6 May 2011.

On Action. Beijing Zaixingdong Guoji Wenhua Zhongxin [Beijing On Action International Culture Center]. http://www.zxd.org.cn/Article/List.asp?lmid=108. Accessed 23 March 2014.

On Action. 2010. On Action 2010 Annual Work Report. http://www.zxd.org.cn/news_view.asp?newsid=2172. Accessed 8 July 2011.

People's Daily. 27 December 2010. Li Liguo: Minzheng Gongzuo Yao Ganyu 'Fangquan', Shanyu Yonghao Shehui Liliang [Li Liguo: In Civil Affairs Work We Must not be Afraid of 'Letting Power Go' and be Adept at Utilizing Social Strengths). http://politics.people.com.cn/GB/13590743.html. Accessed 15 March 2014.

People's Daily. 11 May 2011. Zhicheng Gongyi Shiernian Shehui Guanli Chuangxin Tansuo Zuotanhui Juban [The Zhicheng NGO's Twelfth anniversary Symposium on the Exploration of Social Management Innovation was Convened]. http://society.people.com.cn/GB/42735/14610144.html. Accessed 7 June 2011.

Rural Women. 2010. The 2010 Annual Report of the Rural Women Organization. http://www.nongjianv.org/pdf/seed/2010njnnianbao.pdf. Accessed 8 July 2011.

Saich, Tony. 2000. Negotiating the State: The Development of Social Organizations in China. *The China Quarterly* 161: 124–141.
Saich, Tony. 2003. Social Welfare Provision and Economic Transition: Pluralism and Integration of Service Delivery. International Labor Organization. https://www.ilo.org/ses. Accessed 15 May 2007.
Shang Xiaoyuan. 2007. Funu Gongmin Shehui Zuzhi He Zhili Moshi De Duoyuanhua [Women's Civil Society Organizations and the Pluralism of the Governance Models]. In *Chongji Yu Biange: Duiwai Kaifangzhong De Zhongguo Gongmin Shehui Zuzhi* [*Impact and Change: the Chinese Civil Society Organizations that are Opening Toward the Outside*], ed. Shang Xiaoyuan, 236-297. Beijing, China: Chinese Social Sciences Press.
Simon, Karla W. *Civil Society in China: The Legal Framework from Ancient Times to the "New Reform Era."* New York: Oxford University Press.
Spires, Anthony J., Lin Tao, and Kin-man Chan. 2014. Societal Support for China's Grass-Roots NGOs: Evidence from Yunnan, Guangdong and Beijing. *The China Journal* 71: 65-90.
The Economist. 12 April 2014. Beneath the Glacier. http://www.economist.com/news/china/21600747-spite-political-clampdown-flourishing-civil-society-taking-hold-beneath-glacier. Accessed 15 April 2014.
Tong, James W. 2009. *Revenge of the Forbidden City: The Suppression of the Falungong in China, 1999-2005*. New York: Oxford University Press.
Tong, Lihua (ed.). 2010. *Zhongguo Nongcun Fazhi Redian Wenti Yanjiu* [*Study of the Key Issues Concerning the Rule of Law in Rural China*]. Beijing, China: Law Press China.
US-Asia Law Institute. 2014. Zhang Wenjuan of Zhicheng Public Interest Lawyers Organization Speaks at USALI. https://usali.org/media-entities/zhang-wenjuan-of-zhicheng-public-interest-lawyers-organization-speaks-at-usali-2/. Accessed 23 March 2014.
Wang, Fei-ling. 2010. Conflict, Resistance and the Transformation of the *Hukou* System. In *Chinese Society: Change, Conflict and Resistance (Asia's Transformations)*, eds. Elizabeth J. Perry and Mark Selden, 80-100. New York: Routledge.
Wang, Kan. 2012. The Changing State-NGO Relations and Development of Social Media Strategies of Labor NGOs. Unpublished report.
Wong, Linda. 1998. Marginalization and Social Welfare in China. New York: Routledge.
Wright, Teresa. 2010. *Accepting Authoritarianism: State-Society Relations in China's Reform Era*. Stanford: Stanford University Press.
Xinhua News Agency. 1 March 2011. China's 'Floating Population' Exceeds 221 Million. http://english.cntv.cn/20110301/112148.shtml. Accessed 5 May 2011.
Xu Ying and Zhao Litao. 2010. China's Rapidly Growing Non-Governmental Organizations. *EAI Background Brief No. 514*. http://www.eai.nus.edu.sg/BB514.pdf. Accessed 17 November 2011.
Zhao, Shukai and Andrew Kipnis. 2000. Criminality and the Policing of Migrant Workers. *The China Journal* 43: 101-110.
Zhongguo Shehui Zuzhi Wang. 19 May 2014. Zhongguo Cishan Fazhan Baogao: Shehui Zuzhi Zongliang Da 54.1 Wange, Tongbi Zengzhang 8.4 % [China Charity Development Report: Total Number of Social Organizations Reach 541,000, a Growth of 8.4 % Year-on-Year]. http://www.chinanpo.gov.cn/1938/77052/index.html. Accessed 19 May 2

Chapter 3
State Linkage Has Its Advantages

Abstract State linkages have real benefits for NGOs in China. They can help NGOs build organizational capacity, achieve political legitimacy, and access the state's policymaking apparatus and process. State linkages are also especially valuable for NGOs with aspirations to create national networks because those linkages allow NGOs to access state networks around the country. Organizational linkages to the state begin with the NGO leader's state linkages. In contrast to writings about the significance of membership in the ruling Communist Party of China (CPC), this study has found that the most significant form of personal state linkage is not CPC membership. Instead, the leaders of China's most successful NGOs possess affiliations with well-endowed and authoritative state organizations. Moreover, membership in China's ruling party is not necessary for NGO leaders to establish the proper state affiliations. Armed with state affiliations, China's NGO leaders can establish formal relationships between their respective organizations and state agencies even in the absence of the nonprofit registration status.

Keywords State linkage · State affiliations · Organizational capacity · Policymaking · Lawmaking · Legitimacy · National People's Congress · Chinese People's Political Consultative Conference · All-China Women's Federation · All-China Lawyers' Association

3.1 Linkages, Affiliations, and Effectiveness

What are the benefits of state linkages for NGOs in China? Based on the cases presented in this book and the findings of other researchers, state linkages have benefited Chinese NGOs in three respects: building organizational capacity, enhancing organizational legitimacy, and influencing state policies and practices.

J.W. Tai, *Building Civil Society in Authoritarian China*,
SpringerBriefs in Environment, Security, Development and Peace 20,
DOI 10.1007/978-3-319-03665-6_3

These benefits have had direct bearings on the NGOs' ability to meet organizational goals. Let us first consider organizational capacity.

The issue of organizational capacity is essentially about the organization's ability to carry out activities that support the organization's goals. And material resources are at the foundation of any organization's ability to meet its goals. The most basic form of material resource is monetary in nature. In this respect, the most obvious form of state resource is financial support from the state. A related form of organizational resource is consisted of other material resources, such as office space and personnel income, including the salary of the NGO leader, which is no small matter because if an NGO leader can maintain a livelihood without tapping into the NGO's limited financial resources, then it means that more financial resources can be devoted to building organizational capacity and undertaking activities to further the NGO's mission.

Other types of resources are important to achieve specific organizational purposes, such as the geographical networks of state institutions. In China, no NGO is legally permitted to establish formal geographical branches outside the administrative area in which it is registered. Although this legal requirement may not be applied to unregistered NGOs, logic suggests that unregistered organizations have to be just as cautious, if not more so, about establishing cross-regional networks. In addition, access to state networks means that NGOs do not have to incur the significant cost required to establish networks from scratch.

The second area of interest when it comes to establishing state linkages is the perception that those linkages can confer legitimacy on the Chinese NGOs and organizational legitimacy has advantages. As mentioned earlier in this book, formal linkages to the state can solve the problem of legitimacy for the Chinese NGOs in large part because both the elite and the general public continue to see the state as the most legitimate institution. Therefore, linkages to the state should be able to help the NGO overcome its own legitimacy deficits.

It is important, however, to note that there are different forms of formal linkage to the state. The different forms of nonprofit registration status constitute one type of formal state linkage. However, as mentioned earlier in this book, the attainment of this legal status has been extremely difficult to achieve. In contrast, registration as a commercial enterprise does not require state sponsorship. Therefore, by default, the commercial registration status has become the choice of most Chinese NGOs that require some kind of legal status for organizational purposes, such as the establishment of an organizational bank account.

Although this type of registration provides the NGO with a legal status, it does not resolve the Chinese NGO's legitimacy dilemma because the NGO is operating as a nonprofit enterprise with a commercial license. In this sense, Chinese NGOs with this type of registration are still illegal because they are technically engaging in activities under false pretense. Of course, government officials are aware of this phenomenon and in many instances have chosen to look the other way. However, as the case of Gongmeng has demonstrated, the state can always use this legal technicality as a "legitimate" excuse to crack down on Chinese NGOs.[1]

Therefore, many NGOs with commercial registration status would want to find ways to develop other types of state linkage that would serve as protective covers. In this respect, the case of the Little Bird Migrant Workers Mutual Support Hotline demonstrates that there is an effective alternative to the nonprofit registration status. Although Little Bird has been unable to attain the nonprofit registration status, it has been designated publicly by the local justice bureau as a nongovernmental labor dispute resolution committee (*renmin laodong zongcai xietiao weiyuanhui*) that is affiliated with the justice bureau. As we shall see, this affiliation has established Little Bird's legitimacy in the eyes of employers who owed monies to the NGO's migrant worker clients.

As Little Bird's experience shows, formal state linkages can enhance an NGO's social legitimacy. However, those linkages are especially valuable because they demonstrate the organization's political legitimacy to the state. Put simply, organizational legitimacy derived from formal state linkages serves as a signal to the state that the NGO in question is politically reliable. By extension, we should expect the NGO to derive some material benefits from political legitimacy. To be more precise, we might expect the state to be more willing to provide material resources to a linked NGO and even partner with that NGO, which further solidifies the NGO's legitimacy. In addition, as will be detailed in Chap. 5, organizational legitimacy derived from state linkages can also be valuable in terms of its positive effects on acquiring international resources.[2]

Lastly, as Zhicheng's case will demonstrate, organizational legitimacy facilitates access to the state's policymaking process. In a closed political system like China's, it is virtually inconceivable that individuals and organizations without state linkages can access the policymaking process. Again, the central factor is that political legitimacy signals political reliability. Therefore, for NGOs with the aspiration to influence policy, developing state linkages is an important strategy.

Given the benefits of establishing state linkages, which forms of state association will allow a Chinese NGO leader to establish state linkages that can facilitate access to those benefits? Before we examine the specific forms of state association, we should be explicit about two related premises. The first premise is that there are variations among Chinese state agencies in terms of the nature

---

[1] The Beijing-based Gongmeng constitutional law NGO was registered as a commercial enterprise. In 2009, it was closed down by the Beijing government on charges of tax evasion. Commercial enterprises are subjected to higher tax rates than nonprofit organizations and the charge was that Gongmeng had failed to report its real income. The leader of Gongmeng was temporarily detained in the process. It was widely speculated that the real reason for the state's actions was that Gongmeng had written and circulated a research report on the factors contributing to the unrest in Tibet in March 2008.

[2] According to some experts, foreign donors are often more willing to fund Chinese NGOs with proper state linkages than those without. Interview SHWY010511.

Fig. 3.1 State linkage and organizational effectiveness

and extent of material resources and authority under their control. This premise is also applicable to the policymaking process. Therefore, it follows that we should assume that the best forms of personal state linkage are those that allow the NGO to link up with well-endowed state organizations. *I refer to those forms of personal state linkage that facilitate connections to well-endowed state organizations as state affiliations.*

In a single-party authoritarian political system like China's, there are many valuable forms of state linkage, one of which is membership in the ruling party. With important qualifications, scholars have observed the importance of party membership for employment prospects, job advancements, and wage rates in China, during both the Maoist period and reform era (Walder 1995; Li and Walder 2001; Appleton et al. 2008). However, the same scholars have also observed that this form of state linkage has not guaranteed equal access to state resources and authority, in large part because the Communist Party of China (CPC) is an organization with millions of members, and only a very small percentage of its members have benefited significantly from membership (Walder 1995: 324; Li and Walder 2001: 1374).

Similarly, party membership may not help an NGO leader gain access to state institutions with control over important and desirable resources and authority. Indeed, while the respective leaders of Zhicheng, Rural Women, and On Action have all acquired CPC membership, the three NGOs have demonstrated varying abilities to access state resources and secure organizational legitimacy. Meanwhile, although the leader of Little Bird has not acquired the coveted CPC membership, Little Bird has nonetheless established beneficial state linkages. In other words, as valuable as CPC membership may be, by itself it may be insufficient to ensure the creation of valuable state linkages that can lead to the presence of effective Chinese NGOs. Indeed, instead of the leader's CPC membership, the cases show that affiliations with well-endowed state organizations have distinguished Zhicheng, Rural Women, and Little Bird from On Action with respect to organizational effectiveness.

Figure 3.1 is a diagram of the elements involved in the pathway from state linkages to organizational effectiveness. The dotted arrow between CPC membership and state affiliations reflects the possible, yet unnecessary, connection between the two forms of personal state linkage.

In the next section, the four core cases will be presented to demonstrate the path from the state affiliations of the NGO leader to organizational effectiveness. The first three cases will demonstrate the positive effects of the NGO leader's state affiliations. The fourth case will show what can happen when the NGO leader has not developed the appropriate state affiliations.

3.2 Tong Lihua and the Zhicheng Public Interest Law NGO

Since 1999, Zhicheng has been an advocacy NGO committed to the protection of the rights of minors and migrant workers as well as the establishment of appropriate laws and regulations in these areas. Accessibility to state resources to build organizational capacity, the establishment of organizational legitimacy that has resulted from the development of formal organizational linkages to the state, and participation in the state's lawmaking process have characterized Zhicheng's ability to meet its goals. By most accounts, Zhicheng's achievements have been inextricably connected to its founder and leader, Tong Lihua. And Tong's multiple state affiliations have allowed the NGO to derive benefits from linkages to the state, which have had direct effects on the organization's effectiveness.

Tong's state linkages encompassed both state affiliations and other forms of personal state linkage. For example, Tong has been a member of the CPC. He acquired this form of linkage when he was a student at the Beijing-based China University of Political Science and Law (CUPL), which is home to one of China's most prestigious law schools.

As a CPC member and a graduate of one of China's most prestigious law schools, Tong was probably correct in speculating that he could have easily obtained employment with many party organizations and government ministries at the central level (Tong 2009: 8–9). But he did not choose this career route. Instead, in 1995, after graduation, Tong chose to work at a small law firm that was administratively attached to the justice bureau of Beijng's Fengtai district. Although there is no evidence that Tong obtained this position as a result of his CPC membership, based on empirical evidence and logic, we can reasonably speculate that this form of state linkage certainly did not harm his employment prospects.

Although the law firm position promised neither financial security nor bureaucratic advancements, it provided the freedom and the space for Tong to begin his foray into public interest law, which he officially started in 1998 with the establishment of a legal hotline in collaboration with his alma mater. Meanwhile, Tong was promoted to become a deputy director of the law firm in August 1997. At the end of 1999, he was promoted to the directorship of the law firm, which effectively made him a mid-level bureaucrat in the justice system.[3]

Employment at the law firm also led to Tong's first state affiliation, which was with the local justice bureau. This affiliation has been advantageous for Zhicheng's development in terms of accessing state resources (building capacity), establishing organizational linkages to the state (achieving organizational legitimacy), and participating in the lawmaking process (influencing state policies).

First, in terms of accessing state resources, Tong's affiliation to the justice bureau provided the NGO with free office space. Given the costly nature of

[3] Tong's bureaucratic rank was equivalent to the associate "chu" level. In general, county magistrates are at this level of bureaucracy. See Zhou (2004: 55).

Beijing's real estate, free access to office space is a valuable NGO resource. When I visited the NGO on two occasions in the summer of 2008, all the spaces in the law firm's two-story building were devoted to the NGO's work. In addition, the justice bureau had given the NGO additional space in a building across the street from the law firm to operate the organization's migrant worker arm. In 2010, after a newly-constructed building was allocated to the justice bureau, Zhicheng's staff moved into the new building. Since Tong had used his own funds to start up the NGO enterprise, the free use of the law office has meant that he could apply his limited start-up funding to areas that were directly germane to the NGO's mission.

Second, Tong's personal connection to the justice bureau eventually translated into a formal state linkage for Zhicheng. Zhicheng's organizational linkage to the state was first established in early 1999, when Tong co-founded the "Fengtai District Juvenile Legal Aid Station" in partnership with the justice bureau and the local Communist Youth League (CYL) (Ibid: 12).[4] The organizational linkage was formalized in 2005, when the justice bureau agreed to sponsor the migrant worker arm of the NGO for nonprofit registration status (Ibid: 25).

Third, Tong's affiliation with the justice bureau facilitated the NGO's ability to influence state policies by providing opportunities for Tong to establish connections with policymaking state agencies. In this sense, Tong's institutional affiliation to the justice bureau provided opportunities for Tong to establish affiliations with other state agencies. In the fall of 2002, Tong was recruited by the Beijing municipal CYL to become the research coordinator for the drafting committee to amend the municipal regulations concerning child protection (Ibid: 173). Tong's recruitment was likely a result of the CYL's prior working relationship with Tong in setting up the juvenile legal aid station, thereby making Tong a known quantity to the CYL. This experience marked the first time that Zhicheng had participated in any law-making processes, and it took the form of an agreement between the office of the Beijing Municipal Child Protection Commission and Tong, thereby demonstrating Tong's central role in Zhicheng's access to the state's lawmaking apparatus (Ibid).[5]

It is worthwhile to point out here that Tong had established a reputation as a *state* legal expert on the legal rights of minors just one year prior to his recruitment as the drafting committee's research coordinator. In early 2001, Tong's first book, which was apparently the first published work on Chinese jurisprudence for minors, was published.[6] The book earned Tong the praises of senior officials in the justice system and established his reputation as a legal expert (Ibid: 19–20). Although Tong never suggested any connection between his legal reputation and his recruitment to serve as the drafting committee's research coordinator, we cannot

[4] The legal aid station was administratively connected to the district.

[5] Tong recruited a deputy director of the NGO and another NGO staff lawyer to make up the research team. The commission is staffed by the same set of personnel as municipal CYL's department of rights and interests (*quanyi bu*).

[6] The book is entitled *Weichengnianren Faxue* [*Juvenile Jurisprudence*].

rule out that Tong's expertise in an issue area that was directly connected to the CYL's organizational mission was a factor in his recruitment.

Nonetheless, it would appear that Tong's affiliations with the CYL and state organizations that are connected to the CYL had more significant effects on Tong's recruitment. In addition to the partnership with the Fengtai district CYL (and the district justice bureau) to establish the legal aid station in early 1999, later that year Tong collaborated with the China Juvenile Crime Research Association (CJCRA) to establish the "Juvenile Legal Aid and Research Center," which was the precursor to Zhicheng's children's legal rights arm (Ibid: 13). The CJCRA is another state organization that has strong historical ties to the CYL. The founding president of CJCRA was a political leader who had served as a senior official in the CYL.[7]

The CJCRA affiliation appears to have had a direct impact on the NGO's effort to gain organizational legitimacy for Zhicheng's children's legal rights arm. Since Tong's initial partnership with the CJCRA, he has become a member of the association's standing committee. In 2003, CJCRA became the sponsor of the NGO's successful attempt to attain the nonprofit registration status for its children's legal rights arm (Ibid).

In the meantime, Tong also established institutional affiliation with the All-China Lawyers' Association (ACLA), the Chinese equivalent of the American Bar Association. The closeness of the inter-institutional relationship between ACLA and the Chinese justice system should not be of great surprise. Many former senior officials of the justice system have moved onto ACLA after retiring from their government positions (Ibid: 79).

In 2002, Tong was appointed head of the newly-created professional committee on the protection of children's rights in the Beijing municipal ACLA (Ibid: 78). Not soon after, Tong was appointed the convener of the same committee in the national ACLA (Ibid: 80). Eventually, Tong became the leader of that committee. In 2005, Tong was appointed the deputy director of the newly-created legal aid committee in the national ACLA (Ibid: 26). Since then, Tong has been a member of the ACLA's 35-member standing committee.

Just as Tong's affiliation with the justice bureau has resulted in the NGO's access to valuable state resources, i.e., free office space, his affiliation with ACLA has given Zhicheng access to ACLA's national network of local lawyers associations to create nationwide networks of public interest lawyers and organizations to provide legal aid to children and migrant workers around the country, which has been a goal of the NGO. As the national GONGO that allegedly represents the interests of all Chinese lawyers, ACLA has direct working relationships with sub-national lawyers associations, which are often obligated to accept directives from ACLA. To gain the use of the ACLA network, Tong at first successfully convinced other ACLA leaders to support formally the creation of a nationwide network of lawyers for the protection of children's rights (Ibid: 82). Later Tong was equally successful with

[7] The founding president of the association was Zhang Liqun. For his biographical sketch, see http://baike.baidu.com/view/1514850.htm (accessed 22 December 2011).

respect to promoting the creation of a nationwide network of lawyers and law firms to provide free legal aid to migrant workers (Ibid: 26–29).

Among the three benefits of state linkages, it is worthwhile at this point to examine more closely the significance of formal organizational linkages to the state because Zhicheng's relationship with ACLA illustrates the value of formal state linkages for the NGO. I stated earlier that formal organizational linkages to the state can be valuable because they can confer organizational legitimacy on the NGOs in question, and organizational legitimacy has benefits. One of those benefits for Zhicheng has been enhanced direct collaboration between the NGO and the state, which can be observed through Zhicheng's working relationship with ACLA to operate the nationwide networks of public interest lawyers.

After Tong successfully persuaded the ACLA leadership to support the idea of creating these national networks, the ACLA leadership entrusted the ACLA committees under Tong's leadership with the responsibility to coordinate them. Subsequently, the ACLA, through those committees, entered into formal agreements with Zhicheng to coordinate these networks, which involved thousands of lawyers and dozens of local lawyers associations (Liu 2011). Formally, these networks have operated in the name of the ACLA; however, practically speaking, with Tong at the helm of the relevant ACLA professional committees and with his NGO contracted by ACLA to operate the networks, they have become Zhicheng's networks. To be sure, we do not deny Tong's central role in creating and sustaining Zhicheng's access to the ACLA's network. Yet, at the same time, we cannot overlook the fact that Zhicheng's organizational legitimacy, indeed legality, had facilitated the establishment of formal inter-organizational collaboration between Zhicheng and the ACLA because, as stated earlier, organizational legitimacy is a signal to the state that the NGO is politically reliable.

In a similar fashion, Zhicheng's organizational linkages to the state have sustained and even enhanced the NGO's ability to participate directly in the state's lawmaking processes. Although Zhicheng's initial involvement in the lawmaking process may be directly attributed to Tong's personal state affiliations, the NGO's subsequent involvements in the lawmaking process have been more clearly instances of direct NGO-state collaboration. This is evident in the fact that in nearly all cases of the NGO's participation in the lawmaking processes since 2002, they had taken the form of written contracts between state institutions and the NGO (Tong 2009: 173–212). For instance, in 2003 the national CYL signed a letter of agreement directly with Zhicheng's children's legal aid arm to conduct background research in preparation for the amendment of the national child protection law (Ibid: 181). In other words, while Tong was clearly the reason that Zhicheng was able to gain initial access to the lawmaking process, Zhicheng's organizational legitimacy has at a minimum facilitated the NGO's subsequent involvements.

Furthermore, Zhicheng's organizational linkages facilitated the NGO's access to even more state resources to undertake activities that would meet its organizational goals. As stated earlier, a central organizational goal has been the provision of legal aid to migrant workers. In 2006, after the migrant worker arm of the NGO had acquired the nonprofit registration status, the NGO received financial support from the China Legal Aid Foundation for its work on legal aid to migrant workers (Interview with Zhicheng research director Zhang Wenjuan 2008).[8] The foundation has been one of the three financial supporters that financed Zhicheng's free provision of legal aid (Interview with Zhicheng staff lawyer Wang Fang 2008). Another financial supporter was the Beijing municipal justice bureau, which paid the NGO at the rate of 800 Chinese yuan per legal case (Interview with Wang), which was a significantly higher rate than the national average (Ibid). In fact, it would appear that Zhicheng received this contracted rate as a part of the package that granted nonprofit status to the NGO's migrant worker arm (Tong 2009: 24–25). With these resources, within 6 years of the establishment of the migrant worker arm, Zhicheng was able to handle approximately 35,000 cases involving migrant workers throughout China (People's Daily 2011). And those legal cases have served as the empirical data for Zhicheng's subsequent published reports, thereby contributing to the NGO's expertise that in turn enhanced the NGO's effectiveness as a policy advocacy organization.

In addition to access to more state resources, we should not overlook the fact that after Zhicheng's organizational linkages to the state were established it received substantial support from foreign sources, which will be detailed in Chap. 5.

In sum, Zhicheng's successes as an advocacy NGO have been the consequence of its leader's state affiliations. Tong's state affiliations have exercised direct effects on Zhicheng's organizational capacity, organizational legitimacy, and access to the state's policy apparatus. Zhicheng' s organizational legitimacy has in turn enhanced the NGO's ability to develop inter-organizational collaboration with state organizations and helped the NGO obtain even more monetary resources from the state. In the following section, we will see how Rural Women has benefited similarly from the state affiliations of its leader. However, in contrast to Tong's multiple state affiliations, the ACWF has been Xie Lihua's sole state affiliation. Her case suggests that, when it comes to state affiliations, it is not necessary to emphasize quantity.

3.3 Xie Lihua and Rural Women

Since its establishment in 1993, Rural Women has been the leading Chinese NGO that addresses issues concerning rural Chinese women. The organization is committed to the promotion and protection of the interests and rights of women in China's rural areas, which has included helping female migrant workers acquire the necessary skills and awareness to transition successfully to urban life. To support its mission of developing the potential of rural women, advocating the rights and interests of rural women, and cultivating rural women talents (Rural Women Brief

---

[8] The foundation is a GONGO led by senior central government officials representing the legislative, justice, and public security systems. See China Legal Aid Foundation's website, http://www.claf.com.cn/.

Introduction), the NGO has focused on the following activities: the publication and dissemination of the *Rural Women* magazine; the implementation of the anti-illiteracy network in Chinese villages; and the operation of the Practical Skills Training Center. Rural Women's success in carrying out these activities has been the result of its ability to access state resources and achieve organizational legitimacy through the establishment of organizational linkages to the state. And, as was the case with Zhicheng, Rural Women has been able to derive those advantages from the state as a consequence of the state affiliations of Xie Lihua, one of its two co-founders.

Before we focus our attention on Xie Lihua (no relations to Zhicheng's Tong Lihua), it is necessary to talk briefly about Wu Qing, Rural Women's other founder. Wu's contributions to the NGO, while considerable, have been concentrated primarily in the areas of soliciting international resources and publicizing the NGO, especially the training center, through her personal networks, which have included state officials. Wu was a long-time member of the Beijing Municipal People's Congress and was a popular instructor at the Beijing Foreign Studies University. Wu came from an illustrious family, both intellectually and politically. After the Cultural Revolution, Wu's mother was appointed a delegate to the National People's Congress (NPC), China's national legislature.

Nonetheless, in terms of accessing state resources and building formal organizational linkages to the state, Xie Lihua has arguably exercised stronger influences through her affiliation with the ACWF. However, before establishing her connections to the ACWF, Xie Lihua, like Tong Lihua, was already a member of the ruling CPC. In the early 1980s, she left the military, where she served for approximately a decade, to join the staff of the *Women's News*, which is ACWF's flagship newspaper. The federation is a mass organization directly subordinate to the CPC. When Xie co-founded Rural Women in 1993, she was already a member of the newspaper's editorial committee and the director of the department of family and society for the publication (Shang 2007: 239). Soon after Rural Women was established, Xie was promoted to the position of deputy chief editor.

As was the case with Tong Lihua, we can only speculate about the kind of impact that the Xie's CPC membership has had on the establishment of her state affiliations. Yet, similarly, we can reasonably conclude that her CPC membership would certainly not constitute an impediment to the establishment of those institutional affiliations. Nonetheless, it is clear that Xie's ACWF affiliation has had a more direct impact on the NGO's ability to access state resources and establish formal linkages with the state than her membership in China's ruling party.

Xie Lihua's personal affiliation with the *Women's News* played a direct role in the NGO's ability to access valuable organizational resources from the state, especially during the organization's start-up period. In fact, the NGO's creation was a direct result of the fact that Xie obtained insider's information about the intent of the *Women's News* to sponsor a magazine publication about rural women (Interview with Xie Lihua 2009). Before Xie took over the publication, *Women's News* had difficulties in getting either a state organization or a state employee to take on the task of running the publication (Ibid). Xie approached the *Women's News* about her interest to operate it (Ibid). In this manner, the *Women's News* became the NGO's first benefactor.[9]

The initial agreement between Xie and *Women's News* regarding the operation of the magazine suggested that, in essence, the state newspaper would provide the majority of the resources necessary to run the magazine. Under the agreement, *Women's News* provided a start-up fund of 60,000 Chinese yuan and Xie was responsible for meeting any additional costs (Ibid). In exchange, *Women's News* would receive dividends for its investment when the magazine became profitable (Ibid; Shang 2007: 240). However, if the magazine failed to be profitable within three years of its establishment, Xie would declare bankruptcy and she would not have to pay the newspaper back for its investment (Interview with Xie).[10]

In addition to the one-time financial support, the magazine's operations would also gain use of the newspaper's office space, which eliminated the need for Xie to search for and finance, on her own, other spaces to operate the publication. Moreover, Xie's agreement with *Women's News* was such that Xie would retain her position in the state-run newspaper but allowed to work full-time on the magazine. In other words, Xie would remain on the newspaper's payroll. Therefore, in essence, the newspaper was supporting the magazine by providing its personnel resources to the magazine, free of charge, while supplying the start-up funding and the free office space. Furthermore, *Women's News* would take on the task of disseminating the publication through its own nationwide publication network, again at no charge to Xie.

Over time, Rural Women developed into a full-fledged organization with multiple distinct organizational components. In addition, as a consequence of Xie's personal affiliations with the ACWF, Rural Women has benefited directly from access to state resources that helped the NGO achieve its organizational goals.

One such goal has been to raise public awareness about the interests and rights of rural women throughout China. Another related goal has been to reach out to Chinese rural women to help them improve their personal "qualities" or "levels of sophistication" (*suzhi*). An important organizational project, in this respect, is the anti-illiteracy project, which has been ongoing since 1996, and as of 2010 it has

---

[9] Shang, "Funu gongmin shehui zuzhi," 239–240. The arrangement between Xie and the *Women's News* was a product of China's initial transition into the reform era. In the late 1980s and early 1990s, to relieve the state from the burden of subsidizing numerous state institutions, the state provided limited incentives to individual state officials to "plunge into the sea" of commercial activities (*xiahai*). This phenomenon was most apparent in the commercial sector, but this activity was also encouraged in the nonprofit sector. The arrangement was also a result of a state campaign at the time to encourage rural women to make greater contributions to society. As the state organization that allegedly represented the interests of all Chinese women, ACWF actively sponsored programs and publications that focused on the state of rural women in China.

[10] Xie claimed that ultimately the money from *Women's News* was never used and she promptly returned the investment to the newspaper when she received foreign support to publish the magazine.

affected the lives of nearly 10,000 rural women (Zhang 2011a). The successful achievement of these organizational objectives and the successful implementation of the anti-illiteracy project in particular require access to the state's nationwide networks because Chinese NGOs are legally and practically barred from establishing formal networks of operation beyond their base areas. And politically, local Chinese governments are guarded against suspicious outside influences. These institutional considerations have made it necessary for Chinese NGOs to work with state organizations in order to expand their influences.

The central component of the anti-illiteracy project is the establishment of book clubs for women in villages throughout China. The book club is a physical space that female villagers can access to borrow reading materials and conduct other activities. Given the public nature of the book club program, its sustainability is very much dependent on the political support of village leaders, particularly those with responsibilities for women's affairs. One of those responsibilities was to implement projects sponsored by ACWFs at higher levels. Therefore, any project that bears the ACWF imprints is likely to face little resistance from local village leaders. In this respect, we can see that the network of Rural Women book clubs in rural China was established in large part because Xie was a senior ACWF manager at the central state level (Shang 2007: 245). Xie in fact acknowledged that she has used her affiliation with the central ACWF to get the NGO's projects implemented at the local level (Interview with Xie). In this manner, as of early 2011, Rural Women has been operating 50 book clubs in villages throughout China. On paper, each book club has been a part of the local ACWF. In reality, each book club has been funded by Rural Women and the NGO has maintained "control" over each book club through funding ties. In this respect, each book club is a reflection of the partnership between the NGO and the local ACWF. It is difficult to imagine how Rural Women could have implemented this project without the collaboration of the local ACWFs. That it has been able to do so has been an outcome of Xie's ACWF affiliation.

Although Xie's personal ACWF affiliation clearly had an effect on Rural Women's partnerships with local ACWFs, in all likelihood Rural Women's organizational linkage to the ACWF also contributed significantly to the realization of those inter-organizational partnerships. Because those partnerships were between Rural Women and the local ACWFs, it was important, from the perspective of the local ACWFs, that the partnering NGO was a legitimate entity, especially politically. This illustration shows the importance of organizational linkage to the state. As indicated earlier, organizational linkage to the state has been valuable for Chinese NGOs because it has conferred legitimacy upon the NGO in question. Organizational legitimacy, in turn, has allowed the Chinese NGO to access other state resources. In this respect, Rural Women's organizational legitimacy has played a role in its ability to access local ACWFs, which has been critical to the NGO's ability to meet a key organizational goal.

Xie's affiliation with ACWF has also contributed directly to the Rural Women's success in securing the nonprofit registration status for its training center. Since its establishment, over 10,000 migrant women have received vocational and other training at the center. Many state agencies and state-owned enterprises have contracted with the center to provide training for their employees (Kang et al. 2010: 48). The training center's official sponsor has been the local ACWF in Beijing's Changping district (Kang et al. 2010: 61). Although Wu Qing's connection to the municipal people's congress, her personal network of state officials and intellectuals, and her background as a prominent university instructor might have facilitated the establishment of the training center's nonprofit registration status, the training center's official ACWF sponsorship suggested that Xie's ACWF affiliation has had a more direct and significant impact.

Like the NGO's book club project, the training center has derived benefits from the organizational legitimacy that resulted from the establishment of formal organizational linkages to the state. In particular, the training center has benefited from direct funding by state agencies. In 2010, approximately one-third of the training center's financial resources came from the ACWF and the Changping District Education Commission, which approved the training center's registration application and has acted as one of its supervisory organizations, even though the local ACWF has been on the books as the training center's sole official sponsor (Deng 2000; Kang et al. 2010: 61, 89; Rural Women 2010).

As one can see, Xie has played a central role in accessing valuable state resources for Rural Women. Similarly, her affiliation with *Women's News* has established the political legitimacy of all components of the organization, especially given that the Cultural Development Center of the NGO has been operating as a legally-registered commercial enterprise, a status that is legally inconsistent with the organization's operations as a nonprofit entity. Hence, it is unsurprising that Xie was actively making plans to preserve those institutional affiliations in anticipation of her retirement from *Women's News*. During my 2009 interview with Xie, she indicated that she placed a trusted friend, who she had appointed to be Rural Women's chief executive officer for several years, on the staff of the newspaper (Interview with Xie). In addition to this personnel move, Xie did other things to maintain institutional connections to the state, one of which has been to make sure that the publisher of *Women's News* always occupies a seat on the NGO's board of directors.

In sum, as Tong Lihua had done for Zhicheng, Xie Lihua has used her state affiliation to advance Rural Women's organizational objectives. Moreover, the case of Xie Lihua and Rural Women shows that, depending on the objectives of the NGO, a strong linkage to one state organization that possesses valuable organizational resources can be just as valuable as establishing linkages with multiple state organizations. Furthermore, this case suggested that, to some extent, the ill effects of the absence of formal legal nonprofit registration status could be mitigated by the establishment of inter-organizational partnerships with the state. In the following section, through an examination of the case of Little Bird, we will obtain an even clearer picture about the potential benefits of this type of partnership with the state.

3.4 Wei Wei and the Little Bird Migrant Workers Mutual Support Hotline

Since its establishment in 1999, Little Bird has been committed to the protection of the legal rights of migrant workers in China's major cities. To further this organizational objective, the NGO has been involved in the provision of free legal aid to migrant workers. In particular, Little Bird has become directly involved in the process of resolving disputes between migrant workers and their employers. Over time, from its base in Beijing the NGO has established a physical presence in Shenzhen, which is the catalyst of China's post-Mao economic reforms and which lies across from Hong Kong; in Shenyang, which is the center of China's northeastern rustbelt; and in Shanghai, which is China's financial center.

Wei Wei, the leader of Little Bird, did not begin his NGO career with the personal state linkages enjoyed by Zhicheng's Tong Lihua and Rural Women's Xie Lihua and Wu Qing. He was neither a CPC member nor an employee of a state institution. In addition, unlike Wu, he did not descend from a politically-connected family. Moreover, unlike the three aforementioned NGO leaders, Wei was not a university graduate. Yet, in 2004, under his leadership, Little Bird established an organizational linkage to the Beijing municipal justice system.

How was this possible? The answer is that Wei Wei was able to make up for his "political deficiency" by building extensive connections to the Chinese media. Given the institutional connection between the Chinese media and the ruling CPC, it may be reasonable to argue that Wei's media connections constituted his initial form of state affiliation. However, as will be discussed in Chap. 4, the institutional relationship between the communist state and the media in China is no longer one-dimensional in the sense that the Chinese media remain strictly an instrument of the state. Media organizations and individual journalists in China have attempted to carve out more independent spaces vis-à-vis the state. In fact, there are media organizations that are distinctly non-state in nature. For this reason, connections to the media are not automatically synonymous with state linkages. Therefore, it seems appropriate to distinguish media connections in general from state linkages.

In this section, we focus on Wei's affiliations with the Beijing municipal justice system. This state affiliation has provided the critical resource for Little Bird to become a successful defender of the migrant workers' legal rights through its effectiveness as a dispute resolution mechanism. That critical resource is political legitimacy.

In 2004, five years after the founding of the NGO, a district justice bureau in Beijing authorized Little Bird as a labor dispute resolution committee of the justice bureau (*sifaju renmin xietiao weiyuanhui*) (Liu 2009: 98). In this manner, as Little Bird's leader, Wei Wei's institutional affiliation to the justice bureau was established. As one can see, in Little Bird's case, the NGO leader's state affiliation and the NGO's organization linkage to the state were established simultaneously. But, as will be elaborated in Chap. 4, Wei's central role in creating this linkage is undeniable.

Like Rural Women's Cultural Development Center, Little Bird has been officially registered as a commercial enterprise. Therefore, Little Bird's partnership with a government agency that possesses law-enforcement powers also highlights the possibility for NGOs without proper registration status to establish positive working relationships with the state. More importantly, Little Bird's case shows that the disadvantages of improper registration status could be overcome by the establishment of *professional* relationships with government agencies. An important indicator is the fact that Little Bird was not targeted by the local government in Beijing in the campaigns against Chinese NGOs during the period from 2007 to 2011, despite the organization's focus on the sensitive issue of labor rights.

Of course, Little Bird has derived benefits from this form of state linkage in a more direct manner. Put simply, the partnership allowed the NGO to become highly successful in settling wage disputes between migrant workers and their employers. Since the establishment of the partnership in 2004 and ending in October 2010, the NGO was successful in helping its clients reclaim approximately 90 million Chinese yuan in wage arrears (Wang and Ye 2011). In comparison, in the five-year period prior to the establishment of the government partnership, the NGO was only able to help its clients reclaim approximately 10 million yuan in wage arrears (Ibid). How did the partnership have such a dramatic effect on Little Bird's performance?

The answer is that Wei and his colleagues have used the affiliation with the justice bureau very liberally. As Wei stated matter-of-factly, he often identified himself and his organization to employers as agents of the justice bureau (*yi sifaju de mingyi*) (Interview with Wei Wei 2008). In other words, Wei and his NGO colleagues have advertised themselves not as NGO workers but as government agents. When the employers were told of this identity, they apparently became "fearful" that the state had become actively involved in the process (Wang and Ye 2011). In this manner, Little Bird was able to satisfactorily resolve 50 % of the wage disputes over the telephone (Interview with Wei). As one can see, the political legitimacy associated with the NGO's connection to the justice bureau has served the NGO effectively in its dealings with other non-state actors, which has had direct bearings on organizational effectiveness. The importance of the government affiliation for the NGO's performance is further demonstrated by the fact that Little Bird's branch offices in Shenzhen and Shenyang were not nearly as successful as the Beijing office. Wei admitted that the challenge was that the NGO had yet to secure the same type of partnership with government agencies in Shenzhen and Shenyang (Interview with Wei; Wang and Ye 2011).

In a nutshell, Little Bird's relationship with the justice bureau has conferred political legitimacy on the NGO, even though the NGO has yet to achieve the nonprofit registration status. Moreover, Little Bird's case has provided strong evidence of how advantageous organizational linkages to the state can be achieved even without the formal nonprofit registration status. Furthermore, Little Bird's case is proof that formal state linkages have had direct bearings on the effectiveness of Chinese NGOs.

Lastly, as was the case for Zhicheng and Rural Women, Little Bird's linkage to the justice bureau has made it possible for the NGO to obtain additional financial support from the Beijing municipal government. The additional funding, which has totaled, on an annual basis, hundreds of thousands of Chinese yuan, have been used to support the NGO's other projects (Interview with Wei).

In sum, Little Bird's organizational successes have been the direct outcome of its connections to the justice bureau. That form of state linkage gave Wei and the Little Bird the political legitimacy that was necessary for them to serve their clients effectively because state linkage has acted as a "deterrent" against uncooperative employers (Wang and Ye 2011). Therefore, it was the credible threat that state linkage provided which made Little Bird an effective advocate of the rights of migrant workers.

3.5 Ma Yang and the On Action International Cultural Center

As described in Chap. 2, On Action has been the least effective of the four NGOs with respect to meeting its goals of becoming an advocate for migrant workers and a provider of services to them and their families. The case of Ma Yang and On Action is significant in that it confirms that the state affiliations of the Chinese NGO leader constitute the Chinese NGO's most valuable assets. The absence of institutional affiliations to state organizations with strong resource endowments and authority can impact negatively the NGO's ability to achieve its goals. We begin our account with Ma Yang, the founder of the NGO.

Ma Yang shares the political profile of Rural Women's Xie Lihua. Before arriving in Beijing in the late 1990s, Ma had served in the army. Ma also became a CPC member. However, unlike Xie, Ma did not obtain employment with another state institution upon the end of his military service. Instead, he arrived in China's national capital without clear employment prospects and essentially became one of the millions of migrant workers in the city. During that period of time, he worked as a volunteer at Little Bird.

About two years after On Action's founding, Ma established an affiliation with the China Migrant Worker Social Service Center (CMWSSC) as the center's Beijing office director and director of human resources (Interview with Ma 2008). By Ma's own admission, neither his CPC membership nor his service in the military was a factor in being offered a position in CMWSSC. Instead, Ma claimed that he was hired because no one in the CMWSSC had substantive knowledge about the situations facing the average migrant worker.

The CMWSSC was a creation of eight industry and professional GONGOs, including the All-China Charity Federation (Luo 2007). According to Ma, CMWSSC was created as a part of a state-sponsored effort to engage in poverty alleviation and rural reconstruction throughout China. The leaders of CMWSSC were mostly former executives of business enterprises and they were appointed to their CMWSSC posts by various GONGOs. As the Beijing office director, Ma received no compensation. However, as the center's director of human resources, he received 1,200 RMB a month for "visiting" the headquarters for perhaps two to three days each week. In this sense, Ma's affiliation with CMWSSC has had some benefits for On Action. At a very minimum, like the relationship between Xie Lihua and *Women's News*, the CMWSSC affiliation provided for Ma's personal financial needs, so that Ma did not have to worry about allocating funds from the NGO to stay alive. But this marked the extent to which the connection to CMWSSC was beneficial for On Action.

The reality was that as a low-ranking and poorly-endowed GONGO, CMWSSC was in no position to offer any additional organizational resources to On Action. Moreover, CMWSCC could not provide the sponsorship necessary to help On Action to achieve the nonprofit registration status. This situation had important implications for the NGO's development. First, without access to state resources, it was not certain that On Action would be sustainable after its international resources ran out. As will be detailed in Chap. 5, On Action had been more fortunate than most Chinese NGOs in that it was able to obtain foreign funding at the outset of the organization's establishment. However, there was no guarantee that foreign funding would continue even if the projects had been completed satisfactorily. Indeed, when the international resources ran out in 2009, it had a serious sustainability problem.

Collected donations for victims of the 2008 earthquake in Sichuan stored in On Action's office before they're sent to distribution centers. *Source* The author

We can only imagine that had On Action been able to access state resources at the same time, the situation might have been different.

Even more importantly, the absence of state affiliations has meant that Ma would have a difficult time in fulfilling his original aspirations for the NGO. As described in Chap. 2, Ma had intended the NGO to provide free legal aid to migrant workers and develop policy impact. Moreover, he expressed to me his plan to develop a national influence. But he did not have the state affiliations that Tong Lihua, Xie Lihua, and Wei Wei possessed. The CMWSSC did not possess the organizational networks that were found in ACLA and ACWF, respectively. It also was not in a position to provide On Action access to the policymaking process. Moreover, the CMWSSC could not provide Ma and On Action the same organizational legitimacy that Wei and Little Bird received from the justice bureau, which allowed the latter to become a successful representative of migrant workers in Beijing. In fact, CMWSSC could not even provide free office space to On Action.

In July 2010, Ma was able to establish connection with a state organization that could provide resources and legitimacy to On Action. The "partnership" with the Beijing Municipal Social Work Commission (*Beijing Shi Shehui Gongzuo Weiyuanhui*) provided On Action with the nonprofit registration status and state resources in the form of "government purchases" of the NGO's services. However, the formal linkage also came with a new name and a new organizational mandate. The new full name of the NGO is the Beijing On Action Social Work Office (*Beijing Zaixingdong Shehui Gongzuo Shiwusuo*). The stated mandate of the reconstituted NGO became the promotion of social stability and harmony among the residents of urban communities in Beijing, as well as the training of social workers (China Social Worker Service Platform 2010). One can only speculate that Ma had decided on this path due to his lack of success with CMWSSC. The cost of this new affiliation has been that Ma had to abandon his original goals and adhere to the new mandates of the partnership.

3.6 Corroboration from Additional Cases: Friends of Nature, Yirenping, and the Self-Empowerment Service Center for Disabled Persons

Is the significance of the NGO leader's affiliations limited to those working on issues concerning migrant workers? In addition, some may argue that On Action's level of organizational effectiveness might have been a result of its young organizational age. However, the cases of the Friends of Nature (FON), Yirenping, and the Self-Empowerment Center show that the importance of state affiliations can be found in Chinese NGOs working in different issue areas and geographical regions. Just as importantly, they show that organizational effectiveness is not necessarily limited to "mature" NGOs.

As China's first and most prominent environmental NGO (ENGO), FON has taken the lead to put the issue of environmental degradation in China on the political and public agendas. The NGO's political and public legitimacy have gone a long way in facilitating its ability to achieve organizational objectives. In its illustrious history, FON had led or coordinated several well-publicized public environmental campaigns. In addition, it has provided recommendations to the state. Moreover, it has continued to raise public awareness about environmental issues through various activities, including the publication of an annual edited volume, with essays written by journalists, government officials, and scholars. The NGO's organizational legitimacy and access to the policymaking process were the results of the state affiliations of Liang Congjie, one of its co-founders.

Liang, who died in October 2010, was not a CPC member. However, he was a multi-term delegate to the Chinese People's Political Consultative Conference (CPPCC), which is the Chinese central government's chief advisory body. He was an historian by training and he was a founding member of China's environmental GONGO, the All-China Environment Federation. Ling's family background was even more illustrious than Wu Qing's. Liang's grandfather was the late-Qing political reformer Liang Qichao. His father was a famous U.S.-trained Chinese architect who, after the PRC's founding, strongly urged, without success, Mao Zedong and the other CPC leaders to preserve the old Beijing city walls.

The Friends of Nature is the first Chinese ENGO to attain the nonprofit registration status. And Liang secured this organizational linkage through his affiliation with the Chinese Academy of Culture,[11] where he was one of the three vice presidents. The academy, which was under the direct administration of the Ministry of Culture, was established in 1984. The academy reportedly received the support of top political leaders, notably Deng Xiaoping and his first designated successor Hu Yaobang. The academy's founding faculty included China's top establishment intellectuals. The first two heads of the academy were senior members of the CPPCC and NPC, respectively.

In a similar manner, Liang served as the conduit through which the NGO could access the state's policymaking process. Liang's membership in the CPPCC was the means by which Liang played this role. As an advisory body, CPPCC has no authority to create laws. However, members of the institution are legally permitted to engage in policy deliberations (*yizheng*) and can submit legislative proposals. In March 1997, during the CPPCC annual meeting, Liang submitted a proposal to relocate the state-owned Capital Steel Corporation from Beijing. As the third largest steelmaker in China, Capital Steel has been a major contributor to the poor air quality in the Chinese capital. In February 1997, FON conducted a meeting to discuss the feasibility of relocating Capital Steel's upstream operations from Beijing

[11] The following profile of the academy is based on information garnered from a range of Internet sources, including the websites maintained by the academy. Chinese Academy of Culture, http://www.iafcc.org/ and http://www.guoxue.com/study/iacc/iaccindex.htm (accessed 9 June 2011).

(Friends of Nature 1997). The results of that meeting formed the basis for Liang's proposal to the CPPCC in March.[12]

The example of FON's access to the state's policymaking apparatus has shown that, in the reform era, opportunities have increased for non-state actors to engage the policymaking process in China. At the same time, however, the case of Liang and FON has demonstrated that the NGO leader's state affiliations have remained central to the organization's ability to access the state, especially with respect to accessing the state's policy apparatus. It is telling that over the last several years, since Liang had retreated from the NGO scene due to failing health and eventually, death, FON's direct access to the policymaking apparatus had deteriorated, even though FON is still treated with some degree of reverence because it is contemporary China's first ENGO. In this respect, the case of FON serves as a strong reminder that the leader's membership in the state apparatus is critical to the effectiveness of the Chinese NGO. Nonetheless, the situation is evolving and Yirenping's case has shown the possibility that Chinese NGOs can access the state's policymaking institutions without having a leader with Liang Congjie's political profile.

Yirenping, which was established in 2006, has within a short time become the leading Chinese NGO to advocate the rights of individuals and groups that have been the objects of employment and educational discrimination. In particular, Yirenping has been recognized as the champion of the approximately 120 million Hepatitis-B carriers and patients. It has successfully worked to end employment discrimination against victims of Hepatitis-B by multinational corporations like Nokia. In 2009 Yirenping scored its greatest victory to date when the Ministry of Health issued new rules that abolished mandatory Hepatitis-B testing during physical examinations for prospective college students, factory workers, and government employees (Jacobs 2009).

One of the main factors that affected Yirenping's rise within such a short period of time is its leader's personal connections to sympathetic NPC delegates and CPPCC members. Lu Jun's connections to those delegates have been a reflection of the increased opportunities for ordinary Chinese citizens to engage in lobbying during the reform era (Kennedy 2005; Zhang 2011b). Beginning in 2003, before Yirenping's establishment, Lu Jun has written proposals to NPC delegates and CPPCC members concerning discriminatory practices against Hepatitis-B carriers in the workplace. In 2004, he submitted a proposal to Professor Zhou Hongyu, who was an NPC delegate and the vice chairman of the standing committee of the Hubei

[12] Since then, the steelmaker has taken public measures to reduce its pollution emission and just prior to the 2008 Olympic Games, had been building a new plant in Caofeidian, which is located on the Bohai Bay inlet, about 200 km east of Beijing. In January 2011, the steelmaker shut down its entire operation in Beijing. Xinhua News Agency, "Beijing Steelmaker Halts Production in Capital," http://www.chinadaily.com.cn/china/2011-01/13/content_11847739.htm (accessed 29 December 2011).

provincial people's congress since 2008, regarding this issue (Shidai Zhoubao 2011). In 2003, Zhou established a website that solicited legislative proposals from ordinary citizens and non-state groups.[13] Through this channel, Lu established connections with Zhou and the two have collaborated on many occasions. The basic form of collaboration was as follows. Lu Jun would provide the data and even drafted the legislative proposals. Zhou would then submit those proposals to the NPC. During the March 2007 annual NPC meeting, Zhou was one of the delegates who sponsored proposals to end discriminatory policies and practices against victims of Hepatitis-B (Lu 2007). Between 2004 and 2007, the two men had a third collaborator, the recently-deceased constitutional scholar Cai Dingjian (Shidai Zhoubao 2011).[14]

The persistent efforts of NPC delegates and CPPCC members like Zhou had served as pressures on government ministries, notably the Ministry of Health. In this sense, Zhou's role was very similar to that of Liang Congjie. Therefore, the main difference between FON and Yirenping has been that whereas Liang was the embodiment of two identities—political advisor to the state and NGO leader, Lu had to rely on his personal connections to members of those advisory and legislative bodies to get his message to government officials. Nonetheless, regardless of the nature of the NGO leader's state affiliations, the results have been the same.

Just as the case of FON and Liang Congjie, the case of Yirenping and Lu Jun shows that the value of connections to the right state institutions has applied to NGOs working in issue areas other than migrant workers. In addition, Yirenping's case shows that the NGO leader's personal connections to members of the state apparatus may be just as effective in serving the needs of the Chinese NGO as NGO leaders who are directly affiliated with state institutions. Moreover, Yirenping's case demonstrates that organizational age is not a necessary determinant of organizational performance. Yirenping and On Action were both established in 2006. The fact that there are such differences with respect to the organizational effectiveness of the two organizations should cast doubt on any speculation that organizational "youth" is a necessary cause of reduced organizational effectiveness.

Similarly, the findings presented thus far in this chapter have also been observed in other parts of China, including rural and inland jurisdictions. The case of the Self-Empowerment Center, which is located in a rural county within the jurisdiction of the Chongqing municipality in south-central China, shows that the NGO leader's state affiliations have been equally valuable for NGOs operating in China's hinterland.

Since the Self-Empowerment Center's establishment in 2002, it has focused on servicing the needs of the local population. Approximately one-third of the county's 230,000 residents have either been working outside the county as migrant workers

[13] The name of the website is called "Hongyu Zaixian" [Hongyu Online], http://www.hongyu-online.com/ (accessed 29 December 2011).

[14] Before starting an academic career, Cai was a long-time senior NPC staff member.

or returned to the county after suffering physical disabilities as a result of work-related injuries. As a result of the nature of the NGO's clientele, the center has devoted considerable time, energies, and resources to serving and promoting the needs of the disabled former migrant workers.

In this respect, the NGO has benefited from its connections to the local disabled persons' federation. Like the case of Wei Wei and Little Bird, the center and its leader obtained their institutional affiliations with the disabled persons' federation almost simultaneously. In particular, the center is a fully registered nonprofit institution with the local disabled person's federation as its sponsor. Meanwhile, Chen Yuying, the founder of the NGO, has been the chairperson of the local association of physically disabled persons, which is a subsidiary of the local disabled persons' federation, a GONGO. Although Chen's position is largely honorific, the institutional affiliation has benefited the NGO in both tangible and intangible ways. For instance, the NGO has used the local disabled persons' federation to spread the word about the NGO and its work (Written correspondence with Chen Yuying 2011). Moreover, the local government even mobilized "volunteers" among local government employees to support the center's seasonal effort to distribute leaflets to outgoing migrant workers in the county's main transportation centers (Interview with Chen Yuying 2009). Furthermore, as the chairperson of the local association of physically disabled persons, Chen has been invited to participate in events sponsored by the Chongqing municipal disabled persons' federation and the county disabled persons' federation, where Chen would use the opportunities to submit reports about the state of disabled persons in the county and provide policy recommendations to address their needs (Ibid).

Along the way, the center has been able to effect some policy changes in the local government. For instance, the NGO had organized a forum of handicapped residents and invited a local people's congress delegate to participate in the forum. During that forum, four policy recommendations about handicapped persons were developed. The delegate took the recommendations to the local people's congress, where they were ratified (Interview with Chen Jian 2008). More specific policy recommendations that were implemented included the placement of plaques designating seats for handicapped persons and the elderly on public transportations carriers (Interview with Chen Jian 2009).

As these shadow cases have demonstrated, the value of the state affiliations of the Chinese NGO leader has been observed across issue areas, geographical regions, and organizational age. And all the cases have shown that state linkage is an important precondition for NGO effectiveness in China.

3.7 Conclusion

In this chapter, I show that state linkage has been an important precondition for NGO effectiveness in China. However, the positive impact of state linkage is premised on the presence of several conditions. First and foremost, the personal state linkages of the NGO leader, notably the leader's state affiliations, have constituted the central causal mechanism that determined whether an NGO would be able to acquire state resources, organizational linkages to the state, and access to the policymaking apparatus. Because state resources are scattered among state organizations, it is important for NGO leaders and their NGOs to establish relations with resource-rich organizations. In this sense, it is clear that not all institutional connections to the state have been equally beneficial for Chinese NGOs. Therefore, an important lesson for Chinese NGO leaders is that whenever possible, choose wisely when it comes to the state organizations with which one would like to associate.

The cases also suggest that for advocacy NGOs in China, the opportunities to access the policymaking apparatus have increased in the reform era. This phenomenon is in part the result of the increasing willingness among NPC delegates and CPPCC members to exercise their constitutional responsibilities. Therefore, even NGOs that have not been led by leaders with explicit state identities, such as Yirenping's Lu Jun, have been able to lobby the state successfully through collaboration with likeminded NPC delegates and CPPCC members.

This type of policy collaboration has taken place in an environment in which the Chinese media organizations and journalists have also become more independent and daring. Indeed, as many have discovered, Chinese media organizations and individual journalists have played a valuable role in spreading the influence of the Chinese NGOs. In so doing, the Chinese media have also contributed to the positive interactions between the NGOs and the state and its agents. In the next chapter, we will examine how Chinese NGO leaders have been able to use the media to develop and enhance their state affiliations.

References

Appleton, Simon et al. 2008. The Economics of Communist Party Membership: The Curious Case of Rising Numbers and Wage Premium during China's Transition. IZA Discussion Paper Series, DP No. 3454 (April). http://ftp.iza.org/dp3454.pdf. Accessed 23 December 2011.

China Social Worker Service Platform. 9 November 2010. Beijing Zaixingdong Gongzuo Shiwusuo Zhaoping Qishi [Job Announcement of the Beijing On Action Social Work Office]. http://cncasw.blog.163.com/blog/static/169137968201010111105912974/. Accessed 29 October 2011.

Deng Guosheng. 2000. Nongjianu Baishitong [Rural Women Knowing All]. In *Zhongguo NGO Yanjiu: Yi Anli Wei Zhongxin* [*Study of Chinese NGOs: Focus on the Cases*], 222-237. UNCRD Research Report Series No. 38 (August). Beijing, China: United Nations Centre for Regional Development.

Friends of Nature. Dashiji – 1997 [Timeline – 1997]. http://www.fon.org.cn/channal.php?cid=67. Accessed 29 December 2011.

Jacobs, Andrew. 12 October 2009. China to End Required Testing for Hepatitis B. *The New York Times*. http://www.nytimes.com/2009/10/13/world/asia/13beijing.html. Accessed 20 October 2011.

Kang Xiaoguang et al. 2010. *NGO Yu Zhengfu Hezuo Celue* [*Strategic Study on NGO-Government Collaborations*]. Beijing, China: Social Sciences Academic Press.

Kennedy, Scott. 2005. *The Business of Lobbying in China*. Cambridge: Harvard University Press.
Li, Bobai and Andrew G. Walder. Career Advancement as Party Patronage: Sponsored Mobility into the Chinese Administrative Elite, 1949-1996. *American Journal of Sociology* 106, 5 (March): 1371-1408.
Liu Yaotang. 10 June 2011. Zhicheng Gongyi Lushituandui Maishang Xintaijie [The Zhicheng Public Interest Legal Team has Taken a New Step Toward the Top]. *Zhongguo Lushi Wang* [*All-China Lawyers' Association Online*]. http://www.mzyfz.com/cms/lvshijulebu/lvshiku/gongyilvshi/html/882/2011-06-10/content-79571.html. Accessed 27 July 2011.
Liu Zhihua. 2009. 'Sheng' Yuan Yu Meiti, 'Zhang' Ji Yu Meiti – Cong 'Xiaoxiaoniao Dagong Huzhu Rexian' Fazhan Kan Caogen NGO De Meiti Zhanlue ['Birth' through the Media, 'Growth' on Media's Foundation – Observing the Media Strategy of the Grassroots NGO through the Development of the 'Little Bird Migrant Worker Mutual Support Hotline']. *NGO Anli Ji: Zhongguo Feiyingli Zuzhi Guanli* [*NGO Cases: Management of Chinese Nonprofit Organizations*], 96-102. Shanghai, China: NPI. http://www.npi.org.cn/uploads/magazines/anliji.pdf. Accessed 23 June 2011.
Lu Jun. 29 March 2007. Lianghui Qijian Duowei Daibiao, Weiyuan Tichu Yigan Ti'an, Rang Women Jizhu Tamen Ba [During the Two Meetings, Many [NPC] Delegates and [CPPCC] Members Submitted Proposals Concerning Hepatitis-B, Let Us Remember them]. http://blog.sina.com.cn/s/blog_492aa7be010008dl.html. Accessed 29 December 2011.
Luo Xun. 9 February 2007. 'Xiaochu Pingkun Yu Jibing Cujing Jihua' Qidong [The Launching of the 'Project to Eliminate Poverty and Diseases]. *Guangming Daily*. https://www.gmw.cn/01gmrb/2007-02/09/content_550086.htm. Accessed 15 July 2011.
People's Daily. 11 May 2011. Zhicheng Gongyi Shiernian Shehui Guanli Chuangxin Tansuo Zuotanhui Juban [The Zhicheng NGO's Twelfth anniversary Symposium on the Exploration of Social Management Innovation was Convened]. http://society.people.com.cn/GB/42735/14610144.html. Accessed 7 June 2011.
Rural Women. 2010 Annual Report. http://www.nongjianv.org/web/Files/content/njndzb/2010njnnianbao.pdf. Accessed 26 December 2011.
Shang Xiaoyuan. 2007. Funu Gongmin Shehui Zuzhi He Zhili Moshi De Duoyuanhua [Women's Civil Society Organizations and the Pluralism of the Governance Models]. In *Chongji Yu Biange: Duiwai Kaifangzhong De Zhongguo Gongmin Shehui Zuzhi* [*Impact and Change: the Chinese Civil Society Organizations that are Opening Toward the Outside*], ed. Shang Xiaoyuan, 236-283. Beijing, China: Chinese Social Sciences Press.
Shidai Zhoubao [Time Weekly]. 10 March 2011. Gongmin Jianke Jin 'Lianghui' [Citizen Lobbyists Entered the 'Two Meetings']. http://www.time-weekly.com/story/2011-03-10/112221.html. Accessed 29 December 2011.
Tong Lihua. 2001. *Weichengnianren Faxue* [*Juvenile Jurisprudence*]. Beijing, China: Zhongguo Minzhu Fazhi Chubanshe.
Tong Lihua. 2009. *Weile Zhengyi: Zhicheng Gongyi Shinian* [*For Justice: Ten Years of Zhicheng Public Interest Law NGO*]. Beijing, China: Law Press China.
Walder, Andrew G. 1995. Career Mobility and the Communist Political Order. *American Sociological Review* 60, 3 (June): 309-328.
Wang Qi and Ye Sai. 13 November 2011. 'Xiaoxiaoniao' Dagongzhe Weiquan Lushang de Jixianfeng" ['Little Bird' is the Vanguard of the Road to Defend the Rights of Laborers]. *Jing Bao*. http://www.thefirst.cn/1366/2011-01-13/379845.htm. Accessed 26 December 2011.
Xinhua News Agency. 13 January 2011. Beijing Steelmaker Halts Production in Capital. http://www.chinadaily.com.cn/china/2011-01/13/content_11847739.htm. Accessed 29 December 2011.
Zhang Jianli. 28 February 2011a. The 50th Rural Women Book Club was Established in Zhejiang's Deqing. http://www.nongjianv.org/web/Html/newspaper/2011-2/28/0228221632.html. Accessed 9 June 2011.

Zhang Chen. 5 August 2011b. Huiwai Youshui: Lianghui Baodao Xin Keti [Lobbying Outside the Meeting: New Topics in Reporting about the Two Meetings]. *Qingnian Jizhe* [*Youth Journalists*]. http://media.people.com.cn/GB/22114/42328/227852/15338385.html. Accessed 29 December 2011.

Zhou, Xueguang. 2004. *The State and Life Chances in Urban China: Redistribution and Stratification, 1949-1994*. New York: Cambridge University Press.

Chapter 4
Media Connections: Bridging the State-Society Divide

Abstract The literature on media-state relations has generally emphasized the media's function as a watchdog of the state. According to this line of reasoning, a strong democracy is characterized by the presence of independent media to check the state. By extension, the absence of independent media generally characterizes non-democracies, thereby implying that media in non-democracies do not hold any value for non-state actors. However, this simple distinction of state-media relations by regime type ignores the fact that media in transitioning states like China have retained their connections to the state even as they exercise greater independence. In fact, Chinese media's ability to straddle the state-society divide makes them uniquely positioned to contribute to the work of the NGOs, which can employ media connections to establish state linkages as well as obtain other state resources. Although it is important not to overstate the value of media connections when compared with that of state linkages, the cases presented in this chapter nevertheless show that media connections are extremely valuable as means to acquire state linkages and can help NGOs strengthen those linkages.

Keywords Media connections · State-society divide · Decentralization · Commercialization · Legitimacy · Journalist · Internet

On 5 March 2005, the *People's Daily*, which is the flagship newspaper of the ruling Communist Party of China (CPC), published a report entitled "An anti-leak cover is being installed under the lake in the Old Summer Palace: Is this an act of protection or destruction?"[1] Soon after the report's publication, it was republished on the websites of other Chinese newspapers and online platforms. In the next two days, the *People's Daily* continued its coverage of the issue and other newspapers,

[1] The account of this incident in this section is based completely on its recounting in Jia et al. (2008, 238–241).

J.W. Tai, *Building Civil Society in Authoritarian China*,
SpringerBriefs in Environment, Security, Development and Peace 20,
DOI 10.1007/978-3-319-03665-6_4

including the *Southern Weekend* and the *China Youth Daily*, also published lengthy reports about the anti-leak project that was taking place at the historic Old Summer Palace in Beijing.[2]

The government responded almost immediately to the media reports. On 30 March, the Beijing Municipal Environmental Protection Bureau began an investigation into the project. The following day, the central government's State Environmental Protection Administration (SEPA) issued a statement that the project was undertaken without conducting the required environmental impact assessment; therefore, the project must cease until the assessment has been conducted. On 1 April, the Haidian district government, which had direct administrative jurisdiction over the area in which the palace was located, ordered the complete stoppage of all work relating to the installation of the anti-leak cover.

The media furor over the work project began with an e-mail message from Dr. Li Hao to the *People's Daily* reporter who eventually penned the first news report on the project. Li was the leader of Beijing Earthview Environment Education and Research Center, a Chinese environment NGO (ENGO). Li received the tip-off about the work project from Professor Zhang Zhengchun, who accidentally discovered the project while visiting the palace. Li believed that the best way to get public attention on the project was to inform the media.

After the Haidian district government issued the work-stoppage order, SEPA conducted a public hearing on the anti-leak cover project. The hearing was attended by journalists from dozens of media organizations. After the hearing, media reporting about the project continued. On 5 July, SEPA posted an environmental impact report about the project on its website. The report indicated that the project was illegal and would have a negative impact on the palace's ecology. The entire anti-leak project therefore had to be redone. An even more exciting piece of news for the environmental activists was issued on 22 February 2006. On that day, SEPA promulgated a set of temporary regulations on public involvement in environmental impact assessments (SEPA 2006).

This case illustrates a successful case of collaboration between a Chinese NGO and the Chinese media to change state practices and establish new state policies. It also seems to conform more closely to the media's ideal function, which is to supervise the state. Yet, we can also make the argument that this case is an illustration of how the media have bridged the state-society divide and provided an opportunity for collaboration between one Chinese NGO and a particular government agency that shared an interest in stopping a project that was sponsored, or at least approved, by other state agencies. Even so, the "collaboration" that is illustrated above was indirect. The cases presented in this book, however, demonstrate that the Chinese media may also facilitate the establishment of direct collaborative

---

[2] The Chinese name for the Old Summer Palace is "Yuanmingyuan." The garden was pillaged and plundered in 1860 by British and French troops. The palace has been designated by the Chinese government as a historical monument to China's "century of humiliation," which dates from the mid-19th century to the PRC's founding in 1949.

relationships between Chinese NGOs and China's state agencies. In doing so, the Chinese media can play an invaluable role in facilitating NGO effectiveness.

The Chinese media's positive effects on the NGOs are derived from the various functions of the media. In particular, the media can empower NGOs through the former's roles as (1) publicists of events, issues, organizations, and individuals; (2) legitimating agents; and (3) information collectors for the policymakers. But in order for any NGO to benefit from these media functions, it must gain access to the media. In this respect, the leader of a Chinese NGO plays a critical role. Put simply, a leader with strong media connections is a necessary condition for organizational access to the media. However, it is important to recognize that although the NGO leader's media connections may exert positive effects on the organization's state linkages, the effects of media connections on state linkages may be more indirect when compared to the NGO leader's state affiliations.

4.1 Civil Society, Media, and State

The free flow of information is an important aspect of any free society. Therefore, there should be little disagreement that an "independent and pluralistic media are an essential bulwark of a mature and effective democracy" (Ballentine 2002: 91). As this statement suggests, the role of the state's watchdog has often been championed as the media's most important function. In this manner, the media in a liberal democracy ensure that elected officials will not abuse the powers that they have been entrusted.

By most accounts, the media in non-democracies have not performed this function effectively, if at all. An often-cited reason is that the media in non-democratic states are subjected to a wide range of state measures to control the media (Rugh 2004). Consequently, the media in this type of political environment have become tools of the state. In this role, instead of checking the state, the media in non-democracies have become instruments to weaken the regime's opponents (Ottaway 2003).

When it comes to China, the prevailing opinion is that before the reform era the Chinese media served collectively as the state's mouthpiece. During the Maoist period, the Chinese media's main function was to transmit the state's directives to the masses. Their other function was to serve as an intelligence-gathering agency for the state by collecting information about the Chinese masses and the situations abroad for consumption by the political elite. During periods of intra-elite struggles, the Chinese media could also be counted on to serve the interests of competing factions. To a large extent, the state has sustained its control over the Chinese media into the twenty-first century (Zhao 2008; Stockmann 2013; Young 2013). Therefore, it would appear that the Chinese media, like the media in other non-democracies, have not strayed very far from the authoritarian state. In this sense, we should expect the Chinese media to continue to remain a tool of the state. By extension, it seems inconceivable for the Chinese media to be of any help to Chinese NGOs.

Although there is no doubt that Chinese media organizations do not enjoy the level of freedom that their counterparts in democracies possess, it would be wrong to simply assume that the Chinese media system has remained unchanged over the past 60 years. In fact, the Chinese media system today has become significantly more pluralized, commercialized, and globalized. This development has positive implications for societal actors like NGOs because it ensures that Chinese media organizations are more accessible to non-state entities today than at any other time since the PRC's founding.

To a large extent, the changing Chinese media system is the outcome of its decentralization during the reform era. The process of decentralization has resulted in the creation of numerous media organizations at the sub-national level. In this sense, the decentralization of the Chinese media system has contributed directly to the system's pluralization. Because many of the newly-emerged media organizations are creatures of the local party-state, they reflect local interests to a considerable extent (Zhao 2008). A consequence is the emergence of a wider range of political views, even neo-liberal ones, in the Chinese media landscape.

The decentralization of the media system has undoubtedly contributed to the commercialization of the media industry. Here, it is important to recognize that decentralization is not a process unique to the media system in China. The state has promoted decentralization in other areas of governance. One important reason for promoting decentralization is to relieve the central state of the burden of maintaining an enormous state apparatus. The financial implication of decentralization for Chinese media organizations is thus quite clear. In exchange for greater "freedom," they are now responsible for their own financial well-being. Therefore, in order to maintain and improve their bottom line, Chinese media organizations are motivated to publish contents that will attract more readers, which is important in large part because a large readership will result in increased advertising revenue. In this manner, one can see that just as decentralization has led directly to increases in the number of media outlets in reform-era China, it has also contributed to media commercialization, which have resulted in media contents that are "spiritually polluting," "vulgar," and "pornographic" in nature (Lynch 1999: 6)

Naturally, it would be wrong to suggest that the only consequences of commercialization have been negative in nature. The key implication of commercialization is that Chinese media organizations have become more interested in publishing materials that would be of interest to readers. In this manner, commercialization has affected the editorial process so that Chinese journalists are encouraged to uncover sensitive social issues that would be of interest to the readers. Consequently, just as tabloid journalism has made its way into the Chinese media system, the system has also become a relatively hospitable environment for muckraking journalism.

Lastly, in a related fashion, China's integration with the international community has had profound effects on the Chinese media system, especially in terms of the effects on a new generation of Chinese journalists. As China opened up to the world, its journalists have also had more opportunities to access information about journalistic norms and practices outside China. Nowadays, for those Chinese journalists who have the will and the language proficiency, they often have opportunities to interact with foreign journalists and media experts in China through foreign-sponsored forums and seminars, where they learn about those norms and values (Interview BJ052808). Access to information about media professionals outside China has undoubtedly contributed to the increasingly liberal attitudes of Chinese journalists (Lin 2010).

Collectively, these processes have contributed to the creation of a media environment within which Chinese media organizations are exercising functions that they did not perform during the Maoist days. As shown by the case about the anti-leak cover in Beijing's Old Summer Palace, one of the new functions of the Chinese media has been to check the behavior of the state and its agents. In a similar and related vein, the Chinese media have also become vehicles through which members of the Chinese society have voiced their opinions. Therefore, just as Chinese media organizations have solicited commercial advertisements to bolster their financial conditions, they have become publicists of individuals, organizations, and issues that are of concern to the Chinese public.

However, lest we become too excited about the liberal tendencies of the Chinese media system, we should be aware that the Chinese media organizations have remained keenly cognizant of the political bottom line (Stockmann 2013). In principle, Chinese media's main mandate continues to be that of "[telling] China and the world about the communist party's agenda" (Young 2013: 4). Indeed, although the Chinese political system is no longer a totalitarian system, Chinese media organizations must still adhere to the "party principle," which stipulates that "the news media must accept the [communist] party's guiding ideology as its own; that they must propagate the party's programs, policies, and directives; and that they must accept the party's leadership and stick to the party's organizational principles and press policies" (Zhao 1998). Thus, there are still real consequences for media agencies and individual journalists should they go too far. Similarly, the majority of Chinese media organizations and journalists are fully aware of the need to stay away from issues that are "politically sensitive," which lie within the exclusive domain of the state's official media outlets (Stockmann 2013: 77–103).

But this fact also has important implications for Chinese NGOs. Given that they have retained their close association with the state, Chinese media organizations have thereby maintained their traditional functions, such as transmitters of state policies. The other traditional function of the Chinese media has been that of intelligence collectors and compilers with direct access to policymakers.[3] From the NGO's perspective, Chinese media organizations are thus important conduits to policymakers and the policy apparatus. An important reason that Chinese media

---

[3] This media function has been manifested primarily, if not chiefly, in the presence of the "internal reference" (*neibu cankao*, or *neican* for short) system. The internal reference system is essentially a mechanism through which the Chinese media compile and provide classified reports about domestic and international happenings to top political leaders. Despite its shortcomings, it has been an important portal to China's policymakers and the policymaking process. For criticisms of the mechanism, see Magnier (2004) and The Economist (2010).

organizations are able to access the policy apparatus is that they are *legitimate* state organizations. Therefore, to a certain extent, the media's positive reporting of the Chinese NGO leader and his organization can act as a signal to the state, as well as to the society, about the legitimacy of the leader and his group. In this sense, from the NGO's point of view, Chinese media organizations are valuable because they are *legitimating agents*.

Given that the Chinese media have developed strong connections with members of the society and still retain their close association with the state, the Chinese media have become uniquely positioned to play the role of the bridge across the state-society divide. Moreover, the combination of their new function as publicists of societal organizations and interests with their old function as state intelligence agencies means that Chinese media organizations are very valuable resources for Chinese NGOs that want state recognition and access to the state's policy apparatus. But how can a Chinese NGO acquire these value media commodities? What can a Chinese NGO do to attract the media's attention to itself and its issues of interest?

Analysts have spent much time talking about the importance of issue framing for Chinese NGOs. Indeed, this has been a valuable and effective strategy to establish media connections. At the risk of simplification, issue framing is essentially story-telling. This notion also suggests an intention to create stories. But in many cases, NGOs and their leaders have become media stories without making a conscious effort to create them. At the same time, no matter how compelling the stories may be, there is no guarantee that the media will report on them. In this situation, successful issue framing that garners media attention entails the prior existence of media connections. In this respect, the prior media connections of the Chinese NGO leaders are invaluable.

In sum, just as the Chinese NGO leader's state affiliations to the state have provided her organization with access to the state, her media connections have collectively served as a portal to the world of the Chinese media for her organization. As a portal, the media connections of a Chinese NGO leader need not be exhaustive. The most important thing is that those connections should serve as constant points of media access for the NGOs.

Perhaps the most interesting aspect about the present state of the Chinese media is that Chinese media organizations have become points of state access for Chinese NGOs led by persons without established political ties. Little Bird's experience demonstrate this point clearly. Through connections to media organizations, Chinese NGO leaders are thus able to establish state linkages.

Figure 4.1 is a diagram of the elements involved in the pathway from media connections to organizational effectiveness. As indicated by the solid-line arrow to the NGO leader's state affiliations, the leader's media connections have been more facilitative than direct when it comes to helping the NGO acquire the benefits of state linkage. This means that in all cases the positive effects of media connections have been either contingent on the prior existence of the leader's affiliations to the state or dependent on the creation of state affiliations by those media connections. Most obviously, media connections have not played a direct role in providing the Chinese NGO with access to state resources, such as funding, office space, and personnel.

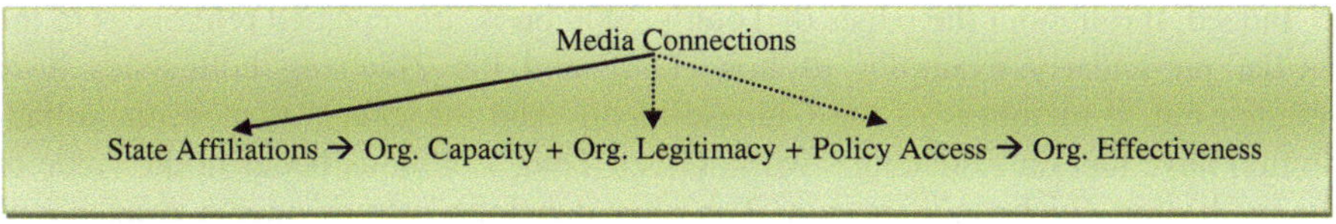

Fig. 4.1 Media connections and organizational effectiveness

As the two dotted arrows to "organizational legitimacy" and "policy access" indicate, the media connections of the NGO leader may help the NGO access certain benefits of state linkage independent of the media connections' impact on the NGO leader's affiliations to the state. But again, in many of those cases, the salutary effects of media connections have been facilitative and involve the prior existence of the leader's state affiliations, thus making it difficult to ascertain the media connections' independent effects. In sum, as valuable as the NGO leader's media connections, we need to be cautious in asserting that the value of media connections can rival that of the leader's state affiliations with respect to obtaining the most important benefits of state linkage.

4.2 Tong Lihua and the Zhicheng Public Interest Law NGO

In Chap. 3, we had seen how Zhicheng benefited from the strong state affiliations of its leader. Tong Lihua has employed his state affiliations to enhance Zhicheng's organizational capacity, develop formal organizational linkages to the state, and established the foundation on which Zhicheng has been able to do what virtually no other Chinese NGOs have been able to accomplish, which is direct participation in the state's lawmaking process. Given the present political situation in China, it would seem that Tong's ties to the state should constitute sufficient resources to support Zhicheng's efforts to achieve its organizational goals.

Instead, both Tong and his Zhicheng colleagues have emphasized how the media have played a significant role in the NGO's achievements (Interview with Zhicheng staff lawyer Wang Fang 2008; interview with Zhicheng research director Zhang Wenjuan 2008). In Tong's 2009 book about his own NGO career and Zhicheng's development, he wrote about the importance of media collaboration for the Chinese NGO:

"[As an NGO, even] if we had shouted ourselves blue in the face, it would seem that, [working] at a time of advanced broadcast media, the simple reliance on speeches and face-to-face [methods] to publicize and popularize [issues] is both wasteful and clumsy. As a public interest law organization, in order to publicize justice, it is necessary to pay attention to the effects of the media; it is necessary to work closely with the media" (Tong 2009: 37; author's translation).

Indeed, throughout the pages of Tong's 2009 book are repeated references to the media, particularly examples of how Tong and his Zhicheng colleagues have reached out to members of the Chinese media, and the effects that media collaboration have had on Zhicheng's work. Given Tong's obvious belief in the value of the media for Zhicheng's success, Tong has developed strong media connections alongside his extensive state affiliations. Tong's media connections have established channels for publicizing Zhicheng and its views on issues of interest to the NGO. Moreover, they have provided additional means through which the NGO could impact state policies and practices, and even establish formal organizational linkage to the state.

Among his media connections, Tong has singled out his professional relationship with Cui Li, a senior journalist at the *China Youth Daily* (CYD) (Tong 2009: 37). Before we describe how Tong and Zhicheng have benefited from his relationship with Cui, it's important to understand the reasons that Cui and CYD might have been willing to establish this connection with Tong. Based on the information that is available, we can infer three reasons.

The first reason was CYD's reputation. Although the CYD is the official newspaper of the Communist Youth League (CYL), it has developed a reputation in recent years as the only outspoken central-level state paper in China (Zhao 2008: 43). True to its reputation, the editors of the CYD have not been afraid to publish writings about controversial subjects and which challenge the dominant state perspectives. In January 2006, for example, the newspaper's weekly published a lengthy article, written by a Chinese university professor, which challenged the dominant state narrative that the Taiping rebels in the nineteenth century and the Boxers in early twentieth century were patriotic anti-imperialists who represented the Chinese peasantry. This historical characterization has important political implications because the CPC has claimed that it is the ideological descendant of the Taiping rebels and the Boxers. For publishing this article, the CYL issued an order to suspend the newspaper's publication and the removal of the weekly's chief editor (Zhao 2008: 57–58).

The second reason was Cui's background. Cui was not simply a CYD journalist. She had been the deputy director of the current affairs department at CYD. Moreover, she earned a law degree and has been the newspaper's senior legal correspondent.[4] She had written many reports about legal matters, including a series of articles about the 2003 Sun Zhigang case, which involved a young university graduate who was beaten to death by the local police in the southeastern city of Guangzhou while in detention for failing to produce a proper working/residential permit. His case became a national and international sensation, and the public outcry generated from the case forced the state to abolish the notorious system of custody and repatriation, which in practice authorized the local police to detain almost anyone who was deemed to be undesirable by the authorities. Although the

[4] For Cui's biography, see the *China Youth Daily* website at http://zqb.cyol.com/node/node_7051.htm (accessed 12 August 2014).

CYD did not break the news about the Sun case, Cui was the first Chinese journalist to write about the petitions made by Chinese legal students and scholars to the National People's Congress (NPC) to authorize special investigations into the case (Cui 2003a, b). Cui's professional background and interest suggest that she would be quite interested in Tong's legal aid operation.

Compared to the two above reasons, the third reason is more speculative. Nonetheless, there is reason to believe that Tong's state affiliations, notably his ties to the CYL, had probably contributed somewhat to his connection to Cui and the CYD. We cannot rule out the possibility that the fact that both Cui and Tong were affiliated with the CYL had played a role in connecting the two individuals'.

Whatever the reasons for their relationship, Tong's ties to Cui have been both long and substantive. The two apparently became acquainted with each other before 2001, which was before Zhicheng's children's legal aid arm attained the nonprofit registration status and before the establishment of Zhicheng's migrant worker arm. Evidently, Cui had developed a high regard for Tong's expertise. In November 2002, she invited Tong to speak to 40 CYD journalists about the NGO's effort to promote the legal protection of children's rights and the inadequacies of the existing legal system in addressing this issue (Tong 2009: 37). And Cui was routinely among the first journalists that Tong would contact if he wanted to publicize the NGO's activities.

From Tong's perspective, the primary benefit of his relationship with Cui has been the latter's role in publicizing Tong and Zhicheng's activities. Over the years, Cui had either written or coordinated the writing of numerous articles that featured Tong and Zhicheng. In most cases, collaboration between Tong and Cui began with Tong contacting Cui about the NGO's work. An illustration of the successful partnership between Tong and Cui is the May 2001 case involving Tong's decision to file a lawsuit against the state tobacco bureau and China's leading tobacco firms. The lawsuit charged the defendants with failure to comply with laws and regulations that required tobacco companies to indicate clearly on their advertisements that smoking was hazardous to health and that minors were prohibited from smoking.

On 11 May 2001, Tong held a press conference to announce his plan to file the lawsuit. He invited journalists he knew to attend the press conference and Cui Li was one of the invitees (Tong 2009: 244). After the press conference, Cui penned a report about the planned lawsuit that was published in the CYD (Cui 2001a). Cui's essay was among the first of dozens of domestic and foreign news writings about the case over the next year and half.

The report by Cui and the writings of other news organizations about the lawsuit produced several observable effects that benefited the NGO. First, the media publicity turned Tong into an instant media sensation (Cui 2001b; Xin 2001). At the same time that Tong became a media personality as a result of the published reports about the lawsuit, the media reporting also put Zhicheng on the map in that nearly all stories about Tong had indicated clearly his association with Zhicheng. Second, the media publicity helped Tong achieve the goal of garnering widespread attention about the negative effects of tobacco on the health of China's minors (Tong 2009: 257).

Third, although the court had refused to accept the lawsuit, the media attention had compelled the defendants to take immediate actions to address the lawsuit's main charges. For example, soon after the plan to file a lawsuit was publicized, the state tobacco bureau, for the first time, made it clear that all advertisements and labels for tobacco companies must include the message that existing regulations prohibited smoking by elementary and secondary students as well as the selling of tobacco to minors (Zhang 2001). Lastly, Tong used all the media interview opportunities resulting from the news reports to talk about broader issues concerning the protection of the rights of minors (Tong 2009: 257). In this manner, Zhicheng became even better known to the media and, by extension, to the state.

In sum, Tong's relationship with Cui in particular and journalists in general have benefited Zhicheng by publicizing the organization and the issues that have concerned the NGO. Furthermore, in getting the state tobacco bureau to make clear its support of the existing regulations and change existing practices, Tong's media connections have provide another means for Tong and Zhicheng to access and impact the state policy apparatus. This example thus illustrates the relatively independent effects of the media on the NGO's ability to influence state policies and practices.

To be sure, Tong's media connections have not been limited to his relationship with Cui Li and other CYD journalists. Journalists associated with the official Xinhua News Agency have also helped to publicize Tong and Zhicheng through its published writings. However, for Zhicheng and any Chinese NGO, it is arguable that Xinhua's greatest value is its access to the pinnacle of China's political system through its writing of the classified "internal references," which are consumed exclusively by senior Chinese political leaders. Tong's connections to Xinhua journalists have often had the effect of getting the ideas that Tong and his Zhicheng colleagues crafted to become featured in internal reference reports. In this fashion, Tong's media connections have provided another means through which Zhicheng was able to affect state policies.

An example of the positive effects of Tong's connections with Xinhua journalists took place in July 2003, not long after the children's arm of the NGO attained the nonprofit registration status. In June of that year, the State Council promulgated the "Management Measures for Supporting the Homeless and Beggars in Urban Life." The new regulations called for government agencies to cease practices that forced the homeless to "accept government assistance." The new regulations would go into effect on 1 August. But Tong and his Zhicheng colleagues felt that the new regulations failed to address the different needs between adult homeless persons and those who were minors. Tong expressed Zhicheng's views on the issue in a 28 July conference. Immediately after the conference, Xinhua journalists approached Tong about conducting an interview with Tong with the intent of submitting the written report to top central government leaders. The report ultimately reached Premier Wen Jiabao, who responded to the report by instructing the State Council's legal office and the Ministry of Civil Affairs to look into the matter (Tong 2009: 230).

This example is illustrative of how a Chinese NGO is able to access the state policy apparatus through connections with the appropriate Chinese media organizations. Yet, it is also suggestive of the contingent nature of the salutary effects of media connections. Although the evidence is not precise, we can speculate that Xinhua journalists' willingness to report about Tong's views was based in part on Tong's established state linkages. As described in Chap. 3, Tong and his colleagues were first invited in 2002 to participate in the Beijing municipal law-making process concerning the revision of the law on child protection. In other words, Tong's political credentials were established before the interview with the Xinhua journalists. Perhaps reflecting his state affiliation, Tong was invited to the state-sponsored conference where he made his 28 July speech. Therefore, to an important extent, we can reasonably infer that at least some of the salutary effects of Tong's media connections have been contingent on his profile as an expert who was politically connected.

Nonetheless, the value of media connections for establishing Zhicheng's state linkages must be recognized. At the very least, positive media coverage has enhanced Zhicheng's reputation, which has made it easier for state agencies to work with Zhicheng. The reason for this speculation is that one of the functions of the Chinese media has been that of a legitimating agent. Positive media coverage of a person, an issue, or an organization is a signal to the political establishment, as well as the general population, that the featured entity is legitimate. Therefore, even if the impact of the media has not been sufficient to create state linkages for the Chinese NGO, the media have nonetheless been facilitative. This function of the Chinese media can be inferred from the example of the establishment of Zhicheng's migrant worker arm.

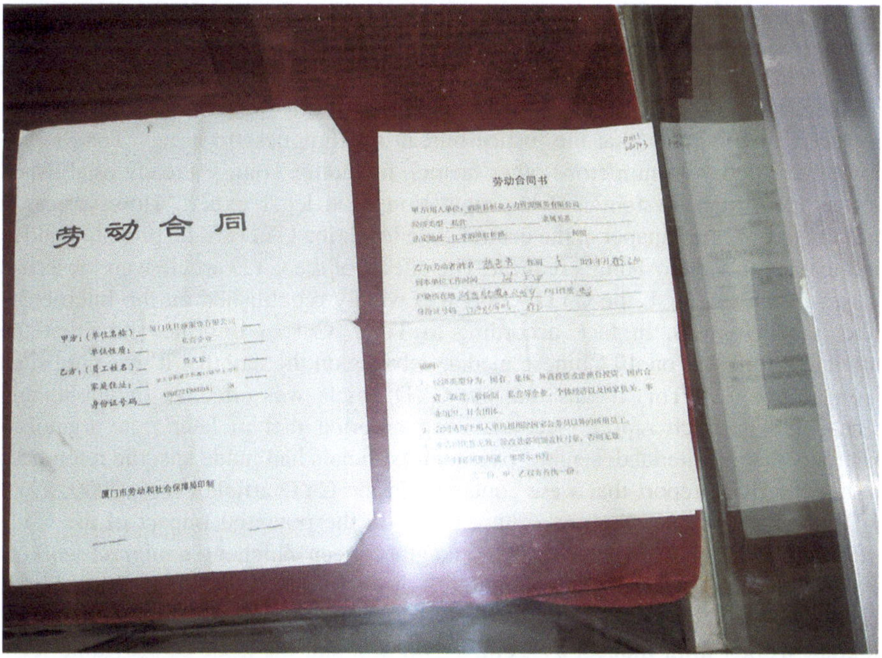

A labor contract agreement between a migrant worker and his employer. *Source* The author

Although Tong established Zhicheng first as an organization promoting the legal rights of children, in 2003 he and his colleagues began to pay attention to the legal conditions of migrant workers. Their interest was spurred in part by a case involving over 60 migrant workers who were unable to obtain wage arrears from their employer. When the leader of the migrant workers could not resolve the case on his own, he sought assistance from Zhicheng with a referral from a journalist. With the help of Zhicheng's lawyers, the leader of the migrant workers filed the first legal case in the Beijing municipality concerning wage arrears for migrant workers (Tong 2009: 347–366).

The case motivated Tong and his colleagues to undertake research about the legal state of the migrant workers, notably their ability to recover wage arrears. The result was a one-year research project that involved extensive interviews with migrant workers, the distribution of 8,000 surveys covering eight provinces, and 17 case studies (Tong 2009: 21). One of the significant findings of the report was that a major challenge to defending the rights of migrant workers was the high cost associated with each case. The cost would be even higher if the government took on those cases directly.

In June 2005, Zhicheng held a forum about the results of the report and Cui Li was the only journalist invited to the forum (Tong 2009: 22). A few days later, CYD published an essay about the report. In August 2005, Tong, on behalf of Zhicheng, submitted a proposal to the Beijing municipal justice bureau, essentially suggesting that the government purchase Zhicheng's services to provide legal aid to migrant workers at a cost that would be significantly cheaper than if the government took on those cases directly. In September 2005, the migrant worker arm of the NGO was registered as a nonprofit organization under the sponsorship of the justice bureau (Tong 2009: 25).

It appears that the CYD report had exercised some influence on the ability of Zhicheng's migrant worker arm to attain the nonprofit registration status. Some, however, may counter that the justice bureau's willingness to accept Tong's proposal had to do with numerous other factors, including Tong's already established ties to the justice system and his reputation as a legal expert. However, as a prominent CPC newspaper at the central state level, the CYD article probably added political legitimacy to Tong's request. The effects of the CYD article were observed in two instances. First, the CYD report was widely republished on the Internet by other media outlets. In fact, according to Tong, Cui Li informed him that the articlewas reposted on 40 Chinese media websites on the day that it was published (Tong 2009: 22). This means that the CYD article was effective in publicizing Zhicheng's research report. Second, Tong reported that in later state-organized meetings that he attended, senior government officials had made specific references to aspects of the report that were contained in the CYD article (Tong 2009: 22).

These facts suggest that we cannot overlook the potential impact of the CYD article on the establishment of formal linkage between Zhicheng's migrant workers arm and the justice bureau. At a minimum, it would not be far-fetched to suggest that the CYD article reduced the political risk for the justice bureau to sponsor the migrant worker arm. Indeed, by Tong's own account, the issue of providing legal aid to migrant workers was still a politically sensitive issue in 2005, so much so that Tong's friends had counseled him against getting involved in this issue area (Tong 2009: 23). It was very likely that the CYD article signaled to the justice bureau that Tong and Zhicheng, as well as Tong's proposal about the government funding a non-state legal aid operation, were politically legitimate.

What can we learn from the case of Tong Lihua and Zhicheng about the importance of the Chinese NGO leader's media connections? First, media connections have indeed played an important role in helping Tong and Zhicheng gain access to the state's policymaking process. Indeed, at times the effects of media connections seem to have exercised independent effects vis-à-vis Tong's state affiliations. Second, as the 2001 tobacco lawsuit and the 2005 CYD article about Zhicheng's survey report on the migrant workers have illustrated, the Chinese NGO leader's media connections can guarantee the leader and his organization access to media outlets to publicize their activities and ideas. To be sure, as these two examples and the 2003 example involving the Xinhua journalists have illustrated, the media's willingness to publicize a Chinese NGO and its leader seemed to also be dependent on the NGO's ability to develop interesting stories. However, the case of Tong and Zhicheng also suggests that successful "story-telling" by the Chinese NGO has often relied on the NGO leader's ability to get the media to report about the story. In this case, strong personal connections to journalists and media organizations have made it easier for Tong and Zhicheng to get their stories publicized. This was demonstrated when Tong was successful in inviting Cui Li to attend the NGO's forum about the report.

However, Tong's relationship with a senior CYD journalist and the 2003 Xinhua example also suggest that Tong's state affiliations might have paved the way for the establishment of media connections that finally led to internal reports to China's top political leaders. Although the evidence cannot be considered robust, we cannot easily dismiss the possibility that Tong's strong state affiliations had exercised a positive effect on his media connections.

For Tong, who was not a professional journalist, having personal ties to top reporters like CYD's Cui Li has been valuable because those personal relationships have allowed Tong to use the Chinese media to promote Zhicheng's agenda. However, for Chinese NGO leaders who are professional journalists, their media connections might simply be their journalist background. In the following section, we will see how Xie Lihua has used the publications under her influence to publicize Rural Women's issues of interest and, by extension, establish collaboration with state organizations.

4.3 Xie Lihua and Rural Women

Xie Lihua has been a journalist for approximately 30 years. As a senior journalist with direct access to multiple national publications, Xie, unlike Tong, did not need to rely on personal connections with journalists to access the Chinese media.

Indeed, Xie's position as a senior editor of multiple national publications has meant that Xie has been her own media connection.[5] And Xie's media connections have provided guaranteed channels to publicize Rural Women and, in the process, facilitated the establishment of collaborative relationships between Rural Women and sub-national state organizations. In fact, the impact of Xie's credentials as a senior state journalist was manifested from the very beginning, starting with the creation of the NGO.

For nearly 10 years before becoming the chief editor of the NGO's flagship publication *Rural Women Knowing All*, which eventually became known simply as *Rural Women*, Xie Lihua was a senior journalist and editor of *Women's News* (Shang 2007: 239),[6] which is the leading women's newspaper in China under the sponsorship of the All-China Women's Federation (ACWF). The newspaper's relationship to the ACWF is quite similar to the relationship between the CYD and the CYL. As a reporter for *Women's News*, Xie had become familiar with the conditions of the rural women during her assignments (Interview with Xie Lihua 2009). In 1994, soon after the establishment of the magazine, Xie was promoted to become a deputy chief editor of *Women's News*.

In Chap. 3, we learned how Xie Lihua's affiliations with ACWF in general and *Women's News* in particular had allowed her to be in a unique position to establish Rural Women. Although it is undeniable that Xie's ACWF affiliation had a decisive impact on the creation of the NGO, Xie's position as a senior journalist at *Women's News* also made her uniquely qualified to do so, particularly since the first manifestation of the NGO was a magazine publication.

According to Xie, the magazine was created by *Women's News* a few years before Xie would take over its rein (Interview with Xie). For two years, *Women's News* had been searching for a qualified person or group to manage the publication. In addition to political reliability, the ideal person or group chosen to operate the publication should possess journalistic/media qualifications. However, for two years, no qualified state employee and no appropriate state organization came forward to take on the responsibility. Xie learned about the publication and submitted her proposal to take over its helm. *Women's News* clearly believed that she possessed the right qualifications and her proposal was approved. In this manner, not only was the NGO created, but its most important state linkage was also established.

---

[5] In her career, Xie has been either a senior editor or the senior editor of three national publications: *Women's News*, *Rural Women*, and *Women Village Heads*, the last publication being the most recent addition to the NGO and was quite short-lived. Xie managed the latter two publications directly. Rural Women had compensated *Women's News* for circulating the latter two publications on behalf of the NGO.

[6] Xie was a reporter, a member of the newspaper's editorial committee, the director of the department of family and social affairs, and was responsible for the newspaper's "marriage and family" columns.

As the case of Tong Lihua and Zhicheng demonstrated, the value of the NGO leader's media connections is due in part to the fact that those connections can facilitate access to media outlets so that the NGO can publicize its activities and issue positions. Because Tong was not employed in a media organization, it was necessary for him to establish connections with journalists and media organizations. As a senior editor for *Women's News* and the chief editor for two publications with national circulations, however, Xie could use the publications under her editorship to publicize her NGO.[7] Although we might expect this to be the case for the two Rural Women publications that have been under her direct control, we might wonder whether Xie would have had similar influence with *Women's News*, in which she was a deputy chief editor and who, according to Xie, had become progressively less involved in the media organization's day-to-day operations (Interview with Xie).

As it turned out, Xie's influence at *Women's News* had not entirely dissipated. In fact, her editorial influence remained quite significant, as we can see in the following example. In August 2011, a group of University of Chicago students made a visit to the NGO's training center. After their visit, *Women's News* published a report about the visit (Pu 2011). The fact that the newspaper had published a report about the visit was suggestive of Xie's influence. But an even stronger indication of Xie's editorial influence was the fact that the news article was penned by a member of the Rural Women training center's staff.

This report was not an isolated incident. In fact, *Women's News* had served as a regular media outlet for publicizing the NGO and its activities. In December 2011, the newspaper published a lengthy report about the NGO's activity to experiment with newly-created materials to teach sex education to children in a county in Hebei, which is the province surrounding Beijing (Ren 2011). In July 2011, the newspaper published on its website a report about a conference on the legal rights of domestic workers, which was co-sponsored by Rural Women, the International Labor Organization, and Oxfam Hong Kong (Funu Wang 2011). Also, in September 2010, the newspaper published a full-length report about the NGO-hosted event for past members of the Migrant Women's Club (Funu Bao 2010).

In the process of publicizing the NGO, *Women's News* had also helped Rural Women establish long-term partnerships with sub-national ACWFs. As indicated earlier, an important ongoing nationwide project of the NGO has been the anti-illiteracy project. The project has been implemented across China through the sponsorship of books clubs by local ACWFs. Again, as described earlier, a major reason that sub-national ACWFs have been willing to sponsor the book clubs has been Xie's affiliation with *Women's News*. However, upon closer inspection, this factor, while critical, might not have been the only reason.

---

[7] Administratively, the two NGO publications were publications of *Women's News*. Rural Women paid annual fees to the ACWF newspaper to "manage" the publications, including their circulation among sub-national ACWFs and their employees.

According to Li Tao, who was a senior staff member of Rural Women in the 1990s, there were important incentives for local ACWFs to partner with Rural Women. According to Li, local ACWFs often have difficulties representing the interests of local women effectively because their financial resources are extremely limited, resulting in heavy dependency on local governments (Shang 2007: 245). Consequently, local ACWFs have had to engage in activities in accordance with the interests of local CPC committees, which have generally emphasized family planning matters when it came to women's issues. Therefore, if local ACWFs wanted to better represent local women's interests and obtain the skills and resources to do so, they needed to find qualified partners. From the perspective of the local ACWFs, Rural Women was an ideal partner because the NGO shared the general objectives of the national ACWF. In addition, Rural Women had access to skills and resources that the national ACWF did not possess.

Li's analysis suggests two things about the local ACWFs' perception of Rural Women. First, the local ACWFs were able to distinguish Rural Women from the national ACWF. Second, the local ACWFs had sufficient knowledge about the NGO that they concluded that the NGO shared similar goals with the national ACWF. But how did the local ACWFs obtain the information about Rural Women?

Based on the data that are available, we can infer that local ACWFs most likely have benefited from access to *Women's News* and the Rural Women publications, to which the local ACWFs have had regular access. Indeed, *Women's News* is a must-read for all local ACWFs and the newspaper, in an agreement with Rural Women, had promised to circulate the NGO's publications to all the local ACWFs. In a nutshell, there has been no shortage of information about Rural Women for local ACWFs. In short, *Women's News* and Rural Women's own publications have served as effective outlets for publicizing the NGO and its activities, and in so doing they have facilitated the establishment of partnerships between the NGO and local ACWFs. And we must not overlook the fact that all the media reporting about Rural Women had taken place under Xie's editorial leadership.

There are some lessons that we can draw from the case of Xie Lihua and Rural Women. First, through this case study, we can see how having a leader who is a journalist can benefit a Chinese NGO. Many writers have observed that successful Chinese NGOs have been led by professional journalists. However, most have not specified how this leadership trait can serve the NGO. The case of Xie Lihua and Rural Women clarifies to some extent the causal connection between a leader with a journalist background and the success of the NGO. Second, along with the case of Tong Lihua and Zhicheng, the case of Xie Lihua and Rural Women demonstrates the media's value as a publicist for an NGO.

Third, as was the case with Tong Lihua and Zhicheng, the potential impact of Xie Lihua's state affiliations on her media connections cannot be easily dismissed. For example, starting with the creation of the *Rural Women* magazine and ending with the implementation of the book club project, one may understandably wonder whether the respective effects of Xie's state affiliations and her media connections could be easily observed. In other words, was it her affiliations to *Women's News* that helped her acquire the magazine or was it her background as a journalist? Similarly, was it her ACWF affiliations that obtained the collaboration of the local ACWFs or was it the media publicity that was generated from the reporting directed by her? Xie is likely to respond that it has been her state affiliations, and that may be the logical answer. If so, then it would offer proof about the contingent nature of the NGO leader's media connections.

Lastly, this case has shown, to a limited extent, that the NGO leader's media connections can facilitate an NGO's formal linkages to state organizations. However, the evidence on this point is significantly stronger and more observable in the case of Wei Wei and the Little Bird Migrant Workers Mutual Support Hotline. And it is to that case that we will now turn.

4.4 Wei Wei and the Little Bird Migrant Workers Mutual-Support Hotline

State linkages are important for NGO successes in China and the NGO leader's state affiliations play a critical role in facilitating organizational linkages to the state. However, this observation presents a dilemma for aspiring NGO entrepreneurs who lacked the necessary political connections. How can these individuals help their NGOs acquire state linkages? Wei Wei's case shows that there is hope for the aspiring NGO entrepreneurs, so long as they acquire media connections.

In Chap. 3, we focused on Wei's affiliations with the Beijing municipal justice bureau. Through that connection, Wei has been able to use the legitimacy associated with state linkage to resolve labor disputes between his migrant worker clients and their employers. In the process, Wei and Little Bird also received state funding to finance their activities. But how was Wei able to do that given his disadvantaged background?

Indeed, in comparison to Tong Lihua and Xie Lihua, Wei Wei stood out in terms of his weak socio-political background. First, Wei did not receive a college education. Second, in a related manner, Wei had never worked as a white-collar professional. Third, he has never attained membership in the CPC. Fourth, unlike Tong and Xie, Wei did not enjoy institutional affiliations to state organizations before he became an NGO leader. Given the absence of these socio-political pedigrees, it would seem difficult, if not impossible, for Wei to develop state linkages.

The changing media system in China, however, has provided a means through which NGO leaders like Wei, who lacked the desired socio-political background, can still establish state linkages. It seems that Wei had recognized the potential advantages of media connections early in his NGO career. That awareness might have been developed when Wei was working in the advertising department of the

China News Agency, which is one of China's two national-level news agencies (the other being the Xinhua News Agency).[8]

In fact, Wei's success in securing a partnership between Little Bird and the municipal justice bureau can be traced back to his media connections. Here again, Wei differed from Tong and Xie in terms of how his media connections were developed. As we had seen from the case of Xie and Rural Women, Xie's media connections were essentially her professional background as a senior journalist. As for Tong, although the evidence is less clear, it would seem that Tong's media connections were developed as a result of Tong's state affiliations and his reputation as a legal expert. Wei did not possess either set of attributes. An important implication is that, unlike Xie and Tong, Wei would have had to do more to initiate contacts with the Chinese media. That was in fact what Wei had done.

In 1999, Wei developed the idea of starting a migrant worker mutual-support group. By his own account, he realized that he needed members and fellow workers in order to bring the idea to fruition. The challenge for Wei was therefore how to attract similar-minded migrant workers to join his initiative. It was at that time that Wei noticed that the *Beijing Youth Daily*[9] published, on a daily basis, a free message board, on which there were approximately 20 messages each day through which migrant workers in Beijing were either looking for new friends or re-establishing old connections (Wei 2005: 37). Wei then wrote a letter to the newspaper, asking that a message be published in the paper for all migrant workers in Beijing who were interested in knowing more about Wei's idea of establishing a mutual-support group to contact him (Wei 2005: 39). The responses to his message resulted in the group's first meeting, which was attended by 29 migrant workers (Liu 2009: 97). Little Bird was thus born.

In the year-and-half after Little Bird's first meeting, Wei continued to initiate contact with the media. For instance, after establishing the group's first office, Wei and his colleagues hosted a lunch with some journalists to introduce the group (Wei 2005: 42). After reading that the Hong Kong-based Phoenix Television was soliciting innovative entrepreneurial ideas for a competition, with the winner receiving 200,000 Chinese yuan, Wei sent materials about his group to the television company's Beijing station, which led to Little Bird's being featured on a nationally-broadcast television program (Wei 2005: 48). Later, when Wei and his colleagues decided to engage in a tree-planting activity, Wei contacted the state-owned Beijing Television Company to inform its staff that a migrant worker group in Beijing would be engaging in a voluntary tree-planting activity in a part of Beijing's Haidian district that has historically been home to many migrant workers in the capital city. The result was that the group became prominently featured in a major nightly television program (Wei 2005: 50–51).

In many ways, these initial successes with "cold-calling" Chinese journalists and media organizations had to do with the fact that the plight of the migrant workers

[8] The chief function of the China News Agency is to disseminate news stories about China to the foreign and overseas Chinese communities.

[9] The *Beijing Youth Daily* is a newspaper of the Beijing municipal CYL.

was gaining increasing media interest by the turn of the twenty-first century. The large number of migrant workers and the social tensions resulting from their increases due to a rising demand for cheap labor to finance China's urbanization and export machine had contributed to the issue's media prominence. It is therefore unsurprising that a majority of China's most prominent migrant worker NGOs, including Little Bird, were established at this time.

In 2000, a year after Little Bird's establishment, Wei established a media connection that allowed the NGO to acquire a "media profile." In that year, at someone's suggestion, Wei contacted the host of the "Life Hotline" radio program at the state-owned Beijing Radio Station (Wei 2005: 65). He spoke to the program host and asked whether the program could publicize the group. The program host agreed and even asked to meet Wei. At the meeting, the program host told Wei that he wanted to give the Thursday night slot to Wei and Little Bird, renaming the Thursday night program as the "Night of the Migrants" (Wei 2005: 66). The new program would be hosted by Wei and his fellow migrant workers. By this time, Wei had already started to move the mutual-support group in the direction of becoming a counseling hotline, through which free legal advice would be given. The radio program advertised Little Bird's organizational goals and activities, and as a result, Little Bird began to gain consistent and increasing publicity. In addition, it solidified the direction of Little Bird's organizational development into a legal aid NGO. Through the radio program, migrant workers in and around Beijing have called Wei and Little Bird, seeking advice about many things, particularly legal issues. The radio program has become a mainstay of the NGO's work and it has been funded by foreign donors, including the United States Embassy in China.

With increasing publicity for Little Bird, more and more legal professionals had joined the organization's volunteer legal aid team. Meanwhile, the increasing publicity also attracted the attention of government agencies. Through a report of Little Bird by the *Beijing Evening News*, the justice bureau in Beijing's Dongcheng district, which is also home to Beijing municipal government, took notice of Little Bird and its legal aid work. The justice bureau contacted Wei and in September 2004 approved the collaboration between the Little Bird and a judicial office in the district to establish the country's first non-governmental labor dispute resolution committee, which would be led and operated by Little Bird (Liu 2009: 98). In this manner, Wei's institutional affiliation to the state and Little Bird's organizational linkage to the state were established.

To be sure, the collaboration between Wei's Little Bird and the justice bureau was a reflection of the general political environment in 2004, which was favorable to migrant worker NGOs. In 2003, as noted earlier, the Sun Zhigang case had caused a media furor that resulted in the government's decision to abolish the custody and repatriation system. Also in that year, Premier Wen Jiabao made a personal pledge to help migrant workers obtain their wage arrears. Consequently, government agencies at all levels, especially Beijing, took greater interest in defending the legal rights of migrant workers. Wei undoubtedly benefited from this change in the political wind.

However, without Wei's media connections that resulted in constant media coverage for Wei and Little Bird, he would not have been able to take advantage of the opportunities afforded by the changing political environment with respect to migrant workers. More specifically, the positive media coverage resulting from Wei's media connections had legitimized Wei and Little Bird in the eyes of some state agencies. Through the positive reporting, media organizations sent a signal to state organizations that Wei and Little Bird were politically reliable. In this respect, Wei's media connections were instrumental in Wei's ability to take advantage of the new political opportunities.

The case of Wei Wei and the Little Bird demonstrates several items of importance concerning the value of media connections for Chinese NGOs. First, media connections can help Chinese NGO leaders who lack political pedigrees to achieve state linkages for their organizations. In this respect, this case shows that in some cases, the leader's media connections may impact his state affiliations in a direct and independent manner. The media can do so by publicizing the NGO and its activities. In the process of publicizing the NGO, the media send signals to state organizations about the NGO's political reliability. However, the salutary effects of media connections in this respect must still depend on the existence of appropriate political opportunities, thereby once again demonstrating the contingent nature of the media's effects. Nonetheless, an important implication of the case of Wei and the Little Bird is that without media connections, the Chinese NGO leader cannot expect media publicity for his organization, which has implications for the NGO's ability to establish state linkages.

4.5 Ma Yang and the On Action International Cultural Center

In comparison to Zhicheng, Rural Women, and Little Bird, On Action has been noticeably less effective since its establishment. What effects, if any, did Ma's lack of media connections have had on On Action's performance? Could On Action have benefited from a leader with stronger media connections? Let us examine the data for possible answers to these questions.

In the previous chapter, we learned that until July 2010, Ma did not possess institutional affiliations to a state organization that could confer organizational resources, state linkage, and access to the state policy apparatus. We also learned that the absence of institutional connections to an appropriate state organization have deprived On Action of state resources and access that might have made a difference in On Action's performance. However, in July 2010 Ma was able to establish a connection to the Beijing Municipal Social Work Commission.

In addition to the lack of appropriate state affiliations, Ma also did not possess strong media connections. This was clearly manifested in the extremely poor media coverage that Ma and On Action had received over the years. Although Ma had professed his belief in the value of media coverage as a means to publicize the NGO and its work, the unfortunate reality was that between 2006, when Ma established On Action, and January 2012, the NGO and its activities have been covered by less than 10 media organizations, most of which were local and specialized in focus.[10] This record certainly pales in comparison to the regular and extensive coverage by major Chinese and international media organizations that Wei and the Little Bird received, which involved over 100 Chinese and international media organizations as of 2012. Moreover, the great majority of the media organizations that reported about On Action were not major outlets, such as the CYD and the *Southern Weekend*. Furthermore, neither Ma nor On Action had been the featured subject in most of those reports.

What can we infer from the existing data about the likely effects of Ma's lack of media connections and the resultant poor media coverage? We can draw some inferences from a comparison between On Action's situation and the situations of the previous three NGO cases, especially that of Little Bird.

As illustrated earlier, the Chinese media have performed three functions that have facilitated state linkages for Chinese NGOs. Two of those functions, or roles, namely publicist and legitimating agent, are closely connected. To some extent, Zhicheng, Rural Women, and Little Bird have all benefited from these media functions. However, Little Bird has arguably derived the greatest benefit from the media publicity, since Wei Wei established the organization without any prior state linkages at either the individual or the organizational level. To recall, it was the media's coverage of Little Bird that attracted the attention of the municipal justice bureau, which led the bureau to contact Wei about the possibility of collaboration.

In 2006, when On Action was established, issues concerning migrant workers had remained a leading subject of media interest. It was during that year that Zhicheng was involved in a highly-publicized legal battle with Kentucky Fried Chicken (KFC) over the latter's practice of employing workers, most of whom were migrant workers, through third-party human resource companies (Tong 2009: 367–396). Meanwhile, in March 2006, partly in response to increasing incidents of labor disputes, especially those involving migrant workers, the NPC began to conduct research to revise the Labor Contract Law, which would go into effect in 2008. In other words, political and media interest in issues concerning migrant workers remained high when On Action was established, thus ensuring a relatively positive environment for Chinese migrant worker NGOs to publicize their activities, if they could gain access to the media. And Little Bird's case has shown that with proper media coverage, there were opportunities for migrant worker NGOs to establish useful state linkages.

---

[10] This observation was based on a review of the published information on On Action's website in January 2012. The URLs were http://www.swcn.org.cn/news_more.asp?page=1&1m=56&1m2= and http://www.zxd.org.cn/news_more.asp?page=1&word=&1m=122&1m2=&1mname=&open=&n=&hot=&tj=. In April 2014, both URLs have apparently been disconnected and the posted information was not observed. However, relevant information posted on the organization's website indicates that On Action's coverage by the media has remained sparse and localized. See Zaixingdong, http://www.zxd.org.cn/Article/List.asp?lmid=114&page=1 (accessed 27 April 2014).

Therefore, it is highly likely that Ma's lack of media connections deprived him of opportunities to publicize the NGO. In turn, the absence of media coverage had limited Ma's chances of establishing valuable connections with interested state organizations, which might have given On Action the resources, legitimacy, and access that were needed to become more effective in meeting its goals. Media connections alone probably could not have helped Ma and On Action, but they might have supplied the necessary push for Ma and On Action to develop connections with state organizations without having to change the NGO's mandate, which as shown in Chap. 3 was the price that Ma had to pay for him to establish state linkage for On Action.

4.6 Corroboration from Additional Cases: Friends of Nature, Yirenping, the Self-Empowerment Service Center for Disabled Persons

The importance of the leader's media connections can also be seen in NGOs that work in different issue areas in different localities, and at different stages of organizational development. The cases of the Friends of Nature (FON), Yirenping, and the Chongqing-based Self-Empowerment Service Center for Disabled Persons provide further evidence that the NGO leader's media connections have contributed to the creation of state linkages for the Chinese NGO and, ultimately, organizational success. Let us begin with the Friends of Nature.

Earlier in this book, we saw how the political connections of Liang Congjie, the principal FON founder, had contributed directly to the establishment of the ENGO's state linkage and its access to the state policy apparatus. Although Liang's media connections might not have had similarly direct effects on FON's linkage to the state, his media connections have allowed him to publicize issues that were of concern to FON and other Chinese environmental activists and groups. The resulting publicity had forced the government to take measures to address those concerns. Before we examine a specific example of the effects of Liang's media connections, let us review Liang's media connections.

Liang's connections to the media, both Chinese and international, were in large part the reflection of the fact that he was the leader of contemporary China's first ENGO. To be sure, even before FON was established, Liang had already developed friendships with Chinese journalists who became interested in environmental issues. One of those journalist friends, for example, was Dai Qing, who actually encouraged Liang to establish an ENGO.[11]

[11] Elizabeth C. Economy (2004, 146). Economy credited Dai, along with two others, with establishing the "intellectual roots of environmental activism" in contemporary China.

Possessing media connections has allowed NGO leaders like Liang to have ready outlets to publicize issues of concern to them. The advertising of those issues, particularly if they involved government agencies, might result in government actions to address those concerns. This theory was put to the test, relatively successfully, early in FON's history. In 1995, a media uproar took place over extensive logging in a Yunnan county, which contributed to the near-extinction of the Yunnan snub-nosed monkey (Economy 2004: 149–151). Xi Zhinong, the man who noticed this problem, was an employee of the Yunnan Forestry Department and his protest to the department's leadership fell on deaf ears. Even the local media refused to report on the issue due to the sensitivity of the matter resulting from the involvement of the local Tibetan minority. A friend of Xi then contacted Tang Xiyang, a former *Beijing Daily* reporter who, along with Dai, was considered the intellectual leader of contemporary Chinese environmental activism. Tang then contacted Xi to write a letter to the then-head of the State Environmental Protection Commission about the linkage between the deforestation and the near-extinction of the snub-nosed monkey. Meanwhile, Tang passed the letter onto Liang, who then shared copies of the letter with international and Chinese media. The letter was first published by the Associated Press and was picked up quickly by Chinese media organizations. The immediate result of the media campaign was a central government investigation. Eventually, the central government ordered the local government to stop the illegal logging.

This example pertaining to Liang and FON is very similar to the incident involving the anti-leaking cover in Beijing's Old Summer Palace. In both cases, media connections involved the passing of relevant information to the media in the "old-fashioned" way. As China moved into the twenty-first century, the use of the Internet technology has become another means by which Chinese NGO leaders can build media connections. In particular, younger enterprising NGO leaders have often established their media presence on the Internet, through which they have developed their connections with traditional media organizations.

Yirenping's Lu Jun is a member of this generation of Chinese NGO leaders and an example of one who has successfully established media connections through the Internet. Lu Jun first became involved in the work to protect the interests of Hepatitis-B carriers when in 2003 he joined the online forum "In the Hepatitis B Camp" (*dangan xiangzhao*). The online forum was created in 2001 and as of 2010 it had 400,000 registered members.[12] The forum allowed members to post information as well as to share their views.

Lu became very active soon after he joined the forum. Consequently, he was elected to become a forum moderator. In that position, Lu started writing commentaries and petitions to government officials. It was in this position that he established connections with NPC delegates and members of the Chinese People's Political Consultative Conference (CPPCC), China's top political advisory body. He also wrote letters to media organizations, supplying them with information about the

[12] Hepatitis B Carriers BBS, http://www.hbvhbv.com/english/ (accessed 18 January 2012).

disease as well as informing them that their writings and advertisements had violated anti-discrimination regulations. He even appeared on television shows to talk about the issue (Hao 2009). In a sense, despite operating on a technologically advanced platform, Lu was nevertheless engaging in relatively traditional media work. In any case, in this manner, Lu established his connections with the media organizations.

When Lu and several colleagues established Yirenping, those media connections that he developed stayed with him, and the effects of the media connections were immediate and clear. According to Lu, in the first two and half years after the establishment of the NGO, there were over 1,900 media reports about Yirenping and its work, or an average of 64 reports per month in a 30-month period (Hao 2009). Yirenping's numerous successes, including the 2009 Ministry of Health order to abolish mandatory Hepatitis-B testing, were testaments of Lu's extensive media connections, which have done much to publicize Yirenping and its issues of concern. The media publicity had pressured government agencies, state institutions such as commercial enterprises and educational institutions, and foreign enterprises to address charges of discriminatory practices. In sum, Lu's media connections, when coupled with his connections to like-minded NPC delegates and CPPCC members, contributed to Yirenping's success as an advocate for victims of discrimination.

Lu had done much to create the media connections that would eventually serve Yirenping's needs. At times, media connections have been developed not as a conscious act on the part of the NGO entrepreneur. Chen Yuying's situation is a case in point. Although the initial circumstances surrounding the development of Chen's media connections were tragic, the outcome ensured that the Self-Empowerment Center would be established with the blessings of the state, which allowed the NGO to achieve state linkage, acquire state resources, and access the state policy apparatus.

Chen gained considerable Chinese and international media attention as a result of her unintended involvement in the 19 November 1993 fire that engulfed a Shenzhen toy factory and which took the lives of over 80 young Chinese female migrant workers. As a result of the fire, Chen suffered severe burns to 75 % of her body, and her left leg had to be amputated while she lost three fingers on her left hand.

As one of the survivors of the factory fire, Chen became the focus of local and overseas media reporting. The nationally well-known muckraking newspaper *Southern Weekend* wrote stories about her (Self-Empowerment Center 2009). Similarly, Hong Kong newspapers, such as *The Sun*, had also published stories about Chen. Media attention continued after she returned to her hometown nearly a year later. Put simply, as tragic as her personal experience was, her identity as a survivor of the incident also made her an instant media celebrity.

As Chen's physical rehabilitation progressed, Chen began to provide free individual counseling and other services to handicapped persons who lived in her neighboring communities (Interview with Chen Jian 2008). In the process, she was motivated to consider establishing a more formal service platform. In early 2002, Chen approached the local civil affairs bureau, which had the responsibility for registering and managing NGOs, and the local handicapped person's association about her idea of establishing a service station for people with physical disabilities (Self-Empowerment Center 2009).

According to Chen's brother, Chen's status as a media celebrity was instrumental in the establishment of her NGO and the state linkages that it has enjoyed (Interview with Chen). As a result of her media celebrity status, Chen was able to obtain from the local civil affairs bureau nonprofit registration for her organization, and the local disabled persons' federation agreed to serve as the NGO's supervisory agency (Ibid). As noted earlier in this book, one persistent complaint by Chinese NGO practitioners and experts has been the challenges associated with attaining this registration status. One of those challenges is the difficulty of getting state agencies to serve as a supervisory agency. The fact that Chen was able to attain this legal status for her organization was impressive enough. Yet, even more impressive than this achievement was the fact that the civil affairs bureau approved the application in one month (Interview with Chen). In the process, all fees and cumbersome administrative procedures, such as a review of the organization's assets by an independent auditing firm, which are associated with an application for nonprofit registration, were waived by the bureau (Interview with Chen). At the same time that the NGO was officially registered, Chen was made the chairperson of the local association of physically disabled persons, which operates under the jurisdiction of the local disabled persons' federation (Interview with Chen). Chen was also appointed a senior manager in the local association of physically disabled persons. In sum, Chen's media connections have helped her and her NGO to achieve legitimacy to work in the county and access to make changes to local policies and practices.

Collectively, the cases of the Friends of Nature, Yirenping, and the Self-Empowerment Center show that, regardless of the issue area, the location, and the stage of organizational development, successful Chinese NGOs have been led by leaders with strong media connections. However, as was the case for the four core cases, the three shadow cases also demonstrate the facilitative nature of the Chinese NGO leader's media connections. In other words, media connections might not be sufficient for Chinese NGOs to establish optimal state linkages. Nonetheless, through the power of media publicity and the fact that the Chinese media occupy a position that straddles the public-private divide, successful Chinese NGOs have been able to enhance and even establish their relationships with the state.

4.7 Conclusion

Scholars have long observed the lack of a genuine civil society in contemporary China because the authoritarian state remains the dominant political actor and it has continued to obstruct any efforts to institutionalize civil society. The cases presented in this book support this long-held perspective. The reality is that state linkages remain important for civil society actors that want to be effective, particularly those working in non-democracies. Of course, in the eyes of most NGO practitioners in China, effectiveness is not defined in terms of successes in challenging and even subverting the state. To those NGO leaders who are featured here, organizational effectiveness has been defined largely in terms of the extent to which the NGO can provide needed services to its "clients." And the state controls important resources that could allow these NGOs to become better service providers. Therefore, it has made perfect sense for NGO leaders to pursue the establishment of positive linkages with the state.

However, this chapter shows that Chinese NGOs do not have to be led by leaders with political pedigrees and connections, which has interesting implications for the Chinese civil society's current state of development. The fact that there are successful NGOs in today's China, which are led by members of the non-elite, or "non-mainstream" (*feizhuliu*) individuals in Chinese parlance, suggests that Chinese civil society may be moving in a direction that is more congruent with the hopes and expectations of Western scholars, policy-makers, and pundits.

Although a Chinese NGO leader's media connections have not helped his organization gain direct access to most state resources, media connections have allowed some NGO leaders to acquire state affiliations, thereby helping those leaders obtain state resources for their organizations in an indirect manner. And if the NGO leader already possesses state affiliations, then his media connections will most likely enhance those affiliations. Therefore, from the organization's perspective, the media connections of the NGO leader should be quite valuable. However, the value of the media has not been limited to its contribution to the development of state linkages for the NGO. The media have also contributed to the NGO's public legitimacy, both domestically and abroad. In this respect, the media have in many cases been instrumental in facilitating the development of international ties for the Chinese NGOs.

On this note, we are left with one final set of leadership ties to examine, i.e., international ties. How do the NGO leader's international ties stack up against his media connections and state affiliations? More importantly, can the leader's international ties facilitate state linkages for his organization? In the next chapter, the cases will provide some tentative answers to these questions.

References

Ballentine, Karen. 2002. International Assistance and the Development of Independent Mass Media in the Zech and Slovak Republics. In *The Power and Limits of NGOs: A Critical Look at Building Democracy in Eastern Europe and Eurasia*, eds. Sarah E. Mendelson and John K. Glenn, 91-125. New York: Columbia University Press.

Cui Li. 16 May 2001 (2001a). Lushi Daili Weichengnianren Tiqi Jituan Susong – Xiang Zhongwai Yancao Shuo 'Bu'! [Lawyers to File Class Action Suit on behalf of Minors – Say 'No' to Chinese and Foreign Tobacco Companies!]. *Zhongguo Qingnian Bao* [*China Youth Daily*]. http://zcpi.org/zhi/mtgz/3txt.asp?id=35&keywords=. Accessed 14 January 2012.

Cui Li. 25 May 2001 (2001b). Xiang Zhongwai Yancaoshang 'Fa Nan' – Lushi Tong Lihua Yiju Chengming [Rebelling' Against Chinese and Foreign Tobacco Businesses – Lawyer Tong Lihua Achieved Fame Overnight]. *Zhongguo Qingnian Bao* [*China Youth Daily*]. http://news.sina.com.cn/c/260917.html. Accessed 14 January 2012.

Cui Li. 16 May 2003 (2003a). Sanwei Zhongguo Gongmin Yifa Shangshu Quanguo Renda Changweihui Jianyi Dui 'Shourong Qiansong Banfa' Jingxing Weixian Shencha [Three Chinese Citizens Petitioned the NPC Standing Committee According to the Laws to Conduct Constitutional Review of the 'Custody and Repatriation Method]. *Zhongguo Qingnian Bao* [*China Youth Daily*]. http://www.chinaelections.org/NewsInfo.asp?NewsID=61853. Accessed 13 January 2012.

Cui Li. 28 May 2003 (2003b). Wuwei Faxuejia Tiqing Renda Qidong Tebie Chengxu Diaocha Sun Zhigang An [Five Legal Experts Made Special Request to the NPC to Initiate Special Procedures to Investigate the Sun Zhigang Case]. *Zhongguo Qingnian Bao* [*China Youth Daily*]. http://www.chinaelections.org/NewsInfo.asp?NewsID=77388. Accessed 13 January 2012.

Economy, Elizabeth C. 2004. *The River Runs Black: The Environmental Challenge to China's Future*. Ithaca: Cornell University Press.

Hao Luli. 30 July 2009. Shehui Qishi Bushi Xiaoshi: Fang Beijing Yirenping Zhongxin Fuzeren Lu Jun [Social Discrimination is not Trivial: Visiting With the Leader of the Beijing Yirenping Center]. *Minjiang* (*NGO*). http://www.360doc.com/content/09/0730/10/84177_4546773.shtml. Accessed 27 April 2014.

Jia Xijin et al. 2008. *Zhongguo Gongmin Canyu: Anli Yu Moshi* [*Citizen Participation in China: Cases and Models*]. Beijing, China: Social Sciences Academic Press.

Lin, Fen. 2010. A Survey Report on Chinese Journalists in China. *The China Quarterly* 202 (June): 421-434.

Liu Zhihua. 2009. 'Sheng' Yuan Yu Meiti, 'Zhang' Ji Yu Meiti – Cong 'Xiaoxiaoniao Dagong Huzhu Rexian' Fazhan Kan Caogen NGO De Meiti Zhanlue ['Birth' through the Media, 'Growth' on Media's Foundation – Observing the Media Strategy of the Grassroots NGO through the Development of the 'Little Bird Migrant Worker Mutual Support Hotline']. *NGO Anli Ji: Zhongguo Feiyingli Zuzhi Guanli* [*NGO Cases: Management of Chinese Nonprofit Organizations*], 96-102. Shanghai, China: NPI. http://www.npi.org.cn/uploads/magazines/anliji.pdf. Accessed 23 June 2011.

Lynch, Daniel C.1999. *After the Propaganda State: Media, Politics, and "Thought Work" in Reformed China*. Stanford: Stanford University Press.

Magnier, Mark. 16 February 2004. Rosy Reports from Underlings Leave Chinese Leaders in the Dark. *Los Angeles Times*. http://articles.latimes.com/2004/feb/16/world/fg-info16. Accessed 11 January 2012.

Ottaway, Marina. 2003. *Democracy Challenged: The Rise of Semi-Authoritarianism*. Washington, DC: Carnegie Endowment for International Peace.

Pu Ruixia. 25 August 2011. Meiguo Zhijiage Daxue Yanjiusheng Canfang Xuexiao [Graduate Students of the University of Chicago in America Visit the School]. *Zhongguo Funu Bao* [*Women's News of China*]. http://www.nongjianv.org/index.php?m=content&c=index&a=show&catid=39&id=2989. Accessed 16 January 2012.

Ren Ran. 4 December 2011. Zhongxiaoxue Qingchunqi Jiaoyu Zai Nongcun De Pobing Zhilu [The Icebreaking Journey of Adolescent Education for Elementary and Secondary Students in a Rural Village]. *Zhongguo Funu Bao* [*Women's News of China*]. http://www.china-woman.com/rp/main?fid=open&fun=show_news&from=view&nid=77386. Accessed 16 January 2012.

Rugh. William A. 2004. *Arab Mass Media: Newspapers, Radio, and Television in Arab Politics*. Westport: Praeger.

Self-Empowerment Center. 28 September 2009. Fenghuang Niepan: Chen Yuying Rang Rensheng Chongtou Zailai – Shenzhen Huozai Shigu Dagongmei Jiehouyusheng De Zhenqing Gushi [Phoenix nirvana: Chen Yuying Started Life Over – The Real-Life Story of the Migrant Woman's Escape from Death after the Shenzhen Fire Incident"]. http://www.cqzq.org/html/zlsj_572_91.html. Accessed 22 June 2011.

Shang Xiaoyuan. 2007. Funu Gongmin Shehui Zuzhi He Zhili Moshi De Duoyuanhua [Women's Civil Society Organizations and the Pluralism of the Governance Models]. In *Chongji Yu Biange: Duiwai Kaifangzhong De Zhongguo Gongmin Shehui Zuzhi* [*Impact and Change: the Chinese Civil Society Organizations that are Opening Toward the Outside*], ed. Shang Xiaoyuan, 236-283. Beijing, China: Chinese Social Sciences Press.

State Environmental Protection Administration. 22 February 2006. Huanjing Yingxiang Pingjia Gongzhong Canyu Zhanxing Banfa [Temporary Method for Public Participation in Environmental Impact Assessments]. http://www.gov.cn/jrzg/2006-02/22/content_207093_2.htm. Accessed 8 January 2012.

Stockmann, Daniela. 2013. *Media Commercialization and Authoritarian Rule in China*. New York: Cambridge University Press.

The Economist. 17 June 2010. Chinese Whispers: China's Secret media. http://www.economist.com/node/16379897?story_id=16379897&source=hptextfeature. Accessed 11 January 2012.

Tong Lihua. 2009. *Weile Zhengyi: Zhicheng Gongyi Shinian* [*For Justice: Ten Years of Zhicheng Public Interest Law NGO*]. Beijing, China: Law Press China.

Wei Wei. 2005. *Woshi Yizhi Xiaoxiao Niao: Yige Dagonzai De Gongyi Shiye Meng* [*I am a Little bird: a Migrant Worker's NGO dream*]. Zhengzhou, China: Henan Wenyi Chubanshe.

Xin Zhoukan [New Weekly]. 23 July 2001. Ti Weichengnianren Zhuanggao 25 Da Yancaoshang: Tong Lihua – Wo Yongbuzhao Chaozuo [Suing the 25 Major Tobacco Firms on Behalf of Minors - Tong Lihua: I Don't Need to Sensationalize the Issue]. http://zcpi.org/zhi/mtgz/3txt.asp?id=37&keywords=. Accessed 14 January 2012.

Young, Doug. 2013. *The Party Line: How the Media Dictates Public Opinion in Modern China*. Singapore: John Wiley & Sons.

Zhang Xiaosong. 28 May 2001. 'Jinzhi Zhongxiaosuesheng Xiyan' Pai Jiang Baishang Juanyan Guitai ['Elementary and Secondary Students Prohibited from Smoking' Label will be Placed on Tobacco Sales Counters]. *Xinhua Wang* [*Xinhua Online*]. http://finance.sina.com.cn/x/64745.html. Accessed 14 January 2012.

Zhao, Yuezhi. 1998. *Media, Market, and Democracy in China: Between the Party Line and the Bottom Line*. Champaign: University of Illinois Press.

Zhao, Yuezhi. 2008. *Communication in China: Political Economy, Power, and Conflict*. Lanham: Rowman & Littlefield Publishers.

Zhongguo Funu Bao [Women's News of China]. 30 September 2010. Weile Shinianqian De Yige Yueding [For a Promise that was Made 10 Years Ago]. http://www.nongjianv.org/index.php?m=content&c=index&a=show&catid=8&id=222. Accessed 16 January 2012.

Zhongguo Funu Wang [Chinese Women Online]. 26 July 2011. Yong Zhidu He Falu Baohu Jiazhenggong Timian Laodong De Quanli [Employ Institutions and Laws to Protect the Right to Decent Work for Domestic Workers]. http://www.nongjianv.org/index.php?m=content&c=index&a=show&catid=8&id=2935. Accessed 16 January 2012.

Chapter 5
Capacity Without Legitimacy: The Limits of International Ties

Abstract International resources have played an invaluable role in the creation of NGOs in contemporary China, mainly through financial support and human resource training. However, contrary to some views, the cases presented in this book show that international resources may not necessarily contribute to the political legitimacy of China's NGOs. In fact, optimal utilization of international resources by China's NGOs may require the prior establishment of state linkages and media connections to ensure an NGO's political legitimacy and its ability to access state resources that are necessary to realize the objectives of the NGO. In addition, the cases presented here show that state linkages and media connections may play a direct role in helping China's NGOs acquire international resources. These observations suggest that aspiring Chinese NGO entrepreneurs should not view international ties as the ultimate solution to their needs, but instead should invest considerable resources to developing state linkages and media connections.

Keywords International ties · INGO · Financial support · Human resource training · State linkages · Media connections

In 2009, the Chinese government's State Administration of Foreign Exchange (SAFE) issued the "Circular of the State Administration of Foreign Exchange on Relevant Issues Concerning the Administration of Donations in Foreign Exchange by Domestic Institutions" (SAFE 2009). Chinese NGO practitioners as well as foreign donors and experts expressed concern about the circular's implications, especially with respect to its stipulation that all foreign grant agreements must be legally notarized and presented to Chinese banks before the funds are released. A central challenge was that the foreign donors and their Chinese beneficiaries, as well as the Chinese banks, seemed to lack the knowledge on how the agreements would be notarized (Zhang 2010; Ford 2010; Davis 2010). What made the situation worse was that SAFE made no effort to clarify the uncertainties.

In general, for Chinese NGOs, the majority of which are dependent on foreign funding, this circular, if implemented, represented a potentially critical blow. In

A version of this chapter was presented at the International Studies Association Annual Conference in April 2013.

J.W. Tai, *Building Civil Society in Authoritarian China*,
SpringerBriefs in Environment, Security, Development and Peace 20,
DOI 10.1007/978-3-319-03665-6_5

addition, the announcement about the new set of regulations was, although not entirely surprising given what seemed to be the state's increasing hostility toward members of the Chinese civil society, was nonetheless disconcerting because it represented another shift in the state's attitude about international funding for Chinese NGOs. There was thus noticeable concern among members of the Chinese NGO community that most, if not all, Chinese NGOs would have their lifelines cut by this new regulation.

The SAFE circular provides an illustration of the power of the Chinese government as well as the constantly shifting nature of its attitude toward NGOs. At the same time, this example underscores the importance of international funding for Chinese NGOs. Indeed, it is difficult to underestimate the importance of foreign financial support for Chinese NGOs. Nearly every NGO in contemporary China has received international funding at some point in its organizational lifetime. All of the most prominent and successful NGOs have been the recipients of foreign largess. And the most successful organizations have consistently and continuously benefited from international support.

The positive effects of international ties, notably foreign funding, has been observed among NGOs operating at different stages of organizational development, working on different issue areas, and locating in different geographical regions. However, the relationship between international ties and NGO effectiveness is not a clear-cut one. There are important conditions that facilitate the positive effects of international ties for NGO effectiveness. In particular, the cases in this book show that the positive effects of international ties on NGO effectiveness are contingent on the presence of the following conditions: (1) the NGO leader's institutional affiliations with state organizations and (2) the leader's media connections.

This finding suggests that the effects of international ties on the Chinese NGO's ability to achieve organizational effectiveness may be less significant than those of state linkages and media connections. In particular, it appears that international ties do not facilitate state linkages, which is a critical element of an effective NGO. Given the importance of international resources for the capacity of Chinese NGOs, the leader's international ties cannot be dismissed. However, that its utility has been dependent on the leader's state affiliations tells us much about the power of state linkage for successful Chinese NGOs. Collectively, the importance of state linkages and the value of media connections vis-à-vis international ties demonstrate the greater significance of domestic factors for NGO effectiveness in contemporary China.

5.1 International Ties: Functions and Limitations

There is no denying that international forces have played a positive role in civil society-building in contemporary China. In particular, international events and actors have contributed to the establishment of individual NGOs in China and the NGO sector's subsequent growth. For example, experts have noted the positive impact of the Fourth World Conference on Women, which was held in Beijing in 1995, in launching many individuals, including Rural Women's Xie Lihua, onto careers in the NGO sector. Through these individuals, the 1995 conference contributed significantly to the formation of "an independent Chinese women's NGO sector" (Kaufman 2012: 594). In addition, international NGOs (INGO) have produced demonstration effects: China's NGO entrepreneurs have established their organizations because they wanted to create Chinese NGOs similar to those found outside the country (Chen 2012: 80). In this respect, many founders of Chinese NGOs benefited from valuable experiences working for prominent INGOs operating in China (Chen 2012: 84). These experiences suggest that, as a type of relationship, international ties can yield tangible benefits for China's NGO leaders, the most significant of which is resources to build up organizational capacity.

In general, international actors have provided two types of resource support to Chinese NGOs. Financial resources have constituted one type of support (Chen 2012: 85–87). As mentioned earlier, most Chinese NGOs are heavily dependent on international funding. The reasons for this phenomenon are many, but they mostly have to do with the fact that the Chinese NGOs have had difficulties accessing local resources. Chinese government and local private enterprises have generally been unwilling to share their resources with NGOs. At the same time, individual giving in China has not been significant. Although this situation is evolving as government purchases of NGO services (Teets 2011, 2013) and private corporate donations have been experiencing incremental rises, there is no clear indication that Chinese NGOs will be able to forego international funding any time soon in the near future.

Human resource training has constituted the second type of support to enhance organizational capacity (Chen 2012: 84). In addition to providing financial resources directly to Chinese NGOs, Chinese NGO workers have also benefited from participation in training programs as part of the efforts of international donors and their Chinese partners to engage in NGO capacity building. Training venues have taken place both inside and outside China. For example, the leaders of Zhicheng, Rural Women, and Yirenping, as well as many full-time staff members of the respective organizations, have gone overseas, including Hong Kong, to participate in training conferences or on short-term fellowships. Chinese NGOs, such as Little Bird, have also conducted in-country training programs on behalf of their international donors.

Notwithstanding some writings noting that international ties have enhanced the legitimacy of local Chinese NGOs (Shieh 2009: 35; Chen 2012: 96), the verdict is still out on this claim. In fact, even though international NGOs and international funding have been operating in China for a long time, the Chinese government has continued to express concern that international support of local NGOs has constituted the key vehicle through which foreign subversive ideas and activities have entered China. Many in fact believed that the SAFE circular illustrated at the beginning of this chapter was a reflection of the state's uneasiness about international forces' potential impact on domestic political situation through their provision of financial resources to Chinese NGOs. Many Chinese NGO leaders have thus expressed the view that international ties, while providing the necessary organizational resources, may negatively affect the political legitimacy of their persons and organizations (interview with Wei Wei 2009; interview with Xie Lihua 2009). In this sense, an important limitation of international ties is its potentially negative impact on a Chinese NGO's ability to achieve state linkage. Therefore, alleviating the state's concerns is very important for unleashing the benefits of international ties.

The situation in China closely resembles the situations in other non-democracies, where the positive effects of international ties have been found to be premised on the presence of domestic factors, notably the attitude of the state. In her study of the effects of foreign assistance on NGOs in post-communist Russia, Lisa McIntosh Sundstrom observed that international aid has had the most positive effects on local NGOs in Russian regions with relatively supportive local states (Sundstrom 2006). With respect to China, analysts have likewise observed that the successes of donor-supported migrant worker NGOs in China have depended a great deal on the support of the local states (Zhang and Smith 2009; Hsu and Hasmath 2013).

To a significantly lesser extent, however, analysts have not written about the impact of the media on the NGO's international support. However, the cases presented in this book show that media connections have in some cases had direct causal effects on Chinese NGOs' access to international support. In this vein, not only had media connections resulted in positive media exposure that led to access to international support; negative media exposure appears to have had similarly strong, albeit negative, impact on access to international support. In fact, negative media exposure has in some cases removed international support from those NGOs that had previously possessed foreign support.

An important reason that state linkages and media connections have acted as catalysts of international ties has to do with the fact that international donors have shown a general proclivity to work with Chinese NGOs that have been deemed politically legitimate. State linkages and media connections serve as signals to the international donors that the Chinese individuals and organizations are legitimate entities, which suggest that such organizations are less likely to attract the ire of the state (Zhang and Smith 2009). Similarly, international donors have adopted the view that NGOs with political legitimacy have been more likely to access the state's policy apparatus and have greater potential to develop national networks (Ibid). Furthermore, international donors have subscribed to the general perception that organizations with formal state linkages, such as the nonprofit registration status, have been more likely to comply with laws and regulations, which has important implications for organizational accountability (Ibid). In short, the importance of organizational legitimacy has mattered not only for the state and the local NGO; it has also acted as a positive signal to the international donors about the reputation, political and otherwise, of the NGO in question, which has implications for its potential as a local partner to international donors.

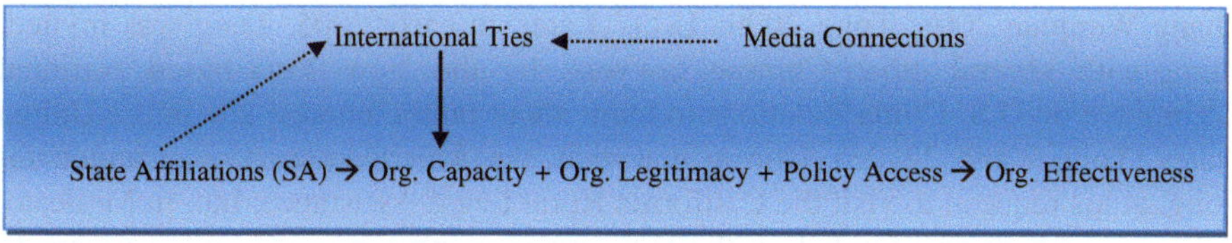

Fig. 5.1 International ties and organizational effectiveness

To a large extent, a Chinese NGO's access to international resources has been linked to its leader's international ties. But those international ties have in most cases been preceded by the leader's possession of state affiliations, media connections, or both. In fact, the cases presented in this book show that a Chinese NGO leader's international ties have often been the direct outcome of the leader's state affiliations and/or media connections. In a nutshell, not only have the Chinese NGO leader's state affiliations and media connections conditioned the positive effects of her international ties; state affiliations and media connections may even *cause* the leader's international ties.

In Fig. 5.1 we can see how the three types of leadership ties interact with each other to impact organizational capacity. As indicated, international ties often have direct effects on a Chinese NGO's organizational capacity. However, enhanced organizational capacity is only one element of organizational effectiveness. In fact, as important as organizational capacity is, the cases in this book show that organizational legitimacy and policy access may be more valuable, especially for advocacy NGOs. The reason is obvious: an advocacy NGO will have limited success in achieving its goals if it cannot access the state's policymaking process, no matter how much money or other material resources it has at its command. Yet, the direct effects of international ties on the latter two elements are insignificant, if non-existent. Instead, the NGO leader must rely on his state affiliations and, to a lesser extent, his media connections to attain organizational legitimacy and access to the policymaking process. More interestingly, as mentioned earlier, international ties may constitute an outcome of the prior existence of state affiliations and media connections.

5.2 Tong Lihua and the Zhicheng Public Interest Law NGO

Just as he has built up strong state affiliations and media connections, Tong Lihua has established extensive international ties. Much of Tong's international ties are connected to the United States. In 2004, Tong served as a PILNET fellow at the Columbia University's School of Law (Interview with Zhicheng research director

Zhang Wenjuan 2008; Tong 2009: 181).[1] Later, in early 2005, Tong was invited, along with several other Chinese lawyers, by the New York-based National Committee on U.S.-China Relations to learn about public interest law in the United States (Tong 2009: 22–23). During that visit, the American sponsor arranged for Tong, at his request, to visit the California Rural Legal Assistance, Inc. in Modesto, California, which was funded by the U.S. government to provide legal services to rural migrant workers. The experience inspired Tong to approach the local Beijing government to fund the provision of free legal aid to migrant workers (Tong 2009: 22–23).

In October 2006 in Beijing, at the annual conference of the Multinational Association of Independent Law Firms (Multilaw), which is an association of law firms in over 70 countries, Tong was given the association's 2006 award in recognition of his work in public interest law. The award reflected Tong's growing international prestige, which has since then been enhanced through his participation in several rounds of human rights dialogue between China and the European Union (EU) as well as visiting fellowships to foreign academic institutions, including a visiting fellowship at the Yale University School of Law in summer 2008.

Befitting the extent of its leader's international ties, Zhicheng has gained access to numerous international sources for organizational resources. Zhicheng has in fact enjoyed strong ties with many benefactors, which cut across countries and regions and which include governments, international organizations, and INGOs. In particular, Zhicheng has received support from the Ford Foundation, the Asia Foundation, the American Bar Association, the United Nations Development Program (UNDP), the EU, and individual European governments. Those connections have provided the material resources for Zhicheng to become a leading advocacy NGO for migrant workers and minors. Through a few illustrations, we can see how international resources have supported Zhicheng's work.

An important aspect of the work of advocacy NGOs around the world is to conduct original research and produce published reports based on the research. Like other advocacy NGOs, and perhaps even more so, Zhicheng's staff has been extremely prolific with respect to the production of published writings. Up until 2009, Zhicheng's staff had produced over 50 publications. And international assistance has made much of the NGO's work in this area possible. For example, Zhicheng's first published report about the rights of migrant workers, which was based on cases that the NGO had handled and nationwide surveys that the NGO had conducted, was supported by the UNDP and the Belgian government (Interview with Zhicheng staff lawyer Wang Fang 2008). These published reports have established Zhicheng's staff as the leading legal experts in China on migrant workers and the rights of minors.

[1] PILNET (http://www.pilnet.org) was originally established in 1997 as a program in the Columbia University's School of Law. The mission of the organization has been in part to promote public interest law and to build civil society in post-communist countries.

Books on migrant workers, legal governance in China's rural areas, and minors by Tong Lihua and his colleagues. *Source* The author

Similarly, international resources helped Tong to establish Zhicheng's national networks of lawyers and legal aid stations to provide free legal aid to migrant workers and minors. Between 2007 and 2009, for example, a Hong Kong philanthropist had contributed 1.5 million Hong Kong dollars to the implementation of a national network of lawyers engaged in the protection of the rights of minors (Tong 2009: 114). In 2007, the Asia Foundation had also allotted funds to support this project (Ibid). In 2008, the Asia Foundation deepened its commitment to Zhicheng by supporting the NGO's project to set up migrant worker legal aid stations in 20 provinces (Ibid). Similarly, in 2007–2008, through the UNDP, the Belgian government provided $500,000 to the NGO to establish migrant worker legal aid stations in 15 provinces (Interview with Zhang).

International support has also built up Zhicheng's human resource capacity. International resources have allowed its staff lawyers, including Tong, to undertake short-term studies overseas, notably in the United States. Tong, as noted above, had been a visiting fellow at Columbia University, had been sponsored by the Committee on U.S.-China Relations to learn about rural legal aid in the United States, and had been a visiting fellow at the Yale University School of Law. However, other members of the NGO staff have also had similar opportunities. When I visited the NGO's migrant worker arm, Wang Fang, the staff lawyer who met with me, informed me that she was preparing to study at Yale University for 6 months (Interview with Wang).[2] Similarly, Zhang Wenjuan, the director of research, had been a visiting fellow at Columbia University and interned at the American Bar

[2] Tong was also a visiting fellow at the school at the same time, but for a shorter period of time.

Association (Interview with Zhang).[3] In 2011, the NGO's manager of the national network of migrant worker legal aid stations also became a PILNET fellow. The overseas experience has helped individual staff lawyers to become better public interest lawyers and together these individual experiences have made Zhicheng the best-staffed public interest law NGO in China.

The NGO staff's overseas experience and the relationships that they had established there, notably in the United States, have also provided the organization with foreign student interns who provided assistance to the NGO in matters concerning the organization's international connections. When I met with Zhang Wenjuan, she introduced me to two Stanford University undergraduate students who were interning at the NGO at the recommendation of their professors. In the summer of 2011, the NGO hosted three American student interns, including a law student from the University of William and Mary. The law student's professor had recommended her to the NGO for a summer internship. And the student stated that one of the projects in which she was asked to be involved was to contribute to a report about the experiences in other countries with respect to pro bono legal work, which was to be submitted to the All-China Lawyers Association (ACLA).[4] In this manner, Zhicheng has utilized foreign intellectual capital to build up its own intellectual capital.

Zhicheng's international support seems to have derived significantly from Tong's international ties. This assumption is logical given the fact that international donors have often based their decisions to provide support to Chinese NGOs on interpersonal networks involving the leaders of those Chinese NGOs. However, we are still left with an even more fundamental question, which is how Tong had acquired his international ties. An examination of the process through which Tong had acquired his international ties shows the importance of Tong's state affiliations. We can begin the process tracing in 2004.

As mentioned earlier, that year Tong became a PILNET fellow. It so happens that during the same year, Zhicheng partnered with the national Communist Youth League (CYL) to implement a project to train state officials involved in the provision of legal aid to minors in 12 mid-western Chinese provinces (Tong 2009: 40). The EU was the benefactor of the project. In particular, the Zhicheng-CYL partnership was one of the projects funded by the China-EU inter-governmental judicial collaboration in 2004. It also represented the first time that Zhicheng's work had received international support. As was shown in Chap. 3, Tong has developed extensive ties to CYL and its affiliated organizations, and those personal ties have

[3] As of February 2014, Zhang, the director of research, was reported to be pursuing an LL.M. at Columbia University. US-Asia Law Institute, "Zhang Wenjuan of Zhicheng Public Interest Lawyers Organization Speaks at USALI," New York University School of Law, http://usali.org/media-entities/zhang-wenjuan-of-zhicheng-public-interest-lawyers-organization-speaks-at-usali-2/ (accessed 10 May 2014).

[4] Laura Bain, "Second Week and Sanlitun," William and Mary Law School, http://law.wm.edu/academics/intellectuallife/researchcenters/postconflictjustice/studentsummerblogs/laurabain/second-week-and-sanlitun.php (accessed 2 October 2011).

benefited Zhicheng. These state ties, including ties to the justice system, have in all likelihood played a significant role in helping Tong gain access to the EU funding, given the fact that the Chinese government almost certainly did not advertise openly the availability of this resource.

One may conclude from this illustration that not only did Tong's state affiliations helped him develop his personal ties to the international community, in fact it was his state affiliations that allowed him to gain access to international resources for his organization. However, one may also argue that the evidence from this example is quite circumstantial. But Tong has provided another example that more clearly indicated the importance of Tong's state affiliations in establishing his international ties and garnering international resources for Zhicheng.

In August 2006, after Zhicheng's migrant worker arm was formally established, Tong was invited to participate in a meeting held by the State Council's Office of Poverty Alleviation. At the meeting, he met a UNDP representative. During the conversation with the UNDP representative, Tong introduced Zhicheng's migrant worker legal aid program (Tong 2009: 27). After the meeting, the UNDP representative contacted Tong to inform him that the Belgian government wanted to provide $500,000 through the UNDP to support a project in China. The UNDP representative suggested that Tong submit an application. In November 2006, UNDP selected Tong's application, which proposed to establish migrant worker legal aid stations in 15 provinces. Because the UNDP/Belgian government fund was allocated through the Chinese government, the project had to be approved by the latter. Sure enough, the Chinese Ministry of Justice approved the project.

This example shows that in some cases the ability to access state resources precede the ability to access international resources if the Chinese NGO leader wants to meet his organizational goals. An important organizational goal for Tong was the development of national networks of legal aid offices and public interest lawyers. To overcome the legal and political barriers to the establishment of such formal networks, Tong has used his state affiliations to serve as the front for the NGO project. In this case, Tong's ACLA affiliations allowed him to do just that. As it turned out, although Tong and his Zhicheng colleagues had drafted the proposal and submitted the funding application to the UNDP, the official beneficiary of the funding was the ACLA (Tong 2009: 28). This fact undoubtedly made it easier for the Ministry of Justice to approve the project. And even though the UNDP had selected Tong's application before the ministry gave its stamp of approval, it would not be inconceivable for the ministry to veto the UNDP selection if the ministry had concerns about any aspect of the project, including Zhicheng. In other words, from the perspective of the Chinese NGO, the utility of international resource is maximized only when the funded project has been given the state's blessings. Indeed, once the ministry gave its approval, ACLA officially commissioned Zhicheng to administer the project. Therefore, although on paper ACLA was the project coordinator and recipient of the UNDP funding, in practice it was Zhicheng that managed the project and the international funding that supported it.

This illustration also highlights the value of state linkages in another way. Tong's state affiliations were probably an important factor in his invitation to attend an official government function at the State Council, which oversees all ministries and ministerial-level government agencies. Tong's attendance at the government meeting gave him the opportunity to establish contact with a representative of an international inter-governmental organization. As a result of that contact, Zhicheng was able to gain access to international funding for a project that was a very important element of the NGO's mission. In a nutshell, it was Tong's state affiliations that gave him the opportunity to develop his international ties (to the UNDP representative), which allowed Zhicheng to be in a position to receive international support.

The central lesson here is that, for the Chinese NGO, the value of international support is highly dependent on the organization's state linkages. Moreover, the Chinese NGO leader's state affiliations can contribute directly to the establishment of the leader's international ties, which can then lead to international support for the NGO.

5.3 Xie Lihua, Wu Qing, and Rural Women

To a large extent, Rural Women's experience with international ties has mirrored that of Zhicheng. However, Rural Women's experience also allows us to see the media's possible impact on international support.

Before we analyze these different factors, let us first gain an understanding about the nature and extent of Rural Women's international support and the international ties of the NGO's two leaders. In the previous chapters, we had focused primarily on Xie's leadership ties. In this chapter, we will pay more attention to Wu Qing, who played a valuable role in the creation of the NGO and was a co-founder, along with Xie, of the Practical Skills Training Center in 1998. Indeed, it is both appropriate and necessary for us to do so given that Wu Qing has played a prominent role in obtaining international resources for Rural Women, especially its training center.

Since its creation in 1993, Rural Women has become the leading Chinese NGO that promotes the interests of Chinese rural women, and international support has played a big role in making this possible. Indeed, international funding has constituted nearly 100 % of the NGO's resource base (Interview with Gao Guanglin, chief executive of Rural Women Cultural Development Center, 2008).[5] And a range of international organizations has constituted that resource base. Approximately 47 % of the NGO funding came from foreign foundations (Ibid). More than 10 % of the funding came from multinational corporations like Microsoft. Nearly 30 % of the funding came from other international groups and individual foreigners. All in all, only 10 % of the NGO's financial base came from domestic sources, namely state agencies.

[5] In 2011, Gao stepped down from this position.

International financial support has been critical for Rural Women to undertake projects that are directly connected to the organization's goals. For example, the Practical Skills Training Center, where the NGO administers training programs for young female migrant workers as well as rural women across China, has received funding from multinational corporations such as J.P. Morgan, Exxon Mobile, and the Barclays Bank, as well as foreign governments and private foreign persons (Rural Women 2010). The Ford Foundation has also been a consistent benefactor of the training center. Similarly, the anti-illiteracy project has been funded by Goldman Sachs and the Tiger Woods Foundation. The book club project, which has been a component of the anti-illiteracy project and which has established 50 book clubs in rural villages across 11 Chinese provinces, was partially funded by Oxfam Hong Kong, which has also supported Rural Women's legal aid project. The United Nations has also provided financial support to the NGO's project to promote the protection of rural girls from sexual predation. Lastly, the Asia Foundation had funded the NGO's involvement in the writing of the NGO shadow report on China for the United Nations' Committee on the Elimination of Discrimination against Women (CEDAW).

Like Zhicheng, international support had also allowed the staff of Rural Women to receive training. In addition to the numerous overseas trips and personal meetings with foreign dignitaries for the NGO's founder and co-founder, other staff members also took part on trips outside China. In 1998, for example, a staff member was sent to George Mason University in Virginia's Fairfax County, a suburb of Washington, D.C., to participate in a one-year training program on strategic management (Shang 2007: 248). Oxfam Hong Kong has been especially active in supporting the NGO staff to receive training and participate in exchanges in Hong Kong (Ibid). Meanwhile, experts and organizations outside China have visited the NGO to provide training to the NGO staff. These international exposures have introduced ideas that impacted Rural Women's organizational goals (Ibid).

The two leaders of Rural Women, especially Wu Qing, possess extensive international ties. To a certain extent, Wu's international ties have been unsurprising given her family history. Wu's parents were educated in the United States. Her father was trained at Dartmouth College and Columbia University after he graduated from China's Tsinghua University. Her mother in turn earned her graduate degree from Wellesley College. Although Wu was born in China, Wu's parents took her and her siblings to Japan in 1945 when Wu's father was appointed a member of the Chinese diplomatic delegation there. In 1951, the family returned to China. In 1961, Wu graduated from the Beijing Foreign Language Institute (later renamed the Beijing Foreign Studies University), where she remained as an English instructor until her retirement in recent years.[6]

---

[6] Wu has an excellent command of the English language. I had an opportunity to listen to her speak to a group of U.S. college students engaging in short-term study at the Beijing Foreign Studies University. Her delivery was smooth and natural. In addition, she deftly handled on-the-spot questions from the students.

A group of college students from the United States visiting a migrant community on the outskirts of Beijing. *Source* The author

In the early 1980s, Wu began to establish her own international connections when she was sent by the foreign language institute to study at the Massachusetts Institute of Technology (MIT) for one year (Phoenix Television 2008). At MIT Wu came into contact with American community activists. By Wu's own account, inspired by what she had seen and learned in the United States, she wanted to promote the rule of law in China (Ibid). Therefore, immediately upon her return to Beijing, she accepted the foreign language institute's nomination to become a delegate to the Haidian district legislature and in 1989 she was appointed to become a delegate to the Beijing municipal legislature, a position that she kept until 2004. While in office, Wu developed a reputation as a consistent and persistent challenger of government practices and the political status quo, as well as a tireless advocate of women's rights. In fact, she is widely known as the first delegate to the Beijing municipal legislature who voted openly against government measures.

Wu's international ties continued to deepen since her time at MIT in the early 1980s. In 1993–1994, Wu once again visited the United States in an academic capacity, this time as a Fulbright Fellow at Stanford University. In 1995, Wu became a member of the Geneva-based Women's World Summit Foundation, which achieved United Nations consultative status. Wu eventually became the president of the foundation. Between 1996 and 2002, she was also a board member of the Global Fund for Women. In 2001, she was awarded the Ramon Magsaysay Award.

A review of Wu's personal history offers some clues about the origins of her international ties. Although we can infer logically that Wu's personal experiences in the United States had exposed her to new ideas and perhaps even established some personal connections, it was probably her political activism after her return to China that truly solidified her international reputation and expanded her extensive international ties. This seems to be a logical connection given that she began to acquire important international recognition in the 1990s, approximately 10 years after she began her activism as a "trouble-making" Chinese legislator. Put simply, Wu's international reputation has been due as much to her state affiliations as to her foreign experiences.

By the 1990s, Wu had already established her international reputation and was in a position to help Chinese NGO entrepreneurs working on women's issues to obtain much-needed international support. One of her first significant contributions to Rural Women was her introduction of Dr. Mary Ann Burris to Xie Lihua. Dr. Burris was the Ford Foundation program officer in Beijing responsible for developing the foundation's programs on reproductive health and women's rights.[7] This connection turned out to be Xie's very first international tie. At the time of the referral, Xie was struggling to operate the *Rural Women Knowing All* magazine with the limited funding that she had received from the *Women's News* of the All-China Women's Federation (ACWF). After the meeting with Xie, Burris offered to fund the magazine for three consecutive years if Xie agreed to include a regular column about reproductive health in the magazine. Ford Foundation funding was in the form of an annual purchase of 10,000 copies of the magazine at an annual cost of 96,000 Chinese yuan. The magazines would then be distributed, free of charge, to women's federations and women living in China's poorest regions (Deng 2000: 222–223).

Xie's relationship with Burris would continue to grow while Burris worked in Beijing until 1995. Between 1993 and 1995, Burris helped Xie gain international experiences and develop new international ties when she arranged for Xie to attend two international women's conferences in 1994 and 1995, respectively (Interview with Xie Lihua 2009). The interesting thing is that in both cases Xie attended as a member of an ACWF delegation (Ibid).[8] Later, during the 1995 UN Women's Conference in Beijing, Burris arranged for a dozen international women's NGOs and journalists to visit Rural Women, further exposing Xie to foreign practices while publicizing Xie and her operation (Deng 2000: 223). Xie again participated in the UN conference as an ACWF staff and an outcome of Xie's participation in the 1995 conference in Beijing was her introduction to Hillary Clinton. Since then, Xie had met the future U.S. senator and secretary of state on five occasions. The most recent meeting took place in 2010, when Clinton was in Beijing to participate in the U.S.-China Strategic and Economic Dialogue. At that time, Xie was invited to meet

[7] Burris's Chinese name was "Bai Mei" (White Plum). She is now the founder and executive director of the Trust for Indigenous Culture and Health (http://www.ticahealthy.org).

[8] Xie told me that attendance at the conferences had to be approved by ACWF and the federation expected the Chinese participants to represent the federation.

Clinton on a Chinese Central Television program in which Clinton was the featured guest.

Here it is worthwhile for us to pause and trace the development of Xie's international ties closely. First, it is clear that Wu Qing played a foundational role in developing Xie's international ties. Second, Xie's relationship with the Ford Foundation in general and Burris in particular had a clear and immediate impact on Rural Women's capacity and the relationship was instrumental in allowing Xie to develop more international ties. Moreover, other than the building of additional international ties, those international experiences served to inspire Xie with ideas about the ideal objectives of a women's NGO and the types of projects that a women's NGO should undertake (Shang 2007: 242). In this manner, Xie would eventually enhance her and Rural Women's international reputations, which would lead to the building of more international ties, because the NGO's projects and organizational goals would become more in line with international practices, thereby gaining the approbation of international donors.

Lastly, and perhaps most significantly, even though Wu Qing was the direct catalyst in Xie's ties with Burris and the Ford Foundation, Xie's ACWF affiliations played a role in allowing her to take advantage of Burris's arrangements, as evidenced by Xie's attendance at the two international conferences between 1993 and 1995 as an ACWF delegate as well as her participation in the UN Women's Conference as an ACWF representative. Although we cannot say with great certainty that it would have been impossible for Xie to take part in these activities if she had not been affiliated with ACWF, I think that it is safe to say that it would have been very difficult for her to take advantage of those opportunities had she not been affiliated with ACWF. The main reason is that as the CPC-sanctioned corporatist body on women's issues ACWF, in essence, monopolizes the issue area of women's affairs in China. In this manner, it determines the overall agenda for the issue area and how other actors, both non-state and foreign, are able to take part in the issue area. Therefore, it should be of no surprise that the most prominent Chinese NGO leaders in the area of women's issues have all possessed some form of connection to ACWF. In addition, we must not forget that the state's decision to crack down on the students in 1989 had just taken place a few years before. The regime was therefore still in a high state of alert about factors that may impact the domestic political situation to the ruling party's detriment. In this context, Xie's affiliations with ACWF provided the political cover that she needed to take advantage of these international opportunities without arousing the state's suspicion.

Just as the international ties of the NGO leader are affected by her state affiliations, the salutary effects of international support for the NGO are conditioned upon the respective leaders' state affiliations. This was clearly evident with both the establishment and operation of the training center and the book club project. As mentioned earlier in this section, Rural Women have received tremendous international support for both the training center and the book club project. However, as demonstrated in Chap. 3, neither the training center nor the book club project could have been established and sustained without Xie's ACWF affiliations because

ACWF has been the official state sponsor of the training center while the book club project involved the cooperation of local ACWFs. In other words, international resources are useful to the extent that they help an NGO engage in projects that serve the NGO's organizational objectives. However, international ties are unlikely to help the NGO establish the state linkages necessary to implement projects that require state collaboration.

In this respect, the book club project also suggests that the salutary effects of Xie's international ties were conditioned upon her media connections. Recalling that in Chap. 4 we had discussed how Rural Women was able to collaborate with the local ACWFs in part because the *Women's News* and the NGO's own publications had served as effective vehicles to advertise the NGO. In this manner, Xie's media connections have facilitated state support for a project that was directly connected to the NGO's objective. With the state support the NGO also achieved legitimacy. As a consequence, as was the case with Tong and Zhicheng, Xie could use international resources freely without being concerned about arousing the state's suspicion. In the process, international resources have helped to ensure the project's successful implementation.

The media connections of both NGO leaders might have exercised more direct effects on their ability to attract international resources for Rural Women. After all, both Xie and Wu have been widely featured in foreign publications, and those international reports have likely contributed to increased media exposure for the NGO, which in all likelihood served as a means to draw international support.[9] But admittedly the evidence for this connection is relatively sketchy.

To a large extent, the case of Rural Women teaches us the same lessons that we have learned from the case of Zhicheng, albeit with some minor twists. As was the case with Zhicheng, the state affiliations of the Rural Women leaders were likely to be catalysts in the development of their respective international ties. However, the means by which Rural Women received the international resources and the nature of those resources were different from Zhicheng's. Whereas Zhicheng had used Tong's state affiliations to obtain international funding that was being distributed by the state, Xie's state affiliations allowed her to obtain approval to use international resources that were not filtered through the state, even though the state most likely had an important say in whether its employees were allowed to access those resources. Moreover, in Rural Women's case, we gained a glimpse of the potential impact of the NGO leader's media connections on international support for the NGO. However, a glimpse is simply not satisfying. Therefore, we now move to the case of the Little Bird, which shows that the effects of the NGO leader's media connections on international ties can be more direct, observable, and immediate.

[9] Both Wu Qing and Xie Lihua had been featured in international press. For example, in January 2008, the *Los Angeles Times* published a story about Xie Lihua: John M. Glionna, "A Voice for Rural Women of China," *Los Angeles Times*, 2 January 2008, http://articles.latimes.com/2008/jan/02/world/fg-women2 (accessed 26 January 2012). Moreover, Xie Lihua and Rural Women have been featured prominently in published studies by foreign scholars.

5.4 Wei Wei and the Little Bird Migrant Workers Mutual Support Hotline

If the evidence for the impact of media connections on international ties was circumstantial and the causal connection was speculative for the case of Rural Women, then the case of Wei Wei and Little Bird shows more clearly the value of media connections. Indeed, among the NGO cases presented in this book, Little Bird offers the best illustration of the power of the media. However, before we analyze the relationship between Wei's media connections and his international ties, let us first get an idea of the nature and extent of Little Bird's international support.

Since 2004, Little Bird has relied extensively on international resources. According to the organization's 2010 annual report, over 70 % of Little Bird's financial resources have originated outside China (Little Bird 2011). Unlike Rural Women, the list of approximately 13 international organizations that have funded the NGO up to 2010 does not include multinational corporations. Nonetheless, the donors have represented a wide range of foreign governments, international governmental organizations, and foreign foundations, all of which have been well-known for their long-term interest in China.

For example, Little Bird has received financial support from the Canadian Embassy in China, the Australian Embassy in China, and the American Embassy in China (Ibid; Liu 2009: 100). In addition, the International Republican Institute (IRI), the Ford Foundation, the Asia Foundation, Oxfam Hong Kong, the American Bar Association, and the German Catholic foundation Misereor have all been Little Bird's benefactors. Moreover, Little Bird has had financial relationships with international government organizations, notably the World Bank and the EU.

These resources have sustained and even enhanced Little Bird's capacity to advocate and serve the migrant workers. For example, the three aforementioned foreign embassies have all supported the NGO's radio programming, which has been an important vehicle for the NGO to advocate the rights and interests of the migrant workers in general (Ibid). The radio programming has also been an important instrument to promote the organization's telephone hotline. And the hosting of the radio program is also important for the NGO to maintain its connections to the media sector and, by extension, the positive state linkages that it has enjoyed.

International resources have even allowed Little Bird to expand its geographical coverage so that it can serve more migrant workers throughout China. In 2006, with the support of the German Embassy in China and the German Catholic foundation Misereor, the NGO opened its hotlines in Shenzhen to the southeast and Shenyang to the northeast (Ibid). In June 2011, through the support of the American Bar Association, the Little Bird telephone hotline was opened in Shanghai (Ibid). With the establishment of the Shanghai branch, the NGO now has a presence in northeastern, northern (Beijing), eastern, and southeastern China. Wei had indicated that his objective was to establish branches in Xi'an, Zhengzhou, and Chongqing, which together would constitute central China.

Furthermore, international support has given Little Bird the capacity to hold training programs to assist Chinese migrant workers. For example, the German government and German foundations have supported the NGO's efforts to hold training sessions for migrant workers on topics that include occupational safety and employment contract signing (Little Bird 2011). Little Bird has also been funded by IRI to hold capacity-building training programs for Chinese NGOs (Ibid). In this vein, international ties have not only supplied Little Bird with material resources; those relationships have also affected the organization's practices and values so that the NGO could meet the standards of the international donors.

Wei had made clear to me that Little Bird's international resources have been the result of his efforts to build international ties (Interview with Wei Wei 2009). In other words, Little Bird's international support has reflected the extent of Wei's international ties. But how did Wei acquire those international ties in the first place? The answer is Wei's media connections.

In Chap. 4, we had seen how Wei's media connections have garnered media exposure for Little Bird, which allowed the NGO to establish a partnership with the local justice bureau. Wei's media connections were also instrumental in the creation of Wei's first international tie. In 2003, Beijing's *Jinghua News*, which was created under the supervision of the *People's Daily*, published a story about Wei and the Little Bird. The *Jinghua* story resulted in a reader telephoning Wei. The reader gave Wei the contact information for an officer in the Canadian Embassy in Beijing. According to Wei, the Canadian Embassy official helped Wei obtain 100,000 yuan in funding from the Canadian government and provided Wei with a letter of reference that he could use to approach other foreign donors, which Wei of course did (Ibid).

As Wei and Little Bird continued to receive extensive media coverage, international resources also continued to pour into the NGO, even during the past few years when Western countries have experienced significant economic downturns. Therefore, it would seem that Wei's media connections have continued to be effective in getting the NGO to access international resources. Even so, we cannot overlook the potential significance of Wei's institutional affiliations to the local justice bureau. As stated earlier, international donors have demonstrated a proclivity to support Chinese NGOs with political and legal legitimacy because of the perception that NGOs with organizational legitimacy have been less likely to arouse state suspicion as well as having greater potential for policy influence and the development of national networks. Although we do not have direct evidence to prove that Wei's state affiliations have had direct causal links to international support for Little Bird, it is safe to conclude that his state affiliations certainly did not harm the interests of the organization. At a minimum, the existence of state linkage has meant that the state would be less concerned about "inappropriate" use of international resources by Wei.[10] In this respect, the state linkages of Wei and the

[10] In this respect, Wei had made sure that he kept the relevant government officials informed about his activities and the sources of support for his activities. Wei believed that it was important to sustain the trust of government officials, even though Wei's behavior has elicited criticisms from other leaders of Chinese migrant worker NGOs.

Little Bird have very likely played an important role in attracting continuous international support.

What lessons can we draw from the case of Wei Wei and the Little Bird? First and foremost, Little Bird shows us that the Chinese NGO leader's media connections can have direct and positive effects on the leader's ability to develop international ties and access international resources for his organization. However, this does not mean that the leader's state affiliations have been unimportant. In fact, the case of Little Bird demonstrates how the combination of media connections and state affiliations can play a valuable role in ensuring a continuous supply of international resources to help the NGO achieve its goals, without causing the ever-suspicious state to be concerned.

5.5 Ma Yang and the On Action International Cultural Center

The focus of this chapter is to show that the maximum value of international ties and the resources that came with those ties can be manifested only when they are coupled with the presence of state affiliations and media connections. Although the three previous cases have illustrated the interactions among the three types of leadership ties, the case of On Action should further clarify the contingent nature of the importance of international ties for the Chinese NGO.

In contrast to Zhicheng, Rural Women, and Little Bird, On Action was established with international resources in hand. Starting in late 2006, when the NGO was established and ending in summer 2009, On Action received approximately 700,000 yuan from three foreign sources (On Action 2008). One of the three donors was the U.S. Embassy in China (Interview with Ma Yang 2008). Funding from the U.S. Embassy had created a reading room for migrant workers in the NGO's office.[11] The other two foreign sources, one of which was the Dutch Labor Union, supplied funds for the NGO to conduct information sessions about occupational health and the 2008 Labor Contract Law (Ibid). According to the report that On Action released in commemoration of the second anniversary of the NGO's establishment, all three projects were completed in 2009.

It seems that these international resources have supported activities that worked toward satisfying an extremely tiny fraction of the organizational goals. For example, none of the projects funded by the international donors dealt with policy research and advocacy and none dealt with the issue of providing legal aid to migrant workers. Moreover, none of the funding seems to have addressed the issue of building up staff capacity that would be important for the NGO to undertake

[11] The reading room was approximately the size of a 10' × 10' room with some shelves along the walls. The collection of readings was a mixture of popular Chinese novels, newspapers, literature from different NGOs, and some state pamphlets about relevant laws and regulations.

activities to address its organizational objectives. Indeed, none of the funded projects required interactions with the state. Yet, in order for On Action to fulfill its mission it was clear that it would have to undertake activities that involve the state.

To be sure, we cannot blame the international donors for this problem, at least not entirely.[12] After all, it was up to the NGO leaders to develop projects that would best address the organization's mission. But On Action's situation once again calls our attention to the contingent nature of international ties. In this case, the fact that Ma did not conceive projects that were more germane to the mission of the NGO has meant that the international resources On Action received were not going to be put to optimal use, at least in terms of meeting the organization's objectives.

However, this raises an interesting and relevant issue, which is that in order for Ma Yang to develop projects that would more directly address the organizational mission he would most likely have to develop better state affiliations because as we have observed, most successful NGO projects have required the support of the state. Yet, as we discovered in Chap. 3, Ma simply did not possess the requisite state affiliations. Similarly, as was shown in Chap. 4, Ma did not possess strong media connections that could help On Action achieve state linkages.

In fact, Ma obtained the international resources independent of any considerations about his state affiliations. The organization's international resources were the result of the efforts and connections one of its original board members, who had worked previously as a program officer for Oxfam Hong Kong and who had studied for an advanced degree in the United Kingdom. Interestingly, the board member's international ties were never converted into Ma Yang's.

Of course, any hope that Ma Yang had of developing his own international ties was dashed, at least for the time being, when Ma was implicated directly in the publicized organizational financial scandal in 2009, which resulted in the departure of the well-connected board member. Through this example we can see the media's "negative" effects on the Chinese NGO leader's ability to acquire international resources. In short, one simply cannot overestimate the independent value of international ties.

[12] Although international donors have been interested in supporting Chinese NGOs that could engage the state, they have also funded other NGO projects with the aim of developing new local NGOs. In fact, international donors like the Ford Foundation have "encouraged" local NGOs to become more self-sufficient by reducing and even cutting off support to local NGOs after a certain number of years of support. Of course, there have been exceptions to this "rule," such as the ongoing support of Rural Women by the Ford Foundation.

5.6 Corroboration from Additional Cases: Friends of Nature, Yirenping, and the Self-Empowerment Service Center for Disabled Persons

The heavy reliance on international sources has not been limited to Chinese NGOs working on issues concerning migrant workers. This phenomenon involves virtually all Chinese NGOs, regardless of the issue areas. It is also a phenomenon that knows no geographical boundaries. Furthermore, it has concerned all Chinese NGOs right from their birth and this concern has been sustained even when the organizations have become more mature in age. Similarly, the contingent nature of international ties has been observed across issue areas, geographical locations, and stages of organizational development.

In the previous chapters, we had seen how the Friends of Nature (FON) had benefited from its chief founder's state affiliations and media connections. Liang Congjie's state affiliations provide FON with political legitimacy while his media connections helped to publicize issues that were of concern to the NGO and, in turn, garnered the attention of the state. In a nutshell, state affiliations and media connections allowed FON to work effectively as an NGO advocate of environmental concerns.

At the same time, Liang Congjie developed a strong international reputation. Between the time when the NGO was founded and his death, Liang had received numerous awards and recognitions from the international community, including the Ramon Magsaysay Award. He was also featured on a wide range of popular and specialized media. Even his death was reported by the world's most important news outlets, including the *New York Times* and *The Economist.*

Since it first received international funding at the end of 1994, which was also the year that the NGO was established, FON has relied heavily on international support for its operations (Interview BJ120908). Undoubtedly, the strength of Liang's international ties was an important factor in FON's ability to access international resources. The NGO's first international funding commitment, for example, was received immediately after Liang and another co-founder visited American foundations and NGOs in the United States at the end of 1994 (Friends of Nature 1994). However, the international community's interest in Liang and FON probably also had to do with his state affiliations, just as they were a point of attraction for those who sought him out to establish China's first legally-registered environmental NGO (Economy 2004: 146).

Furthermore, international funding would lose its value if it could not be employed to further the NGO's goals. As stated earlier, the primary function of international funding has been the enhancement of an NGO's organizational capacity. But the effectiveness of the NGO, especially one with an advocacy focus, has depended on organizational legitimacy and access to the state policy apparatus. Both of these items required some degree of state linkage. International funding alone is therefore insufficient to meet the Chinese NGO's mission. In this respect, the state connections of Liang and FON as well as their media connections have put them in a great position to utilize their international resources to further their objectives. Here, we can use the example of FON's *China Environmental Yearbook* project, which has involved the publication of an annual edited volume since 2005, to illustrate the importance of the NGO's prior media and state connections for the optimal use of international resources.

The yearbook series have been funded by the Delta Foundation, which was established by a Taiwanese-American energy entrepreneur. In essence, the yearbook series are edited volumes containing essays written by Chinese journalists, academic experts, and, to a lesser extent, government officials about various aspects of China's environmental conditions. These writings have included rigorous analyses of environmental issues concerning China. The yearbooks have been published by a reputable Chinese publisher of academic writings and have been made available to the public through purchases. Moreover, the findings of the yearbooks were compressed to be submitted eventually to National People's Congress (NPC) delegates and members of the Chinese People's Political Consultative Conference (CPPCCP) who were connected to FON so that those delegates and members could use the materials to make legislative proposals.

We can see that Delta's money has been able to help Liang and FON achieve the goal of publicizing environmental concerns and making policy impact. But Delta's money was able to make an impact because Liang and FON could enlist journalists, government officials, and other experts to contribute their analyses. The impact of Delta's money would be further felt when the NGO submitted the findings to sympathetic legislators and government advisors. In a nutshell, the effects of Delta's funding on FON were contingent on FON's state linkages and media connections. And Liang played a big role in helping FON achieve those linkages and connections.

The case of FON shows that the general principles concerning the effects of international ties on Chinese NGOs could be applied to NGOs working in different issue areas. Yirenping's case offers further proof of this point. Moreover, the case of Yirenping shows that those principles may also be applied to *young* NGOs.

In the preceding chapters, we have seen how Lu Jun, the chief founder of Yirenping, had developed state affiliations and media connections. We have also seen how those leadership ties have helped Yirenping become the effective voice of 400 million Chinese Hepatitis-B patients and the most successful Chinese NGO advocate against institutionalized discrimination. And the remarkable thing is that those achievements have been made within 5 years of the organization's establishment.

In addition to his state affiliations and media connections, Lu Jun has become a world-famous NGO activist. In 2009, Lu was awarded by the Italian government with a human rights award that was established in commemoration of a Roman Catholic priest who had been gunned down by members of the Italian mafia (Yale University Law School 2010). A year later, Lu was invited to serve as a visiting scholar at the Yale University's China Law Center. Past and present visiting scholars at the China Law Center have included government officials and notable public interest lawyers, such as Zhicheng's Tong Lihua. And Lu has employed his international ties to obtain international resources that would help Yirenping conduct research, write reports, and submit proposals to government officials and state organizations.

However, without their state linkages and media connections, Lu and Yirenping could not have made their impact no matter how much international resources they had obtained. Indeed, in many ways, Lu's media connections and successes in influencing state policies were probably the factors that attracted international attention and helped build up Lu's international ties, which would result in more international resources for the NGO. On this note, it should be unsurprising that Lu received the Italian award on the same year that Yirenping scored its greatest victory to date, when the Ministry of Health, in response to pressures from activists like Lu and Yirenping, issued new rules to abolish mandatory Hepatitis-B testing during physical examinations for prospective college students, factory workers, and government employees (Jacobs 2009). In sum, Lu's state affiliations and media connections have acted as facilitators of his international ties and the conditions under which international resources to Yirenping have helped the NGO to succeed.

Outside Beijing, other Chinese NGOs have also benefited from the interactive effects of international ties, state affiliations, and media connections. Although the Self-Empowerment Center has not received nearly as much international resources as most of the NGOs that are featured in this book, it has nonetheless relied completely on international resources to establish an office space, create a telephone hotline, hire staff, provide services, and print information pamphlets that would be distributed to outgoing migrant workers.

But international support would have been less effective if Chen Yuying, the founder of the center, had not received the media attention that led to the creation of her affiliations with the local handicapped persons' federation. Those media connections and the resulting state affiliations essentially created the platform on which Chen could pursue her goals for the NGO. Those connections, especially the media connections, have legitimized Chen and her organization and, in the process, those connections have allowed Chen to even influence local policies. In important respects, Chen's state affiliations have held the local state in check so that it would not interfere in the center's activities. In fact, those affiliations have even resulted in getting the state to assist the NGO's activities, such as when the local government mobilized local schoolteachers to serve as "volunteers" to help the center distribute its information pamphlets to outgoing migrant workers in the rural county's main transportation portals. Coupled with state connections, Chen's international ties have helped her to overcome many personal and organizational obstacles, and in the process paved the way for the center to become effective providers of much-needed services to under-served residents of Chen's home county. But it's unlikely that Chen's international ties could have been as effective without her state affiliations and media connections.

5.7 Conclusion

In this chapter, I have made the argument that a Chinese NGO leader's international ties are not sufficient to help her NGO achieve its organizational objectives. The reason is that international ties have not had direct and independent effects on the NGO's accessibility to the state, which is important, indeed critical, to the NGO's efforts to achieve its goals because the successful implementation of NGO activities requires the support of the state. In fact, international ties may be a source of concern to the NGO because the Chinese state has remained suspicious of the purpose of international ties. Therefore, in order for international ties to serve the interests of Chinese NGOs, the leaders of the most successful Chinese NGOs have also acquired state affiliations and developed media connections. These two characteristics have been valuable because, among other things, they serve as signals to the state about the legitimacy, i.e., reliability, of the NGO in question. And, of course, as the cases have demonstrated, state affiliations and media connections have often acted in a causal role vis-à-vis the leader's international ties.

Therefore, in an important respect, although a leader's international ties can help her NGO obtain much-needed resources to enhance organizational capacity, the importance of those ties must necessarily take a backseat to the leader's state affiliations and media connections. These findings are generally consistent with observations made by an increasing number of analysts that the effects of international support on the effectiveness of local NGOs have been highly contingent on other factors, especially the attitude of the state. Fewer writers, however, have observed similar effects by the media. Nonetheless, as I tried to demonstrate in this chapter, we should not overlook the importance of the media as a condition for the effective utilization of international ties.

A key lesson to be drawn from this chapter is, therefore, that prospective Chinese NGO leaders need to be mindful about the single-minded pursuit of international ties. In fact, aspiring Chinese NGO entrepreneurs can go a long way in helping their organizations if they first develop their state linkages and media connections. In this respect, Wei Wei and the Little Bird can serve as a useful model for aspiring Chinese NGO entrepreneurs.

References

Chen Jie. 2012. Transnational Civil Society in China: Intrusion and Impact. Cheltenham, UK: Edward Elgar.

Davis, Meg. 14 June 2010. China's New Nonprofit Regulations: Season of Instability. *Asia Catalyst*. http://asiacatalyst.org/blog/2010/06/chinas-new-nonprofit-regulations-season-of-instability.html. Accessed 8 February 2011.

Deng Guosheng. 2000. Nongjianu Baishitong [Rural Women Knowing All]. In *Zhongguo NGO Yanjiu: Yi Anli Wei Zhongxin* [*Study of Chinese NGOs: Focus on the Cases*], 222-237. UNCRD Research Report Series No. 38 (August). Beijing, China: United Nations Centre for Regional Development.

Economy, Elizabeth C. 2004. The River Runs Black: The Environmental Challenge to China's Future. Ithaca: Cornell University Press.

Friends of Nature. 1994. Dashiji – 1994 [Timeline – 1994]. http://www.fon.org.cn/channal.php?cid=64. Accessed 27 January 2012.

Ford, Peter. 20 May 2010. Law Chokes Chinese NGOs' Foreign Funding. *The Christian Science Monitor*. http://www.csmonitor.com/World/Asia-Pacific/2010/0520/Law-chokes-Chinese-NGOs-foreign-funding. Accessed 16 February 2011.

Hsu, Jennifer Y.J. and Reza Hasmath. 2013. The Rise and Impact of the Local State on the NGO Sector. In *The Chinese Corporatist State: Adaptation, Survival and Resistance*, eds. Jennifer Y. J. Hsu and Reza Smith, 120-135. New York: Routledge.

Jacobs, Andrew. 12 October 2009. China to End Required Testing for Hepatitis B. *The New York Times*. http://www.nytimes.com/2009/10/13/world/asia/13beijing.html. Accessed 20 October 2011.

Kaufman, Joan. 2012. The Global Women's Movement and Chinese Women's Rights. *Journal of Contemporary China* 21, 76 (July): 585-602.

Little Bird. 29 June 2011. Work Report in 2010. http://www.xiaoxiaoniao.org.cn/html/aboutus/2011-04-06/174639.html. Accessed 1 October 2011.

Liu Zhihua. 2009. 'Sheng' Yuan Yu Meiti, 'Zhang' Ji Yu Meiti – Cong 'Xiaoxiaoniao Dagong Huzhu Rexian' Fazhan Kan Caogen NGO De Meiti Zhanlue ['Birth' through the Media, 'Growth' on Media's Foundation – Observing the Media Strategy of the Grassroots NGO through the Development of the 'Little Bird Migrant Worker Mutual Support Hotline']. *NGO Anli Ji: Zhongguo Feiyingli Zuzhi Guanli* [*NGO Cases: Management of Chinese Nonprofit Organizations*], 96-102. Shanghai, China: NPI. http://www.npi.org.cn/uploads/magazines/anliji.pdf. Accessed 23 June 2011.

On Action. 2008. Zaixingdong Tongxun: Zaixingdong Liangzhounian Fazhan (Tekan) [On Action Communiqué: (Special Issue) of the Second Anniversary of On Action's Development]. http://www.chinadevelopmentbrief.org. Accessed 15 October 2011.

Phoenix Television (program transcript). 21 March 2008. Bing Xin Zhi Nu Wu Qing – 'Shuo Zhenhua De Renda Daibiao'" [Wu Qing the Daughter of Bing Xin – 'the People's Congress Delegate Who Speaks the Truth']. http://phtv.ifeng.com/program/mrmdm/200803/0321_1597_453302.shtml. Accessed 27 January 2012.

Rural Women. 2010 Annual Report. http://www.nongjianv.org/web/Files/content/njndzb/2010njnnianbao.pdf. Accessed 26 December 2011.

Shang Xiaoyuan. 2007. Funu Gongmin Shehui Zuzhi He Zhili Moshi De Duoyuanhua [Women's Civil Society Organizations and the Pluralism of the Governance Models]. In Chongji Yu Biange: Duiwai Kaifangzhong De Zhongguo Gongmin Shehui Zuzhi [Impact and Change: the Chinese Civil Society Organizations that are Opening Toward the Outside], ed. Shang Xiaoyuan, 236-283. Beijing, China: Chinese Social Sciences Press.

Shieh, Shawn. 2009. Beyond Corporatism and Civil Society: Three Modes of State-NGO Interaction in China. In State and Society Responses to Social Welfare Needs in China: Serving the people, eds. Jonathan Schwartz and Shawn Shieh, 22-41. New York: Routledge.

State Administration of Foreign Exchange. 30 December 2009. Circular of the State Administration of Foreign Exchange on Relevant Issues Concerning the Administration of Donations in Foreign Exchange by Domestic Institutions. http://www.safe.gov.cn/model_safe_en/laws_en/laws_detail_en.jsp?ID=30600000000000000,58. Accessed 14 September 2011.

Sundstrom, Lisa McIntosh. 2006. Funding Civil Society: Foreign Assistance and NGO Development in Russia. Stanford: Stanford University Press.

Teets, Jessica C. 2011. Reforming Service Delivery in China: The Emergence of a Social Innovation Model. Journal of Chinese Political Science. doi:10.1007/s11366-011-9176-9.

Teets, Jessica C. 2013. Let Many Civil Societies Bloom: The Rise of Consultative Authoritarianism in China. The China Quarterly. doi:10.1017/S0305741012001269.

Tong Lihua. 2009. Weile Zhengyi: Zhicheng Gongyi Shinian [For Justice: Ten Years of Zhicheng Public Interest Law NGO]. Beijing, China: Law Press China.

Yale University Law School. 22 April 2010. Visiting Scholar Lu Jun to Speak at Law School. http://www.law.yale.edu/intellectuallife/11566.htm. Accessed 20 October 2011.

Zhang Han. 19 May 2010. NGOs Feel the Squeeze. *Global Times*. http://special.globaltimes.cn/2010-05/533426.html. Accessed 16 February 2011.

Zhang, Hong and Marsha Smith. 2009. Navigating a Space for Labor Activism: Labor NGOs in the Pearl River Delta of South China. In State and Society Responses to Social Welfare Needs in China: Serving the people, eds. Jonathan Schwartz and Shawn Shieh, 66-88. New York: Routledge.

Chapter 6
Understanding the Present, Looking to the Future

Abstract The impressive growth of China's NGO sector belies the fact that China's civil society is still a work in progress. Nonetheless, NGO leaders who are involved in that building process have established effective organizations that are making important contributions to social governance. In this respect, they defy the understanding that effective NGOs cannot be found in non-democracies and that effectiveness can only be defined in terms of effecting revolutionary political changes. As a work in progress, China's NGO sector is under-institutionalized, which means that the NGO leader is especially important for organizational effectiveness. In this respect, contemporary China offers a unique opportunity to assess the importance of NGO leadership and its associated skills. This study's focus on leadership ties in general and state linkages in particular reflects the empirical reality in China. However, it is important to continue monitoring the emergence of other types of leadership ties and a new generation of leaders because such changes may point to important developments in China's civil society and, by extension, changes to China's political landscape. Similarly, the continuing importance of the NGO leader must also be subjected to regular observation. In this vein, the greatest challenges to the development of China's NGO sector in general and NGOs in particular may not be their dependence on linkages with the state. Instead, elitism and individualism may constitute a greater threat to their development. Both Chinese and foreign actors can play an important role in addressing these challenges.

Keywords Organizational effectiveness · Performance benchmark · Revolutionary change · Governance · Leadership ties · Elitism · Individualism

6.1 Leadership Ties and NGO Effectiveness

China's civil society is growing, as evidenced by the impressive growth of China's NGO sector in such a short period of time. However, the fact remains that it is still a work in progress, as reflected in the NGO sector's lack of legitimacy. The leaders of

J.W. Tai, *Building Civil Society in Authoritarian China*,
SpringerBriefs in Environment, Security, Development and Peace 20,
DOI 10.1007/978-3-319-03665-6_6

China's NGOs must be credited for the growth, but they will also be the first to say that many challenges remain and much remains to be done. In any case, whether they realize it or not, these men and women are engaging in a major construction effort and already they are making history by creating effective organizations that are actively taking part in social governance.

China serves as an interesting case to examine the factors that contribute to NGO effectiveness in non-democracies. First, given the dramatic growth in the number of NGOs in China over the past two decades, China is a fertile ground for conducting research about NGOs operating in authoritarian systems. Second, as an authoritarian state, many might not expect China to be home to a large number of NGOs, let alone effective ones. In this respect, China's situation has shown the inadequacy of the argument that effective NGOs and non-democratic political systems are generally incompatible. Yet, it is also true that the Chinese NGO sector has remained under-institutionalized, a consequence of which has been the prominence of NGO leaders. But this phenomenon is valuable because it means that there are numerous observations from which the analyst can draw to understand how NGO leadership can pave the way for organizational effectiveness.

The ability to effect revolutionary political change has been set as the performance benchmark for NGOs in non-democracies. One implication of this perspective is that NGOs in non-democratic settings are unlikely to perform well; in fact, in all likelihood they will under-perform significantly. Yet, it can be unfair and unrealistic to set such a standard of performance for Chinese NGOs, especially given the fact that the communist state in China is a powerful organization that enjoys greater legitimacy than any other group in China and the state controls critical organizational resources. Perhaps more importantly, the leaders of China's NGOs are for the most part uninterested in promoting radical political transformations. They are more interested in finding ways to improve the lives of the Chinese people. Therefore, beyond the issue of the fairness of establishing such a performance standard, focusing on whether China's NGOs have facilitated revolutionary political change has had the effect of ensuring that the visions of China's NGO leaders are ignored and, as a result, analysts would overlook the important contributions that NGO leaders in China have made and are making to public governance.

The cases presented in this book show that Chinese NGOs are indeed contributing to the governance of this vast country. As shown in the preceding chapters, many of these organizations are effective advocates of migrant workers through the provision of free legal aid to migrant workers, the supply of vocational training to migrants, and the sponsorship of publications to raise public and government awareness about the needs of this important population segment. They have also had access, albeit limited, to the state's policy-making process. In the process, these organizations have helped millions of Chinese people who have been otherwise helpless and defenseless. Organizations working in other issue areas are also making similar contributions. Collectively, they show that by defining NGO success in terms of revolutionary political change, the analysts would have overlooked the NGOs' other important political functions.

But this study has also made clear that not all Chinese NGOs have been equally effective. So what factors have contributed to the success of some NGOs? Institutional factors have generally been given to answer this question. For political scientists, the most commonly-cited institutional factor is regime type. According to this line of argument, the type of state that governs society has much to do with whether NGOs are able to achieve effectiveness. Indeed, this argument stipulates that the effectiveness of NGOs, indeed the presence of NGOs, is the consequence of whether the state in question is democratic or authoritarian. In a nutshell, analysts who hold this perspective will argue that NGOs thrive in democracies and these organizations will thus most likely fail in non-democracies.

Unfortunately, empirical evidence does not support this generalization. China, notably, has witnessed an incredible growth of its NGO sector. A clear indication is the fact that there has been a 100 % increase in the number of registered NGOs over the past two decades. And this number does not include the great majority of the millions of grassroots NGOs that either are registered as commercial enterprises or are not registered at all. Moreover, as the preceding chapters have demonstrated, there are many successful Chinese NGOs, in spite of the non-democratic political system with which they have had to contend.

In the face of unfriendly, and often hostile, authoritarian governments, striving for greater organizational autonomy has been advocated by many experts and practitioners as the central precondition for achieving NGO effectiveness. This strategy also has normative value if one views the NGO's *raison d'être* as the authoritarian state's perennial challenger with the ultimate goal of radically transforming the non-democratic political system. Although this strategy seems to make a great deal of sense on the surface, the cases presented in this book suggest that organizational autonomy is certainly not the only precondition for NGO effectiveness. In fact, state linkages arguably constitute a similarly important precondition, if not an even more important precondition, for NGOs operating in non-democratic political systems. And the importance of state linkage seems obvious if one recognizes that the non-democratic state can be an all-powerful state that not only controls critical resources but which is also widely perceived as the most legitimate institution in society.

As the last two statements suggest, state linkages have been important for Chinese NGOs not necessarily because they are inherently valuable. They are valuable because they confer specific benefits to the Chinese NGO. However, contrary to the general impression obtained from the existing literature, the unfortunate reality is that not all state linkages have conferred these advantages on the Chinese NGO, and an important reason is that not all state institutions have been similarly endowed.

The key question therefore is how can a Chinese NGO establish the most advantageous state linkages? What are the mechanisms create those state linkages? The cases presented in this book show that the Chinese NGO leader *is* the mechanism that has created state linkages for the Chinese NGO. In particular, the NGO leader's social ties pave the way for his organization to achieve its goals.

Even though analysts have identified the Chinese NGO leader as an important factor for organizational success, they have not examined in-depth the aspects of the NGO leader that make some more successful than others.

Although some writers have noted the importance of "internal" attributes of the NGO leader, such as personalities, predispositions, and motivations, there are reasons to believe that the leader's social ties exercise a more direct and visible impact on organizational performance. One of those reasons is that, in most cases, the success of the NGO has relied on its ability to manage the organization's external environment, notably in terms of resources. If we adopt this perspective, we should then expect that the most valuable NGO leaders, from the perspective of organizational success, are those who possess the relationships that allow them to deal effectively with the external environment, i.e., obtain external resources.

With this mind, in this study, I have taken three commonly-identified characteristics of the successful Chinese NGO—state linkages, media connections, and international ties—and examined them as leadership ties to understand how they have facilitated organizational linkages to the state, how they have interacted with each other, and, ultimately, how they have contributed to organizational effectiveness. In this manner, I tried to understand the importance of each type of leadership ties relative to the others.

In Chap. 3, I explored the Chinese NGO leader's state linkages in general and institutional affiliations to the state in particular to understand how those individual-level factors have contributed to state linkages for the Chinese NGO and, ultimately, to the NGO's ability to achieve its goals. The cases showed that even though common forms of state linkage, notably CPC membership, might have facilitated the establishment of the Chinese NGO leaders' state affiliations, ultimately state affiliations have had more direct bearing on NGO effectiveness. The reason is that state affiliations give NGOs direct access to state resources that are scattered among state institutions. Through these state affiliations, NGOs acquire material resources, organizational legitimacy, and policy access, which together allow NGOs to fulfill their respective organizational missions. In this conceptualization, the key ingredient is the extent to which an NGO leader is able to form linkages with resource-rich and authoritative state organizations.

In Chap. 4, the Chinese NGO leader's media connections were examined to understand how this type of leadership ties has paved the way for organizational linkages to the state and, ultimately, organizational effectiveness. Most analyses have treated the media either as exogenous to civil society (Ballentine 2002) or as a structural framework of civil society (Butsch 2007). Moreover, the media's relationship to the state has often been portrayed as either for or against the authoritarian state and, by extension, as either for or against civil society actors. But the reality is that in an evolving media system like China's, the media have actually achieved a position that has straddled the state-society divide, putting them in a unique position to bridge the gulf between the authoritarian state and the emerging civil society. In this respect, media connections are extremely valuable for Chinese NGO leaders.

Nonetheless, the relationship between the NGO leader's media connections and the organization's access to the state has been facilitative, especially in comparison to the effects of the NGO leader's state affiliations. Media connections, for example, have not had direct effects on the NGO's ability to acquire material resources from the state. The cases presented in this book show that those resources were obtained as the direct consequence of the leader's state affiliations. However, media connections may facilitate resource acquisition by helping the NGO leader acquire state affiliations. Indeed, this may be the most interesting and valuable aspect of the leader's media connections, especially from the perspective of those NGO leaders who do not possess prior state affiliations. In other cases where state affiliations existed, media connections have helped NGOs establish organizational linkages to the state and access to the state policy apparatus. Yet, in those cases media connections have acted more as enhancers of the NGO leader's existing state affiliations rather than as direct causal agents. In other words, the leader's media connections serve more as lubricants than as catalysts, when compared with his state affiliations.

Lastly, in Chap. 5, I examined the effects of the NGO leader's international ties. Like situations in other non-democracies, the international ties of the Chinese NGO leader have been valuable in that those ties have allowed the leader to acquire international financial resources for the NGO. Given the difficulties in accessing state financial resources and the poor local fundraising environment, international resources have been indispensable to the Chinese NGOs. In other words, the primary function of international ties has been to enhance the Chinese NGO's organizational capacity. However, international ties act, in a sense, as a double-edged sword. The Chinese state remains highly suspicious of international ties for fear that they could serve as vehicles to import subversive ideas from abroad into China. In this respect, international ties can have a negative impact on the legitimacy of Chinese NGOs, which is already in short supply. Moreover, as the cases have demonstrated, many projects that have been necessary for the Chinese NGO to achieve its goals must have the support of the state. This means that the Chinese NGO leader must find other means to establish organizational linkages to the state so as to achieve legitimacy.

Quite often the case, the NGO leader has relied on his state affiliations and media connections to achieve the requisite state linkages so as to undertake those projects that were germane to the NGO's mission. Only then would the NGO leader's international ties be optimized. The implication seems clear: the salutary effects of international ties are conditioned upon the prior existence of the leader's state affiliations and/or his media connections. In fact, in some cases, these two sets of connections have been instrumental in the development of the Chinese NGO leader's international ties.

Based on the cases that have been presented, we can see how the three types leadership ties have worked together to address the needs of the Chinese NGO. In the course of this analysis, we have also seen the importance of state linkages in helping Chinese NGOs achieve their organizational goals. Lastly, by examining the interactions among the three types of leadership ties, we have gained a sense of their weight in importance relative to each other. In this respect, what is especially clear is that the NGO leader's international ties, albeit seemingly indispensable, have not been able to rival the importance of his state affiliations and, to a lesser extent, his media connections.

6.2 Future Research

The state of China's NGO sector is highly fluid. Therefore, it is important to anticipate changes. In this vein, I propose three related avenues for further research. The first avenue involves the exploration of other important leadership ties and qualities. The second avenue involves a close examination of the attributes and skills of the younger generation of NGO leaders. The third avenue involves maintaining a close eye on the coming leadership transitions for many prominent Chinese NGOs. All three research agendas are likely to yield interesting insights not just about the development of the individual Chinese NGO, but more importantly, the development trajectories of the Chinese civil society and even those of the Chinese political system.

The focus on state affiliations, media connections, and international ties reflects current realities. As the state of China's NGO sector continues to evolve, other leadership skills and qualities will surely emerge to exert impact on the Chinese NGO's ability to achieve organizational success. As mentioned in Chap. 2, one type of potential leadership ties is the relations among NGO leaders. As the Chinese NGO sector develops, interactions among NGO leaders have also increased in both frequency and intensity. However, those inter-NGO interactions have not yet yielded noticeable benefits for individual Chinese NGOs. But this does not mean that things will always remain the way it is today. However, an exploration of other leadership ties and qualities is a valuable endeavor not only because of the inherent value in such an exercise. In fact, an exploration of other leadership qualities may tell us much about the evolution of China's civil society and China's overall political trajectory. In this respect, whether or not inter-NGO relationships exercise increasingly important influences on the performance of the Chinese NGO should tell us much about the extent to which a Tocquevillian civil society characterized by interpersonal trust and cooperation as well as public participation independent of the state is being formed in China.

Similarly, researchers should pay attention to the attributes and skills of the new generation of Chinese NGO leaders, namely those who were born after 1980. Unlike previous generations, this generation of NGO leaders has grown up in a China that has been for the most part devoid of domestic political turmoil along with rising prosperity. They have also become more exposed to developments outside China. In addition, they are accustomed to greater individual freedom and initiatives. Consequently, their perceptions of the state and social activism may be very different from previous generations. It will be interesting to see how their generational background shapes their perceptions and practices as NGO leaders, which collectively should have important implications on how they engage the state.

Lastly, this book's emphasis on the organizational impact of the Chinese NGO leader is reflective of the under-institutionalized nature of the Chinese NGO sector. To a large extent, the finding about the importance of the leadership can only be tentative because the majority of the Chinese NGOs are still being led by their founder(s).[1] Therefore, we do not yet have opportunities to observe the full impact of the founder-leader on the Chinese NGO. In addition to the value of gaining a fuller picture of the organizational impact of the NGO leader, whether and how the Chinese NGO will survive the coming leadership transition can tell us much about the extent to which the Chinese NGO sector in particular and the Chinese civil society in general have become institutionalized, which again may have important implications for China's political development.

6.3 Implications and Recommendations: NGO Effectiveness and Civil Society Development

Depending on how the proper political role of civil society is defined, different people may draw different implications from the observations presented in this book. On the one hand, for those who believe that civil society ought to be in perpetual conflict with authoritarian states, the presentation here may be downright depressing. From this perspective, if Chinese NGOs, which are both promoters and constituents of any civil society that is defined as anti-state, need to develop strong relations with the state in order to be effective, then such NGOs are unlikely to become effective agents of civil society in China. Unfortunately, Chinese civil society groups have been simply no match for the Chinese authoritarian state's primacy, both in terms of legitimacy and resources. Therefore, civil society as defined in terms of opposition to the state is unlikely to be found in China, as long as the communist party-state remains in power.

However, even if Chinese NGOs have not demonstrated clear intentions for political subversion, the NGO leaders have not been reluctant to criticize state policies and practices, whenever it's warranted. Even those NGO leaders with strong political ties have not been reticent in expressing their critical views. For example, the leader of the Zhicheng Public Interest Law NGO has possessed the most substantive state linkages, and yet he has been extremely critical of state policies and practices concerning the rights of children and the rights of migrant workers. A substantial portion of his 2009 book-length review of Zhicheng's NGO work since the organization's establishment has included strong criticisms of the state.

[1] To my understanding, the only Chinese NGO that has experienced the departure of its chief founder is the Friends of Nature, which is featured in this book as a shadow case. But in the next five to 10 years, we are likely to see more leadership shifts among Chinese NGOs as the first generation of Chinese NGO entrepreneurs pass from the scene due to age.

Nonetheless, criticisms of the state are not tantamount to open calls for overthrowing the state. In this respect, the observations here provide further evidence for the argument that the concept of civil society as commonly derived from modern Western experiences is not entirely applicable to the reality of state-society relations in China. Hence, many China specialists have favored the concept of corporatism to describe the nature of state-society relations in contemporary China. The problem with this approach, however, is that just as the concept of civil society has been unable to capture the prominence of the Chinese state in managing state-society relations, the concept of corporatism has downplayed the declining role of the state and the increasing degree of independence for Chinese civil society groups. The disjuncture between theoretical concepts and the reality in China has forced later writers to adopt analytical frameworks that emphasized less about the dichotomy between state and society and more about the interactive nature of their relationship.

In adopting the framework of state linkage, I believe that the interactive approach corresponds better to reality. The interactive approach does not suggest that the power relations between state and society in China are symmetrical. But what it does highlight is that civil society groups like NGOs may have more to gain by working with the state than against it. In this sense, the notion of state linkage suggests that the future of the Chinese civil society is not necessarily bleak. Let's take as example the issue of organizational legitimacy to illustrate the potential value of state linkage for Chinese NGOs.

As illustrated in this study, NGOs in China have yet to achieve political and popular legitimacy, especially in comparison to the state. The lack of popular legitimacy has been raised consistently by leading Chinese NGO experts and practitioners as among the greatest challenges for Chinese NGOs. Indeed, the great majority of ordinary Chinese people have either little or no knowledge about civil society and NGOs. Many of those who have heard of the concepts are unable to explain what they mean. It made sense, therefore, for many, if not most, Chinese NGO leaders to emphasize the building of their respective organizations' "brand names" (*pinpai*). What better way to achieve name recognition than through affiliations with the most legitimate institution in the country, namely the state? To be sure, the ultimate goal should be to achieve legitimacy for the entire NGO sector and, in turn, popularize and legitimize the notion of civil society. But the process has to start at the organizational level.

Arguably, a greater threat to the future development of the Chinese civil society in general and the NGO sector in particular is the fact that the Chinese NGOs are characterized by elitism. This book's conceptual focus is a testament to this fact. If the strength of a civil society involves the extent to which ordinary citizens are involved, as many writers have argued, then the elitist nature of the Chinese NGO sector is a cause for concern. However, it would be unfair to attribute the blame for this phenomenon to the Chinese NGO leaders. There are important cultural and institutional factors that have contributed to this reality. Moreover, China is not unique in this respect among non-democratic countries and developing societies. In fact, it is arguable that NGO sector worldwide has become increasingly elitist with the increasing involvement of highly-educated professionals in the sector. In this sense, the Chinese situation may simply be a reflection of the global trend.

Similarly, individualism among Chinese NGO leaders poses another challenge to the future development of the Chinese civil society. In the scramble to build organizational brand names, Chinese NGO leaders have essentially done so alone. With very few exceptions, such as public environmental campaigns and networking among the leaders of elite Chinese NGOs, there has been little noticeable cooperation among Chinese NGOs. Competition, rather than cooperation, has generally characterized the Chinese NGO sector. Again, it would be unfair to cast all the blame on the NGO leaders for this phenomenon given that, in most cases, they have simply been responding to the existing environment of limited funding and weak legitimacy. Nonetheless, if a critical sign of a strong civil society is the presence of cooperation and mutual trust, then the Chinese civil society has a long way to go.

The situation is not entirely hopeless, but it does call for a concerted effort from all the key stakeholders. With respect to the problem of elitism, private international donors can do much more to cultivate and support non-elite NGO entrepreneurs. Since international donors have gained access to China, they have for the most part interacted with members of the Chinese elite, notably Chinese intellectuals. This has been a reasonable strategy since international donors have understandably wanted to have greater and more direct influences on political decision-making and on shaping societal norms and practices. But after over two decades of presence in China, it is equally reasonable to expect that international donors should do and have done more to promote greater popular participation in the Chinese civil society. In this respect, the observations in this book suggest that one thing that international donors, including foreign governments, can do to promote greater public participation in NGO activities and cultivate more non-elite NGO entrepreneurs is to provide more support and opportunities to empower the Chinese media sector and develop strong ties between the Chinese media and the Chinese NGO entrepreneurs.

To be sure, the prevailing political environment has continued to limit the sphere of activities of the international donors, which continue to be based in the major coastal metropolitan areas in China, namely Beijing and Shanghai. Although this situation should not automatically preclude international donors from accessing the general population to identify potential Chinese NGO entrepreneurs, it has greatly hampered the mobility of the agents of international donors. Therefore, if international donors are to do more to cultivate non-elite NGO entrepreneurs, then they must inevitably rely on others. In this respect, leading Chinese NGO entrepreneurs can aid in this endeavor by working to identify the next generation of Chinese NGO leaders, particularly those who are not members of the elite. It is therefore encouraging that leading Chinese NGOs, such as the Narada Foundation and NPI, which are led by Chinese NGO entrepreneurs with extensive state linkages, have initiated this process through programs like Narada's Gingko Fellow Program and NPI's Incubation Program. Similarly, Chinese NGOs have begun to pay more attention to cultivating connections with the Chinese media.

Even more importantly, leading Chinese NGO entrepreneurs can do more to promote collaboration among Chinese NGOs. At present, inter-NGO collaboration either at the level of the NGO leaders or at the level of the overall organization remains is limited. Other than environmental campaigns and the post-2008 Sichuan earthquake NGO alliance, known inter-NGO collaboration defined in terms of regular and close interactions has been confined to a network of the leaders of China's leading women's NGOs; AIDS/homosexual NGO networks; and looser networks of environmental NGOs led by leading organizations, such as the Friends of Nature, the Global Village, and the Green Earth Volunteers. In a nutshell, there is still much room for inter-NGO cooperation.

Lastly, it would seem that Chinese NGOs are unlikely to become agents of revolutionary bottom-up political changes in China, at least not in the foreseeable future. But for the time being they can help build a civil society that will become even more effective and influential as a participant in governance in China. Robert Putnam had argued that a strong civil society can contribute to better democratic governance. There is no reason to believe that a strong civil society cannot do the same in a non-democratic political system. Although this statement will surely disappoint many who would like to see civil society act in more revolutionary ways, the ordinary Chinese people will likely be impressed, and perhaps even grateful, if Chinese NGOs and other civil society groups can do more to get the Chinese state to meet the needs and demands of the masses.

References

Ballentine, Karen. 2002. International Assistance and the Development of Independent Mass Media in the Czech and Slovak Republics. In *The Power and Limits of NGOs: A Critical Look at Building Democracy in Eastern Europe and Eurasia*, eds. Sarah E. Mendelson and John K. Glenn, 91-125. New York: Columbia University Press.

Butsch, Richard, ed. 2007. *Media and Public Spheres*. New York: Palgrave Macmillan.

Appendix
Scope, Data, Methodology, and Limitations

The field work for this study took place between autumn 2007 and summer 2011 and most of the data for this study were collected during this time period.

In this study, the label NGO is use to describe all social organizations that are not "creatures of the state" in the sense that they enjoy the power to make decisions with respect to personnel, financial, and other management matters. Nongovernmental organizations may, however, be created by state employees. In addition, NGOs may include those that possess some administrative connections to the state at some point during organizational development.[1] Moreover, NGOs may, of course, receive financial and other resource support from state agencies. Furthermore, NGOs include those that have achieved the nonprofit registration status and those that have not. This study distinguishes NGOs from GONGOs, such as the All-China Women's Federation, which is an appendage of the Communist Party of China, and the All-China Lawyers Association, which is a product of state legislation.

In the existing literature on the Chinese NGO sector, scholars have generally included both GONGOs and NGOs in their samples. Some writers have in fact justified this analytical strategy by arguing that both types of organizations face similar environmental constraints and have therefore behaved in similar fashions. Even so, I contend that in doing so we underestimate the differences between the two types of organizations with respect to organizational needs and access to organizational resources. In the process, we are unlikely to capture the different factors that contribute to effectiveness for different organizational types.

[1] Although some may understandably question whether this type of "hybrid" NGO should be included in the sample, I believe that the decision to include such organizations conforms to the reality that organizational origins have not predestined the NGOs to pursue certain paths of development. In other words, how the NGO got started is less important than what it eventually becomes. As long as the organizational leadership can make independent decisions regarding the organization's development, then we should recognize its independent character. Hybrid NGOs in China have enjoyed strong reputations as "nongovernmental" organizations, and one of China's most prominent hybrid NGOs is Rural Women, which is featured in this book.

J.W. Tai, *Building Civil Society in Authoritarian China*,
SpringerBriefs in Environment, Security, Development and Peace 20,
DOI 10.1007/978-3-319-03665-6

To understand the processes through which certain leadership ties are likely to contribute to NGO effectiveness, this research project adopts the case-study approach, which is best suited to analyzing causal processes and mechanisms (Yin 2002; Ragin 2007). In particular, this study is a multiple-case study (Yin 2002: 45–46). In order to explore the effects of leadership qualities on NGO effectiveness, I compare four core NGO cases that display variations with respect to organizational effectiveness. More specifically, three of the core cases have displayed similar levels of effectiveness while the fourth case has displayed a level of effectiveness different from the first three cases. In addition, the leaders of the four cases vary in terms of key personal attributes, notably level of education, occupation, and gender. In this respect, I employ both the "Mill's Method of Agreement" and the "Mill's Indirect Method of Difference" for this study (Ragin 1989: 36–42).

The four cases share certain environmental and organizational characteristics, however. Those characteristics are issue area and geographical location. All four NGO core cases are headquartered in Beijing, which is China's capital city, even though all of them have varying degrees of organizational reach outside Beijing. In addition, the four organizations work primarily in the same issue area, namely issues concerning migrant workers, even though one organization is also prominent in the field of protecting minors.

In an effort to illustrate that the value of NGO leadership, notably the ties of the NGO leader, has also been found in organizations working in other geographical locations and on different issues, an additional three cases that exhibit variations in these two organizational characteristics are examined.[2] The three cases also serve another function. The NGO displaying the lowest level of organizational effectiveness also happens to be the youngest of the four organizations featured as core cases. One may thus come to the conclusion that organizational age may be an unaccounted-for factor to explain variations in organizational effectiveness. One of the three additional cases thus serves to show that organizational effectiveness has also been found in similarly young organizations.

Although NGOs serving migrant workers and their dependents constitute the focus of this study, my knowledge of China's NGO landscape has benefited immensely from personal interviews, informal conversations, participant observation sessions, and written correspondence with the leaders, staff, and volunteers of NGOs that work in other issue areas, including the environment, public health, education, women's issues, NGO capacity building, and disabled persons, as well as donor representatives. Similarly, while Beijing-based NGOs have constituted the bulk of the organizations with which I have interacted, I have

[2] The three additional cases are "shadow cases," which are distinguished from core cases in that they are introduced into the analysis with the specific purposes of corroborating the argument and refuting rival hypotheses. For an example of the use of shadow cases, see Eva Bellin, "Contingent Democrats: Industrialists, Labor, and Democratization in Late-Developing Countries," *World Politics* 52, 2 (January 2000): 175–205. I thank Bruce Dickson for introducing me to this methodological approach.

also engaged NGOs working in other parts of China. Interviews were also conducted with the staff of two GONGOs (Tables A.1, A.2, A.3 and A.4).

Given the difficulty of ascertaining the true universe of grassroots NGOs in China and accessing individual NGOs, I have adopted the well-known technique of snowball sampling. As a form of the non-probability sampling method, the main criticism of the method is that there is no assurance that the sample is representative of the population. The researcher has little control over the selection of the sample and unknown biases may be introduced into the sample, which can adversely impact the reliability of the findings. Yet, the method has important advantages. Other than the fact that it is easier to execute this method than to conduct probability sampling, the adoption of this method allows the researcher to reach difficult-to-access populations.

In addition to field research, I have studied English-language and Chinese-language publications for case studies of Chinese NGOs. The publications include both traditional and Internet forms. They include books, essays in news magazines, online blogs, academic journal articles, and internal NGO publications. To enhance my understanding of particular NGOs and the Chinese NGO sector in general, I have also interviewed and corresponded with scholars, practitioners, and government officials who are experts on the subject.

Despite these efforts, I am fully aware that I can make, at best, very limited claims about the representativeness of the sample. One obvious reason, which has been mentioned earlier, is that the population of NGOs in China remains a mystery. A second reason is that there remains substantial disagreement among China scholars, especially those who operate in different disciplines, about the appropriate defining qualities of a Chinese NGO. Moreover, given that this project is essentially a small-N study that focuses on NGOs operating in major urban centers, very little generalizations can be made about the state of effective NGOs in contemporary China. In order to evaluate all the possible factors that impact NGO effectiveness, large-N surveys of NGOs across different Chinese provinces and municipalities must be conducted. Nonetheless, through a detailed examination of several core cases and a more cursory examination of other cases, I have confidence that this study can yield insights about the mechanisms that facilitate NGO effectiveness and the processes through which those mechanisms might contribute to organizational success.

Another issue concerns the applicability of my findings to situations found in democracies. After all, the importance of leadership for NGOs anywhere has indeed been identified by writers. However, differences between democracies and non-democracies with respect to the state's policies and practices toward NGOs, the relative openness of the state to non-state entities, the nature of the media's relationship vis-à-vis the government, the extent of the NGOs' reliance on international support, the state and history of private philanthropy, and the importance of cultural considerations, such as religious factors, suggest that there would be differences in terms of the types of leadership qualities and connections that are most valuable for NGOs. Even if the same qualities and skills are found present among the leaders of successful NGOs working in democracies, the nature of those qualities, the relationships among them, as well as the relative weight of each quality would in all likelihood be different than those found in non-democracies. Therefore, it is important to keep these considerations in mind when deciding how the findings in this book may be applied to situations in democracies.

Lastly, an additional word about the finding concerning the causal importance of NGO leadership in general and leadership qualities in particular. The emphasis on the NGO leader in this study is reflective of the current reality in China that NGOs are under-institutionalized in terms of organizational legitimacy. In this situation, NGOs have relied a great deal on their respective leaders. Even so, my research suggests that most Chinese NGO leaders would like for their organizations to outlive them. In this sense, the findings from this study may ultimately be applicable only for a limited period of time. It will, therefore, be interesting to see whether and how the value of the NGO leadership changes over time. However, at present, the great majority of the Chinese NGO leaders are alive and well. In fact, only one well-known Chinese NGO, the Friends of Nature, which is featured in this study, has thus far survived, with relative success, the passing of its founding leader. Therefore, at this time, we can only speculate about the extent to which the organizational value of the NGO leader will evolve (or decline) over time.

Table A.1 Interviews

BJ090607	September 6, 2007	Beijing, China
BJMY101107	October 11, 2007	Beijing, China
BJ112307	November 23, 2007	Beijing, China
BJ121207	December 12, 2007	Beijing, China
BJ012908	January 29, 2008	Beijing, China
BJ022108	February 21, 2008	Beijing, China
BJWK022108	February 21, 2008	Beijing, China
BJ022708	February 27, 2008	Beijing, China
NPO Shang Yusheng	April 1, 2008	Beijing, China
BJWK042808	April 28, 2008	Beijing, China
BJ052708	May 27, 2008	Beijing, China
On Action Ma Yang	May 27, 2008	Beijing, China
BJ052808	May 28, 2008	Beijing, China
Rural Women Gao Guanglin	June 12, 2008	Beijing, China
Zhicheng Wang Fang	June 13, 2008	Beijing, China
BJWK061608	June 16, 2008	Beijing, China
BJ061908	June 19, 2008	Beijing, China
BJHD062508	June 25, 2008	Beijing, China
BJWK062708	June 27, 2008	Beijing, China
BJ070308	July 3, 2008	Beijing, China
BJ070408	July 4, 2008	Beijing, China
Zhicheng Zhang Wenjuan	July 17, 2008	Beijing, China
BJWK081208	August 12, 2008	Beijing, China
BJ091608	September 16, 2008	Beijing, China
BJ100108	October 1, 2008	Beijing, China
BJMY100808	October 8, 2008	Beijing, China

continued

Table A.1 (continued)

BJ111908	November 19, 2008	Beijing, China
BJWK111908	November 19, 2008	Beijing, China
BJ112508	November 25, 2008	Beijing, China
BJ112608	November 26, 2008	Beijing, China
BJ120608	December 6, 2008	Beijing, China
BJ120908	December 9, 2008	Beijing, China
BJ121808	December 18, 2008	Beijing, China
BJ122108	December 21, 2008	Beijing, China
BJ122308	December 23, 2008	Beijing, China
BJ021609	February 16, 2009	Beijing, China
BJCD021909	February 19, 2009	Beijing, China
BJ022309	February 23, 2009	Beijing, China
BJ032209	March 22, 2009	Beijing, China
BJ032509	March 25, 2009	Beijing, China
BJWK032509	March 25, 2009	Beijing, China
BJAZ040109	April 1, 2009	Beijing, China
Maple Women Wang Xingjuan	April 1, 2009	Beijing, China
BJWK040809	April 8, 2009	Beijing, China
BJ041409	April, 14, 2009	Beijing, China
BJWK041509	April 15, 2009	Beijing, China
BJZX042209	April 22, 2009	Beijing, China
BJ042609	April 26, 2009	Beijing, China
BJHD052709	May 27, 2009	Beijing, China
BJWK060109	June 1, 2009	Beijing, China
Rural Women Xie Lihua	June 3, 2009	Beijing, China
BJWK072609	July 26, 2009	Beijing, China
BJTZ072709	July 27, 2009	Beijing, China
Little Bird Wei Wei	August 6, 2009	Beijing, China
Self-Empowerment Chen Jian	April 6, 2008	Chongqing, China
CQ040808	April 8, 2008	Chongqing, China
CQ040908	April 9, 2008	Chongqing, China
Self-Empowerment Chen Jian	July 31, 2009	Chongqing, China
JN071008	July 10, 2008	Jinan, China
QD070708	July 7, 2008	Qingdao, China
SH041808	April 18, 2008	Shanghai, China
SH092808	September 28, 2008	Shanghai, China
SH031010	March 10, 2010	Shanghai, China
SHWY010511	January 5, 2011	Shanghai, China

continued

Table A.1 (continued)

SHZJ010511	January 5, 2011	Shanghai, China
SH012510	January 25, 2010	Shanghai, China
SHWY011011	January 10, 2011	Shanghai, China
SHWY031011	March 10, 2011	Shanghai, China
XI030308	March 3, 2008	Xian, China
XI030709	March 7, 2009	Xian, China
ZZ072108	July 21, 2008	Zhengzhou, China
Self-Empowerment Chen Yuying and Chen Jian	May 13, 2009	Zhongxian, China
TPWK020811	February 8, 2011	Taipei, Taiwan
EJIC Huang Xuetao	December 29, 2011	Washington, DC (telephone)

Table A.2 Written correspondence

BJYYSK020909	February 9, 2009	Beijing, China
BJYY04052009	April 5, 2009	Beijing, China
Self-Empowerment Chen Yuying	January 10, 2011	Shanghai, China
Self-Empowerment Chen Yuying	January 15, 2011	Shanghai, China
Self-Empowerment Chen Yuying	July 14, 2011	Shanghai, China

Table A.3 Participant observation sessions

NGO forum on taxation	September 18, 2008	Beijing, China
NGO accountability forum	October 24, 2008	Beijing, China
International Volunteers Day	December 5, 2008	Beijing, China
On action anniversary celebration	December 20, 2008	Beijing, China
China Doll Sunday movie program	February 22, 2009	Beijing, China
Unirule bi-weekly seminar	April 10, 2009	Beijing, China
Conference on Sichuan earthquake relief	June 9, 2009	Beijing, China
Visit to cerebral palsy NGO	April 18, 2008	Shanghai, China

Table A.4 Informal conversation

Foreign NGO officer	March 8, 2011	Shanghai, China

References

Ragin, Charles C. 1989. *The Comparative Method: Moving Beyond Qualitative and Quantitative Strategies*. Berkeley: The University of California Press.

Ragin, Charles C. 2007. Comparative Methods. In *The Sage Handbook of Social Science Methodology*, eds. William Outhwaite and Stephen P. Turner, 67-81. London: Sage Publications.

Yin, Robert K. 2002. *Case Study Research: Design and Methods, Third Edition*. Thousand Oaks: Sage Publications.

About the Author

John W. Tai, Ph.D., is an adjunct professor at the George Washington University's Elliott School of International Affairs and a consultant to the U.S. government. John has written about China's civil society, religious freedom and human rights in China, and US-China-Taiwan relations. He has also conducted public talks, private briefings, and media interviews on these subjects in both the United States and China. From 2007 to 2011, John taught at the Beijing Foreign Studies University and Shanghai's East China Normal University, where he managed study-abroad programs for American university students. Most recently, John authored a lengthy book review essay entitled "Civil Society: The Search Continues," which appeared in the *Fudan Journal of the Humanities and Social Sciences*.

Address John W. Tai, Ph.D., Elliott School of International Affairs, George Washington University, 1957 E Street, NW, Washington, DC 20052
Email johnctai2@gmail.com
Website http://www.linkedin.com/pub/john-w-tai/9/16a/4a9

More information on this author and book is at:
http://afes-press-books.de/html/SpringerBriefs_ESDP_Tai.htm.

J.W. Tai, *Building Civil Society in Authoritarian China*,
SpringerBriefs in Environment, Security, Development and Peace 20,
DOI 10.1007/978-3-319-03665-6

About the Book

How is modern civil society created? There are few contemporary studies on this important question. When it is addressed, scholars tend to emphasize the institutional environment that facilitates a modern civil society. However, there is a need for a new perspective on this issue. Contemporary China, where a modern civil society remains at its nascent stage, offers a valuable site to seek new answers. Through a comparative analysis of nongovernmental organizations (NGOs) in contemporary China, this study shows the importance of the human factor, notably the NGO leadership, in the building of a modern civil society. In particular, in recognition of the social nature of NGOs, this study engages in a comparative examination of the state linkage, media connection, and international tie of Chinese NGO leaders to understand how each contributes to an effective NGO.

More information on this book is at:
http://afes-press-books.de/html/SpringerBriefs_ESDP_Tai.htm.

J.W. Tai, *Building Civil Society in Authoritarian China*,
SpringerBriefs in Environment, Security, Development and Peace 20,
DOI 10.1007/978-3-319-03665-6

MIX
Papier aus verantwortungsvollen Quellen
Paper from responsible sources
FSC® C105338

If you have any concerns about our products,
you can contact us on
ProductSafety@springernature.com

In case Publisher is established outside the EU,
the EU authorized representative is:
Springer Nature Customer Service Center GmbH
Europaplatz 3, 69115 Heidelberg, Germany

Printed by Libri Plureos GmbH
in Hamburg, Germany